Understanding
Understanding

T0180534

Springer
New York
Berlin
Heidelberg
Hong Kong
London
Milan
Paris
Tokyo

Heinz von Foerster

Understanding Understanding

Essays on Cybernetics and Cognition

With 122 Illustrations

Springer

Heinz von Foerster
Biological Computer Laboratory, Emeritus
University of Illinois
Urbana, IL 61801

Library of Congress Cataloging-in-Publication Data
von Foerster, Heinz, 1911–
    Understanding understanding: essays on cybernetics and cognition / Heinz von Foerster.
        p. cm.
    Includes bibliographical references and index.

    1. Cybernetics. 2. Cognition. I. Title.
Q315.5 .V64 2002
003'.5—dc21

                                                                                        2001057676

    ISBN 978-1-4419-2982-2   e-ISBN 978-0-387-21722-2

Printed in the United States of America.

www.springer-ny.com

Springer-Verlag New York Berlin Heidelberg
A member of BertelsmannSpringer Science+Business Media GmbH

Preface

In school I always had difficulties remembering facts, data, lists of events: Was Cleopatra the girlfriend of Lincoln or Charlemagne or Caesar? History and geography thus became forbidden lands, surrounded by insurmountable obstacles. Relationships, on the other hand, I found easy to visualize, so that I felt at home in mathematics and physics.

To overcome my deficiency in history, I invented a device (unknown in the schools I attended) by which I could see at a glance whose girlfriend Cleopatra could possibly be: a timeline, where each memorable event corresponds to a point on a line, with older dates arranged toward the left, newer dates toward the right. I soon found that the farther one went to the left, the scarcer the entries became, while the closer one came to the present at the right, the denser the entries became. Having learned my mathematics, I used a logarithmic scale for my timeline, so that the past year has the same amount of space as the past decade, and the past century, and so forth. I could thus fit the demise of the dinosaurs on the same crib sheet as the conquest of Gaul and the death of the Emperor Franz Josef.

When I had filled in my timeline carefully, I found that my dates were far more nearly uniformly arranged than one should have expected. And so I had the idea that this logarithmic arrangement of dates corresponded to some sort of logarithmic decay of memory: one forgets an amount of data proportional to the amount of data one has in store at any one time. Unfortunately I could not find at that time any references that even discussed such a hypothesis, let alone confirmed or refuted it.

Not too long after the end of World War II, I returned to Vienna, ravaged by the war, occupied by Russian and Allied troops, but trying hard to recreate an appearance of a once-civilized life.

At one point, browsing through a bookstore, I came across an introductory textbook on psychology by a well-known Austrian professor, Hans Rohracher. Opening it up to flip through it, I found a graph showing a decaying line labeled "Ebbinghaus's Forgetting Curve." I immediately bought the book and took it home to see if the curve would fit my logarithmic hypothesis.

At first I had no luck: No matter what I tried, the curve was not a decaying exponential. Finally I turned to the text to see how Ebbinghaus had obtained his data. What he had done was to ask a group of volunteers (that means graduate students) to memorize a large number of nonsense syllables and then periodically, over several days, asked them to reproduce the list. I immediately realized that each test would of course also provide a prompt to relearn a part of the list, so that the "Forgetting Curve" was actually a forgetting-and-remembering curve. When I added the effect of the relearning process to my earlier equation, I obtained a result that fit Ebbinghaus's data very well.

Having thus found that pure forgetting is indeed an exponential decay, I thought of possible physical analogues. To my delight, it turned out that changes in state of macromolecules have about the same time constant as Ebbinghaus's forgetful subjects. It seemed that one could blame the volunteers' memory loss on some forgetful molecules in their brains.

I told this story to the psychiatrist Viktor Frankl, a truly remarkable man, with whom I was fortunate to have had some professional connection. Frankl urged me to publish it, and sent my paper to his publisher, Franz Deuticke. Now, Deuticke had a high esteem for Frankl the psychotherapist, but he was less sure about Frankl's mathematics, and Heinz von Foerster was a completely unknown quantity. However, Deuticke was also the publisher of Erwin Schrödinger, so he sent my paper to Schrödinger in Ireland for review. Soon the response came back: he did not believe a word of it, but he could not find any error in the mathematics. Deuticke decided that he did not care what Schrödinger believed, only whether there are errors, and so he published the story under the title, "Das Gedächtnis: Eine Quantenmechanische Abhandlung" (Memory: A quantum mechanical treatise).

The next year, in 1949, I was invited to come to the United States by Ilse Nelson, the best friend of my wife, who had escaped the Nazis and settled in New York. After some searching, I found a job at the University of Illinois in Urbana, in the Electron Tube Research Laboratory, where I was able to do some interesting work on electronics. We had two main interests. One was the production and detection of very short electromagnetic waves, in the millimeter and sub-millimeter region. The other was measuring events of very short time duration, on the order of tens of nanoseconds down to nanoseconds.

In the late 1940s a group of people had begun meeting every year in New York under the auspices of the Josiah Macy Foundation to discuss "circular causal and feedback mechanisms.". For these meetings the foundation had collected some of the most interesting people at the American scientific scene: The group included Norbert Wiener, who had coined the term "cybernetics"; Claude Shannon, the inventor of information theory; Warren McCullough, one of the leading neuropsychiatrists—he called himself an "experimental epistemologist"; Gregory Bateson, the philosopher and anthropologist; his wife Margaret Mead, the anthropologist who made Samoa famous; John von Neuman, one of the people who started the computer revolution; and many others of this caliber. I came into this group, by pure coincidence.

McCullough, then working at the University of Illinois in Chicago, had also been interested in memory, and when he came across my paper, he found that it seemed to be the only one that matched experimental results with theoretical predictions. I met him during the time of my job search, and through him I had the opportunity to join these meetings. So I participated with these people who had an absolutely fascinating approach to biology, to life, to work, to theory, to epistemology, and edited the resulting volumes. Meanwhile, of course, my main activity was still with electron tubes. But I was thinking, I would like to join that group in a full-time professional way later on.

In 1957, I had a wonderful opportunity to have a sabbatical leave from the university. I used it to visit two laboratories. For one semester I joined McCullough, who was by then at the Massachusetts Institute of Technology, and for the second I went to the laboratory of Arturo Rosenbluth, who was professor of neuropsychiatry in Mexico. After that year, I thought I can dare to start a new laboratory.

I called it the Biological Computer Laboratory. Initially it consisted of just a little seed money from the people who supported the Electron Tube Research Lab. But as soon as I could, I invited people from all over the world who could help me develop this laboratory: McCullough and Jerome Lettvin visited from MIT; Gordon Pask, from England, who developed some of the first true teaching machines, became a regular visitor; W. Ross Ashby, a British neurophysiologist, joined the laboratory and the faculty at Illinois; and many others visited for longer or shorter periods. We dealt with many interesting topics.

One of the notions that was then just coming up was that of self-organization. For example, we know that as the nervous system develops, it becomes more reliable, more refined. But the components, the neurons, remain the same. They fire sometimes; they don't fire sometimes. McCullough asked, how is such a system organized or how does it organize itself, that it becomes a highly reliable system? In the same vein, Pask considered the teaching and learning situation: Here come two people, one knows something, the other one doesn't know something. They go together, sit in a room for 2 or 3 days at Harvard or some other fancy place, and suddenly the one who doesn't know anything knows something. The change does not come from the outside. The system—the two people—changes in such a way as to distribute the knowledge between the two. The system organizes itself.

What is order? Order was usually considered as a wonderful building, a loss of uncertainty. Typically it means that if a system is so constructed that if you know the location or the property of one element, you can make conclusions about the other elements. So order is essentially the arrival of redundancy in a system, a reduction of possibilities.

Another notion we concerned ourselves with was, of course, memory. But at that time, in the 1950s, I was at odds with most of my colleagues who were dealing with memory. As I see it, memory for biological systems cannot be dead storage of isolated data but must be a dynamic process that involves the whole system in computing what is going on at the moment and what may happen in the future. The mind does not have a particular section for memory, another for counting, and another for language. The whole system has memory, can count, can add, can write papers, and so forth.

The notion of memory as simply a dead storage will not work for biological systems for two reasons.

Assume we have stored everything that we have experienced, in a picture catalog or a book or whatever, how do you find the experiences when you need them? When you see Uncle Joe, is there a little demon which zaps through the storage system in your head finding the proper picture from six months ago so you can say: "Hi, Uncle Joe"? The demon itself would have to have a memory of where it had stored things and what Uncle Joe means. So the problem becomes the memory of the demon. This demon, or any biological system, should have no need to store the past. It will never meet what happened in the past; instead, it needs to know what it can expect in the future. You need to judge at the moment what your actions should be so that you don't drop off a cliff or get eaten by a tiger, or get poisoned by putting the wrong food in your mouth. You need memory, but you need hindsight and foresight as well.

A second problem with the idea of memory as storage is that it ignores the whole notion of semantics. If you deliver things into storage, you would like to have exactly the same thing coming out. You bring a fur coat to the furrier in the summer and say: "Here is my mink coat. I would like to pick it up in the fall." If it's stored nicely, then you will get the mink coat when you come back in the fall. You will not get a sheet of paper that says: "Your mink coat." You get a mink coat. On the other hand, if somebody asks me: "What did you eat on your flight from New York?" I don't present scrambled eggs, I say "scrambled eggs," and everybody knows what I had for lunch. But a storage memory would have to produce the scrambled eggs, which I would not like to do at all in a conversation.

Another important theme for me has ben the clarification of terms. We have to make a clear distinction between the language with which we speak about computers and the language we are speaking about neurobiological systems. For instance, there is a distinction between input and stimulus in biological systems that is important to point out. For example, consider Pavlov with his famous experiment: you show a dog a piece of meat while you ring a bell, and the dog salivates; after a week, you only ring the bell and the dog salivates because he has been conditioned, as one says, to take

the independent stimulus for the meat. But some time later, a Polish experimental psychologist repeated the Pavlov experiments with a slight variation: he took the clapper off of the bell so it could not ring, but the assistant
did not know that, so he stepped forward, took the bell ringing it-silence,
and the dog salivated. One could say that the ringing of the bell was a stimulus for Pavlov, but not for the dog.

Or consider the difference between subject and object, a theme that ultimately developed into my notions on constructivism. I am unhappy with
this discrimination between objective and subjective: How do I know the
objects? Where are they? Of course, I can reconfirm or establish a rich connection with an object by touching or by smelling it or talking about it,
and so I had the idea to make the object a representation of the activity or
behavior of the observer, instead of the passive being looked or just sitting
there.

These ideas and questions have been the stimulus for many investigations, both at the Biological Computer Laboratory and later, and the papers
in this collection are built on them.

Heinz von Foerster
Pescadero, California
December 2001

Contents

1
On Self-Organizing Systems and Their Environments*

H. VON FOERSTER

Department of Electrical Engineering, University of Illinois, Urbana, Illinois

I AM somewhat hesitant to make the introductory remarks of my presenta-tion, because I am afraid I may hurt the feelings of those who so generously sponsored this conference on self-organizing systems. On the other hand, I believe, I may have a suggestion on how to answer Dr. Weyl's question which he asked in his pertinent and thought-provoking introduction: "What makes a self-organizing system?" Thus, I hope you will forgive me if I open my paper by presenting the following thesis: "There are no such things as self-organizing systems!"

In the face of the title of this conference I have to give a rather strong proof of this thesis, a task which may not be at all too difficult, if there is not a secret purpose behind this meeting to promote a conspiracy to dispose of the Second Law of Thermodynamics. I shall now prove the non-existence of self-organizing systems by *reductio ad absurdum* of the assumption that there is such a thing as a self-organizing system.

Assume a finite universe, U_0, as small or as large as you wish (see Fig. 1*a*), which is enclosed in an adiabatic shell which separates this finite uni-verse from any "meta-universe" in which it may be immersed. Assume, fur-thermore, that in this universe, U_0, there is a closed surface which divides this universe into two mutually exclusive parts: the one part is completely occupied with a self-organizing system S_0, while the other part we may call the environment E_0 of this self-organizing system: S_0 & $E_0 = U_0$.

I may add that it is irrelevant whether we have our self-organizing system inside or outside the closed surface. However, in Fig. 1 the system is assumed to occupy the interior of the dividing surface.

Undoubtedly, if this self-organizing system is permitted to do its job of organizing itself for a little while, its entropy must have decreased during this time:

* This article is an adaptation of an address given at The Interdisciplinary Sympo-sium on Self-Organizing Systems, on May 5, 1959, in Chicago, Illinois; originally pub-lished in *Self-Organizing Systems*. M.C. Yovits and S. Cameron (eds.), Pergamon Press, London, pp. 31–50 (1960).

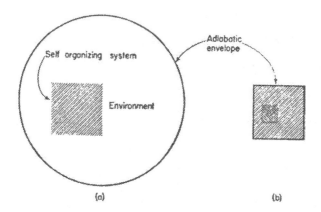

FIGURE 1.

$$\frac{\delta S_s}{\delta t} < 0,$$

otherwise we would not call it a self-organizing system, but just a mechanical $\delta S_s/\delta t = 0$, or a thermodynamical $\delta S_s/\delta t > 0$ system. In order to accomplish this, the entropy in the remaining part of our finite universe, i.e. the entropy in the environment must have increased

$$\frac{\delta S_E}{\delta t} > 0,$$

otherwise the Second Law of Thermodynamics is violated. If now some of the processes which contributed to the decrease of entropy of the system are irreversible we will find the entropy of the universe U_0 at a higher level than before our system started to organize itself, hence the state of the universe will be more disorganized than before $\delta S_U/\delta t > 0$, in other words, the activity of the system was a disorganizing one, and we may justly call such a system a "disorganizing system."

However, it may be argued that it is unfair to the system to make it responsible for changes in the whole universe and that this apparent inconsistency came about by not only paying attention to the system proper but also including into the consideration the environment of the system. By drawing too large an adiabatic envelope one may include processes not at all relevant to this argument. All right then, let us have the adiabatic envelope coincide with the closed surface which previously separated the system from its environment (Fig. 1b). This step will not only invalidate the above argument, but will also enable me to show that if one assumes that this envelope contains the self-organizing system proper, this system turns out to be not only just a disorganizing system but even a self-disorganizing system.

It is clear from my previous example with the large envelope, that here too—if irreversible processes should occur—the entropy of the system now within the envelope must increase, hence, as time goes on, the system would disorganize itself, although in certain regions the entropy may indeed have decreased. One may now insist that we should have wrapped our envelope just around this region, since it appears to be the proper self-organizing part of our system. But again, I could employ that same argument as before, only to a smaller region, and so we could go on for ever, until our would-be self-organizing system has vanished into the eternal happy hunting grounds of the infinitesimal.

In spite of this suggested proof of the non-existence of self-organizing systems, I propose to continue the use of the term "self-organizing system," whilst being aware of the fact that this term becomes meaningless, unless the system is in close contact with an environment, *which posseses available energy and order*, and with which our system is in a state of perpetual interaction, such that it somehow manages to "live" on the expenses of this environment.

Although I shall not go into the details of the interesting discussion of the energy flow from the environment into the system and out again, I may briefly mention the two different schools of thought associated with this problem, namely, the one which considers energy flow and signal flow as a strongly linked, single-channel affair (i.e. the message carries also the food, or, if you wish, signal and food are synonymous) while the other viewpoint carefully separates these two, although there exists in this theory a significant interdependence between signal flow and energy availability.

I confess that I do belong to the latter school of thought and I am particularly happy that later in this meeting Mr. Pask, in his paper *The Natural History of Networks*,[2] will make this point of view much clearer than I will ever be able to do.

What interests me particularly at this moment is not so much the energy from the environment which is digested by the system, but its utilization of environmental order. In other words, the question I would like to answer is: "How much order can our system assimilate from its environment, if any at all?"

Before tackling this question, I have to take two more hurdles, both of which represent problems concerned with the environment. Since you have undoubtedly observed that in my philosophy about self-organizing systems the environment of such systems is a *conditio sine qua non* I am first of all obliged to show in which sense we may talk about the existence of such an environment. Second, I have to show that, if there exists such an environment, it must possess structure.

The first problem I am going to eliminate is perhaps one of the oldest philosophical problems with which mankind has had to live. This problem arises when we, men, consider ourselves to be self-organizing systems. We may insist that introspection does not permit us to decide whether the world as we see it is "real," or just a phantasmagory, a dream, an illusion of our fancy. A decision in this dilemma is in so far pertinent to my discussion, since—if the latter alternative should hold true—my original thesis asserting the nonsensicality of the conception of an isolated self-organizing system would pitiably collapse.

I shall now proceed to show the reality of the world as we see it, by *reductio ad absurdum* of the thesis: this world is only in our imagination and the only reality is the imagining "I".

Thanks to the artistic assistance of Mr. Pask who so beautifully illustrated this and some of my later assertions,* it will be easy for me to develop my argument.

Assume for the moment that I am the successful business man with the bowler hat in Fig. 2, and I insist that I am the sole reality, while everything else appears only in my imagination. I cannot deny that in my imagination there will appear people, scientists, other successful businessmen, etc., as for instance in this conference. Since I find these apparitions in many respects similar to myself, I have to grant them the privilege that they themselves may insist that they are the sole reality and everything else is only a concoction of their imagination. On the other hand, they cannot deny that their fantasies will be populated by people—and one of them may be I, with bowler hat and everything!

With this we have closed the circle of our contradiction: If I assume that I am the sole reality, it turns out that I am the imagination of somebody else, who in turn assumes that *he* is the sole reality. Of course, this paradox is easily resolved, by postulating the reality of the world in which we happily thrive.

Having re-established reality, it may be interesting to note that reality appears as a consistent reference frame for at least two observers. This becomes particularly transparent, if it is realized that my "proof" was exactly modeled after the "Principle of Relativity," which roughly states that, if a hypothesis which is applicable to a set of objects holds for one object and it holds for another object, then it holds for both objects simultaneously, the hypothesis is acceptable for all objects of the set. Written in terms of symbolic logic, we have:

$$(Ex)[H(a) \& H(x) \rightarrow H(a+x)] \rightarrow (x)H(x) \qquad (1)$$

Copernicus could have used this argument to his advantage, by pointing out that if we insist on a geocentric system, $[H(a)]$, the Venusians, e.g. could insist on a venucentric system $[(Hx)]$. But since we cannot be both, center and epicycloid at the same time $[H(a + x)]$, something must be wrong with a planetocentric system.

---

* Figures 2, 5 and 6.

FIGURE 2.

However, one should not overlook that the above expression, $R(H)$ is not a tautology, hence it must be a meaningful statement.* What it does, is to establish a way in which we may talk about the existence of an environment.

___

* This was observed by Wittgenstein,[6] although he applied this consideration to the principle of mathematical induction. However, the close relation between the induction and the relativity principle seems to be quite evident. I would even venture to say that the principle of mathematical induction is the relativity principle in number theory.

Before I can return to my original question of how much order a self-organizing system may assimilate from its environment, I have to show that there is some structure in our environment. This can be done very easily indeed, by pointing out that we are obviously not yet in the dreadful state of Boltzmann's "Heat-Death." Hence, presently still the entropy increases, which means that there must be some order—at least now—otherwise we could not lose it.

Let me briefly summarize the points I have made until now:

(1) By a self-organizing system I mean that part of a system that eats energy and order from its environment.
(2) There is a reality of the environment in a sense suggested by the acceptance of the principle of relativity.
(3) The environment has structure.

Let us now turn to our self-organzing systems. What we expect is that the systems are increasing their internal order. In order to describe this process, first, it would be nice if we would be able to define what we mean by "internal," and second, if we would have some measure of order.

The first problem arises whenever we have to deal with systems which do not come wrapped in a skin. In such cases, it is up to us to define the closed boundary of our system. But this may cause some trouble, because, if we specify a certain region in space as being intuitively the proper place to look for our self-organizing system, it may turn out that this region does not show self-organizing properties at all, and we are forced to make another choice, hoping for more luck this time. It is this kind of difficulty which is encountered, e.g., in connection with the problem of the "localization of functions" in the cerebral cortex.

Of course, we may turn the argument the other way around by saying that we define our boundary at any instant of time as being the envelope of that region in space which shows the desired increase in order. But here we run into some trouble again; because I do not know of any gad get which would indicate whether it is plugged into a self-*dis*organizing or self-organizing region, thus providing us with a sound operational definition.

Another difficulty may arise from the possibility that these self-organizing regions may not only constantly move in space and change in shape, they may appear and disappear spontaneously here and there, requiring the "ordometer" not only to follow these all-elusive systems, but also to sense the location of their formation.

With this little digression I only wanted to point out that we have to be very cautious in applying the word "inside" in this context, because, even if the position of the observer has been stated, he may have a tough time saying what he sees.

Let us now turn to the other point I mentioned before, namely, trying to find an adequate measure of order. It is my personal feeling that we wish to describe by this term two states of affairs. First, we may wish to account for apparent relationships between elements of a set which would impose some constraints as to the possible arrangements of the elements of this system. As the organization of the system grows, more and more of these relations should become apparent. Second, it seems to me that order has a relative connotation, rather than an absolute one, namely, with respect to the maximum disorder the elements of the set may be able to display. This suggests that it would be convenient if the measure of order would assume values between zero and unity, accounting in the first case for maximum disorder and, in the second case, for maximum order. This eliminates the choice of "neg-entropy" for a measure of order, because neg-entropy always assumes finite values for systems being in complete disorder. However, what Shannon[3] has defined as "redundancy" seems to be tailor-made for describing order as I like to think of it. Using Shannon's definition for redundancy we have:

$$R = 1 - \frac{H}{H_m} \tag{2}$$

whereby H/H_m is the ratio of the entropy H of an information source to the maximum value, H_m, it could have while still restricted to the same symbols. Shannon calls this ratio the "relative entropy." Clearly, this expression fulfills the requirements for a measure of order as I have listed them before. If the system is in its maximum disorder $H = H_m$, R becomes zero; while, if the elements of the system are arranged such that, given one element, the position of all other elements are determined, the entropy— or the degree of uncertainty—vanishes, and R becomes unity, indicating perfect order.

What we expect from a self-organizing system is, of course, that, given some initial value of order in the system, this order is going to increase as time goes on. With our expression (2) we can at once state the criterion for a system to be self-organizing, namely, that the rate of change of R should be positive:

$$\frac{\delta R}{\delta t} > 0 \tag{3}$$

Differentiating eq. (2) with respect to time and using the inequality (3) we have:

$$\frac{\delta R}{\delta t} = -\frac{H_m(\delta H/\delta t) - H(\delta H_m/\delta t)}{H_m^2} \tag{4}$$

Since $H_m^2 > 0$, under all conditions (unless we start out with systems which can only be thought of as being always in perfect order: $H_m = 0$), we find the condition for a system to be self-organzing expressed in terms of entropies:

$$H \frac{\delta H_m}{\delta t} > H_m \frac{\delta H}{\delta t} \qquad (5)$$

In order to see the significance of this equation let me first briefly discuss two special cases, namely those, where in each case one of the two terms H, H_m is assumed to remain constant.

(a) $H_m = $ const.

Let us first consider the case, where H_m, the maximum possible entropy of the system remains constant, because it is the case which is usually visualized when we talk about self-organzing systems. If H_m is supposed to be constant the time derivative of H_m vanishes, and we have from eq. (5):

for $\qquad\qquad \frac{\delta H_m}{\delta t} = 0 \cdots \cdots \frac{\delta H}{\delta t} < 0 \qquad (6)$

This equation simply says that, when time goes on, the entropy of the system should decrease. We knew this already—but now we may ask, how can this be accomplished? Since the entropy of the system is dependent upon the probability distribution of the elements to be found in certain distinguishable states, it is clear that this probability distribution must change such that H is reduced. We may visualize this, and how this can be accomplished, by paying attention to the factors which determine the probability distribution. One of these factors could be that our elements possess certain properties which would make it more or less likely that an element is found to be in a certain state. Assume, for instance, the state under consideration is "to be in a hole of a certain size." The probability of elements with sizes larger than the hole to be found in this state is clearly zero. Hence, if the elements are slowly blown up like little balloons, the probability distribution will constantly change. Another factor influencing the probability distribution could be that our elements possess some other properties which determine the conditional probabilities of an elements to be found in certain states, given the state of other elements in this system. Again, a change in these conditional probabilities will change the probability distribution, hence the entropy of the system. Since all these changes take place internally I'm going to make an "internal demon" responsible for these changes. He is the one, e.g. being busy blowing up the little balloons and thus changing the probability distribution, or shifting conditional probabilities by establishing ties between elements such that H is going to decrease. Since we have some familiarity with the task of this demon, I shall leave him for a moment and turn now to another one, by discussing the second special case I mentioned before, namely, where H is supposed to remain constant.

(b) H = const.

If the entropy of the system is supposed to remain constant, its time derivative will vanish and we will have from eq. (5)

$$\text{for} \qquad \frac{\delta H}{\delta t} = 0 \cdots\cdots \frac{\delta H_m}{\delta t} > 0 \qquad\qquad (7)$$

Thus, we obtain the peculiar result that, according to our previous definition of order, we may have a self-organizing system before us, if its possible maximum disorder is increasing. At first glance, it seems that to achieve this may turn out to be a rather trivial affair, because one can easily imagine simple processes where this condition is fulfilled. Take as a simple example a system composed of N elements which are capable of assuming certain observable states. In most cases a probability distribution for the number of elements in these states can be worked out such that H is maximized and an expression for H_m is obtained. Due to the fact that entropy (or, amount of information) is linked with the logarithm of the probabilities, it is not too difficult to show that expressions for H_m usually follow the general form*:

$$H_m = C_1 + C_2 \log_2 N.$$

This suggests immediately a way of increasing H_m, namely, by just increasing the number of elements constituting the system; in other words a system that grows by incorporating new elements will increase its maximum entropy and, since this fulfills the criterion for a system to be self-organizing (eq. 7), we must, by all fairness, recognize this system as a member of the distinguished family of self-organizing systems.

It may be argued that if just adding elements to a system makes this a self-organizing system, pouring sand into a bucket would make the bucket a self-organizing system. Somehow—to put it mildly—this does not seem to comply with our intuitive esteem for members of our distinguished family. And rightly so, because this argument ignores the premise under which this statement was derived, namely, that during the process of adding new elements to the system the entropy H of the system is to be kept constant. In the case of the bucket full of sand, this might be a ticklish task, which may conceivably be accomplished, e.g. by placing the newly admitted particles precisely in the same order with respect to some distinguishable states, say position, direction, etc. as those present at the instant of admission of the newcomers. Clearly, this task of increasing H_m by keeping H constant asks for superhuman skills and thus we may employ another demon whom I shall call the "external demon," and whose business it is to admit to the system only those elements, the state of which complies with the conditions of, at least, constant internal entropy. As you certainly have noticed, this demon is a close relative of Maxwell's demon, only that to-day

---

* See also Appendix.

these fellows don't come as good as they used to come, because before 1927[4] they could watch an arbitrary small hole through which the newcomer had to pass and could test with arbitrary high accuracy his momentum. Today, however, demons watching closely a given hole would be unable to make a reliable momentum test, and vice versa. They are, alas, restricted by Heisenberg's uncertainty principle.

Having discussed the two special cases where in each case only one demon is at work while the other one is chained, I shall now briefly describe the general situation where both demons are free to move, thus turning to our general eq. (5) which expressed the criterion for a system to be self-organizing in terms of the two entropies H and H_m. For convenience this equation may be repeated here, indicating at the same time the assignments for the two demons D_i and D_e:

$$H \times \frac{\delta H_m}{\delta t} > H_m \times \frac{\delta H}{\delta t} \tag{5}$$

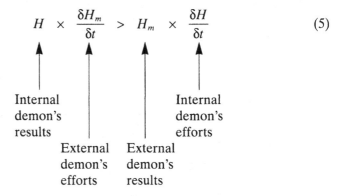

| Internal demon's results | | | Internal demon's efforts |
| | External demon's efforts | External demon's results | |

From this equation we can now easily see that, if the two demons are permitted to work together, they will have a disproportionately easier life compared to when they were forced to work alone. First, it is not necessary that D_i is always decreasing the instantaneous entropy H, or D_e is always increasing the maximum possible entropy H_m; it is only necessary that the product of D_i's results with D_e's efforts is larger than the product of D_e's results with D_i's efforts. Second, if either H or H_m is large, D_e or D_i respectively can take it easy, because their efforts will be multiplied by the appropriate factors. This shows, in a relevant way, the interdependence of these demons. Because, if D_i was very busy in building up a large H, D_e can afford to be lazy, because his efforts will be multiplied by D_i's results, and vice versa. On the other hand, if D_e remains lazy too long, D_i will have nothing to build on and his output will diminish, forcing D_e to resume his activity lest the system ceases to be a self-organizing system.

In addition to this entropic coupling of the two demons, there is also an energetic interaction between the two which is caused by the energy requirements of the internal demon who is supposed to accomplish the shifts in the probability distribution of the elements comprising the system. This requires some energy, as we may remember from our previous example, where somebody has to blow up the little balloons. Since this energy has been taken from the environment, it will affect the activities of the external demon who may be confronted with a problem when he attempts to supply the system with choice-entropy he must gather from an energetically depleted environment.

In concluding the brief exposition of my demonology, a simple diagram may illustrate the double linkage between the internal and the external demon which makes them entropically (H) and energetically (E) interdependent.

For anyone who wants to approach this subject from the point of view of a physicist, and who is conditioned to think in terms of thermodynamics and statistical mechanics, it is impossible not to refer to the beautiful little monograph by Erwin Schrodinger *What is Life*.[5] Those of you who are familiar with this book may remember that Schrodinger admires particularly two remarkable features of living organisms. One is the incredible high order of the genes, the "hereditary code-scripts" as he calls them, and the other one is the marvelous stability of these organized units whose delicate structures remain almost untouched despite their exposure to thermal agitation by being immersed—e.g. in the case of mammals—into a thermostat, set to about 310°K.

In the course of his absorbing discussion, Schrodinger draws our attention to two different basic "mechanisms" by which orderly events can be produced: "The statistical mechanism which produces order from disorder and the . . . [other] one producing 'order from order'."

While the former mechanism, the "order from disorder" principle is merely referring to "statistical laws" or, as Schrodinger puts it, to "the magnificent order of exact physical law coming forth from atomic and molecular disorder," the latter mechanism, the "order from order" principle is, again in his words: "the real clue to the understanding of life." Already earlier in his book Schrodinger develops this principle very clearly and states: "What an organism feeds upon is negative entropy." I think my demons would agree with this, and I do too.

However, by reading recently through Schrodinger's booklet I wondered how it could happen that his keen eyes escaped what I would consider a "second clue" to the understanding of life, or—if it is fair to say—of self-organizing systems. Although the principle I have in mind may, at first glance, be mistaken for Schrodinger's "order from disorder" principle, it has in fact nothing in common with it. Hence, in order to stress the difference between the two, I shall call the principle I am going to introduce to you presently the "order from noise" principle. Thus, in my restaurant self-organizing systems do not only feed upon order, they will also find noise on the menu.

Let me briefly explain what I mean by saying that a self-organizing system feeds upon noise by using an almost trivial, but nevertheless amusing example.

Assume I get myself a large sheet of permanent magnetic material which is strongly magnetized perpendicular to the surface, and I cut from this sheet a large number of little squares (Fig. 3a). These little squares I glue

FIGURE 3. (*a*) Magnetized square. (*b*) Cube, family I.

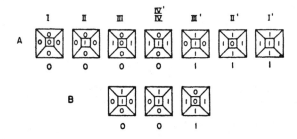

FIGURE 4. Ten different families of cubes (see text).

to all the surfaces of small cubes made of light, unmagnetic material, having the same size as my squares (Fig. 3*b*). Depending upon the choice of which sides of the cubes have the magnetic north pole pointing to the outside (Family I), one can produce precisely ten different families of cubes as indicated in Fig. 4.

Suppose now I take a large number of cubes, say, of family I, which is characterized by all sides having north poles pointing to the outside (or family I' with all south poles), put them into a large box which is also filled with tiny glass pebbles in order to make these cubes float under friction and start shaking this box. Certainly, nothing very striking is going to happen: since the cubes are all repelling each other, they will tend to distribute themselves in the available space such that none of them will come too close to its fellow-cube. If, by putting the cubes into the box, no particular ordering principle was observed, the entropy of the system will remain constant, or, at worst, increase a small amount.

In order to make this game a little more amusing, suppose now I collect a population of cubes where only half of the elements are again members belonging to family I (or I') while the other half are members of family II (or II′) which is characterized by having only one side of different magnetism pointing to the outside. If this population is put into my box and I go on shaking, clearly, those cubes with the single different pole pointing to the outside will tend, with overwhelming probability, to mate with members of the other family, until my cubes have almost all paired up. Since the conditional probabilities of finding a member of family II, given the locus of a member of family I, has very much increased, the entropy of the system has gone down, hence we have more order after the shaking than before. It is easy to show* that in this case the amount of order in our system went up from zero to

$$R_\infty = \frac{1}{\log_2(en)},$$

if one started out with a population density of n cubes per unit volume.

I grant you, that this increase in orderliness is not impressive at all, particularly if the population density is high. All right then, let's take a population made up entirely of members belonging to family IVB, which is characterized by opposite polarity of the two pairs of those three sides which join in two opposite corners. I put these cubes into my box and you shake it. After some time we open the box and, instead of seeing a heap of cubes piled up somewhere in the box (Fig. 5), you may not believe your eyes, but an incredibly ordered structure will emerge, which, I fancy, may pass the grade to be displayed in an exhibition of surrealistic art (Fig. 6).

If I would have left you ignorant with respect to my magnetic-surface trick and you would ask me, what is it that put these cubes into this remarkable order, I would keep a straight face and would answer: The shaking, of course—and some little demons in the box.

With this example, I hope, I have sufficiently illustrated the principle I called "order from noise," because no order was fed to the system, just cheap undirected energy; however, thanks to the little demons in the box, in the long run only those components of the noise were selected which contributed to the increase of order in the system. The occurrence of a mutation e.g. would be a pertinent analogy in the case of gametes being the systems of consideration.

Hence, I would name two mechanisms as important clues to the understanding of self-organizing systems, one we may call the "order from order" principle as Schrodinger suggested, and the other one the "order from noise" principle, both of which require the co-operation of our demons who are created along with the elements of our system, being manifest in some of the intrinsic structural properties of these elements.

I may be accused of having presented an almost trivial case in the attempt to develop my order from noise principle. I agree. However, I am convinced

---

* See Appendix.

FIGURE 5. Before.

that I would maintain a much stronger position, if I would not have given away my neat little trick with the magnetized surfaces. Thus, I am very grateful to the sponsors of this conference that they invited Dr. Auerbach[6] who later in this meeting will tell us about his beautiful experiments *in vitro* of the reorganization of cells into predetermined organs after the cells have been completely separated and mixed. If Dr. Auerbach happens to know the trick by which this is accomplished, I hope he does not give it away. Because, if he would remain silent, I could recover my thesis that without having some knowledge of the mechanisms involved, my example was not too trivial after all, and self-organizing systems still remain miraculous things.

Appendix

The entropy of a system of given size consisting of N indistinguishable elements will be computed taking only the spatial distribution of elements into consideration. We start by subdividing the space into Z cells of equal size

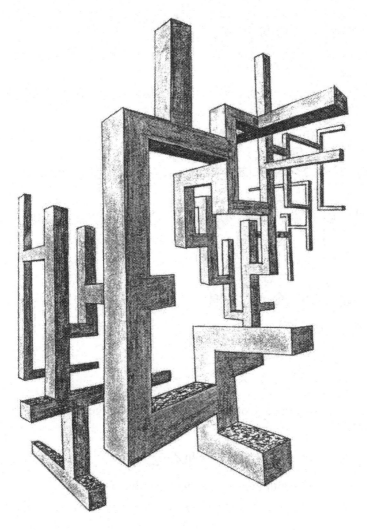

FIGURE 6. After.

and count the number of cells Z_i lodging i elements (see Fig. 7a). Clearly we have

$$\sum Z_i = Z \qquad \text{(i)}$$

$$\sum i Z_i = N \qquad \text{(ii)}$$

The number of distinguishable variations of having a different number of elements in the cells is

$$P = \frac{Z!}{\prod Z_i!} \qquad \text{(iii)}$$

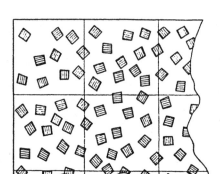

FIGURE 7.

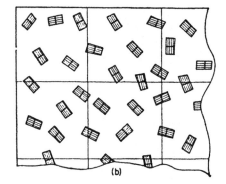

whence we obtain the entropy of the system for a large number of cells and elements:

$$H = \ln P = Z \ln Z - \sum Z_i \ln Z_i \qquad \text{(iv)}$$

In the case of maximum entropy $\overline{H}$ we must have

$$\delta H = 0 \qquad \text{(v)}$$

observing also the conditions expressed in eqs. (i) and (ii). Applying the method of the Lagrange multipliers we have from (iv) and (v) with (i) and (ii):

$$\sum (\ln Z_i + 1)\delta Z_i = 0$$
$$\sum i\delta Z_i = 0 \qquad \qquad \left| \begin{array}{l} \beta \\ \end{array} \right.$$
$$\sum \delta Z_i = 0 \qquad \qquad \left| \begin{array}{l} -(1 + \ln \alpha) \end{array} \right.$$

multiplying with the factors indicated and summing up the three equations we note that this sum vanishes if each term vanishes identically. Hence:

$$\ln Z_i + 1 + i\beta - 1 - \ln \alpha = 0 \qquad \text{(vi)}$$

whence we obtain that distribution which maximizes H:

$$Z_i = \alpha \, e^{-i\beta} \tag{vii}$$

The two undetermined multipliers α and β can be evaluated from eqs. (i) and (ii):

$$\alpha \sum e^{-i\beta} = Z \tag{viii}$$

$$\alpha \sum i e^{-i\beta} = N \tag{ix}$$

Remembering that

$$-\frac{\delta}{\delta\beta} \sum e^{-i\beta} = \sum i e^{-i\beta}$$

we obtain from (viii) and (ix) after some manipulation:

$$\alpha = Z(1 - e^{-1/n}) \approx \frac{Z}{n} \tag{x}$$

$$\beta = \ln\left(1 + \frac{1}{n}\right) \approx \frac{1}{n} \tag{xi}$$

where n, the mean cell population or density N/Z is assumed to be large in order to obtain the simple approximations. In other words, cells are assumed to be large enough to lodge plenty of elements.

Having determined the multipliers α and β, we have arrived at the most probable distribution which, after eq. (vii) now reads:

$$Z_i = \frac{Z}{n} e^{-i/n} \tag{xii}$$

From eq. (iv) we obtain at once the maximum entropy:

$$\overline{H} = Z \ln(en). \tag{xiii}$$

Clearly, if the elements are supposed to be able to fuse into pairs (Fig. 7b), we have

$$\overline{H}' = Z \ln(en/2). \tag{xiv}$$

Equating $\overline{H}$ with H_m and $\overline{H}'$ with H, we have for the amount of order after fusion:

$$R = 1 - \frac{Z\ln(en)}{Z\ln(en/2)} = \frac{1}{\log_2(en)} \tag{xv}$$

References

1. L. Wittgenstein, *Tractatus Logico-Philosophicus*, paragraph 6.31, Humanities Publishing House, New York (1956).

2. G. A. Pask, The natural history of networks. This volume, p. 232.
3. C. Shannon and W. Weaver, *The Mathematical Theory of Communication*, p. 25, University of Illinois Press, Urbana, Illinois (1949).
4. W. Heisenberg, *Z. Phys.* **43**, 172 (1972).
5. E. Shrodinger, *What is Life?* pp. 72, 80, 80, 82, Macmillan, New York (1947).
6. R. Auerbach, Organization and reorganization of embryonic cells. This volume, p. 101.

Discussion

Lederman (*University of Chicago*): I wonder if it is true that in your definition of order you are really aiming at conditional probabilities rather than just an order in a given system, because for a given number of elements in your system, under your definition of order, the order would be higher in a system in which the information content was actually smaller than for other systems.

von Foerster: Perfectly right. What I tried to do here in setting a measure of order, was by suggesting redundancy as a measure. It is easy to handle. From this I can derive two statements with respect to H_{max} and with respect to H. Of course, I don't mean this is a universal interpretation of order in general. It is only a suggestion which may be useful or may not be useful.

Lederman: I think it is a good suggestion but it is an especially good suggestion if you think of it in terms of some sort of conditional probability. It would be more meaningful if you think of the conditional probabilities as changing so that one of the elements is singled out for a given environmental state as a high probability.

von Foerster: Yes, if you change H, there are several ways one can do it. One can change the conditional probability. One can change also the probability distribution which is perhaps easier. That is perfectly correct.

Now the question is, of course, in which way can this be achieved? It can be achieved, I think, if there is some internal structure of those entities which are to be organized.

Lederman: I believe you can achieve that result from your original mathematical statement of the problem in terms of H and H_{max}, in the sense that you can increase the order of your system by decreasing the noise in the system which increases H_{max}.

Won Foerster: That is right. But there is the possibility that we will not be able to go beyond a certain level. On the other hand, I think it is favorable to have some noise in the system. If a system is going to freeze into a particular state, it is inadaptable and this final state may be altogether wrong. It will be incapable of adjusting itself to something that is a more appropriate situation.

Lederman: That is right, but I think the parallelism between your mathematical approach and the model you gave in terms of the magnets organizing themselves, that in the mathematical approach you can increase the information content of the system by decreasing the noise and similarly in your system where you saw the magnets organizing themselves into some sort of structure you were also decreasing the noise in the system before you reached the point where you could say ah ha, there is order in that system.

von Foerster: Yes, that is right.

Mayo (*Loyola University*): How can noise contribute to human learning? Isn't noise equivalent to nonsense?

von Foerster: Oh, absolutely, yes. (Laughter). Well, the distinction between noise and nonsense, of course, is a very interesting one. It is referring usually to a reference frame. I believe that, for instance, if you would like to teach a dog, it would be advisable not only to do one and the same thing over and over again. I think what should be done in teaching or training, say, an animal, is to allow the system to remain adaptable, to ingrain the information in a way where the system has to test in every particular situation a hypothesis whether it is working or not. This can only be obtained if the nature into which the system is immersed is not absolutely deterministic but has some fluctuations. These fluctuations can be interpreted in many different forms. They can be interpreted as noise, as nonsense, as things depending upon the particular frame of reference we talk about.

For instance, when I am teaching a class, and I want to have something remembered by the students particularly well, I usually come up with an error and they point out, "You made an error, sir." I say, "Oh yes, I made an error," but they remember this much better than if I would not have made an error. And that is why I am convinced that an environment with a reasonable amount of noise may not be too bad if you would really like to achieve learning.

Reid (*Montreal Neurological Institute*): I would like to hear Dr. von Foerster's comment on the thermodynamics of self-organizing systems.

von Foerster: You didn't say open or closed systems. This is an extremely important question and a very interesting one and probably there should be a two-year course on the thermodynamics of self-organizing systems. I think Prigogin and others have approached the open system problem. I myself am very interested in many different angles of the thermodynamics of self-organizing systems because it is a completely new field.

If your system contains only a thousand, ten thousand or a hundred thousand particles, one runs into difficulties with the definition of temperature. For instance, in a chromosome or a gene, you may have a complex molecule involving about 10^6 particles. Now, how valid is the thermodynamics of 10^6 particles or the theory which was originally developed for 10^{23} particles? If this reduction of about 10^{17} is valid in the sense that you can still talk about "temperature" there is one way you may talk about it. There is, of course, the approach to which you may switch, and that is information theory. However, there is one problem left and that is, you don't have a Boltzmann's constant in information theory and that is, alas, a major trouble.

2
Computation in Neural Nets*

HEINZ VON FOERSTER
Department of Electrical Engineering and Department of Biophysics, University of Illinois, Urbana, Illinois, USA

A mathematical apparatus is developed that deals with networks of elements which are connected to each other by well defined connection rules and which perform well defined operations on their inputs. The output of these elements either is transmitted to other elements in the network or—should they be terminal elements—represents the outcome of the computation of the network. The discussion is confined to such rules of connection between elements and their operational modalities as they appear to have anatomical and physiological counter parts in neural tissue. The great latitude given today in the interpretation of nervous activity with regard to what constitutes the "signal" is accounted for by giving the mathematical apparatus the necessary and sufficient latitude to cope with various interpretations. Special attention is given to a mathematical formulation of structural and functional properties of networks that compute invariants in the distribution of their stimuli.

1. Introduction

Ten neurons can be interconnected in precisely 1,267,650,500,228,229,401, 703,205,376 different ways. This count excludes the various ways in which each particular neuron may react to its afferent stimuli. Considering this fact, it will be appreciated that today we do not yet possess a general theory of neural nets of even modest complexity.

It is clear that any progress in our understanding of functional and structural properties of nerve nets must be based on the introduction of constraints into potentially hyper-astronomical variations of connecting pathways. These constraints may be introduced from a theoretical point of view, for reasons purely esthetic, in order to develop an "elegant" mathe-

* Reprinted from *BIOSYSTEMS* (formerly *Currents in Modern Biology*), Vol. 1, No. 1, 1967, pp. 47–93, with permission from Elsevier Science.

matical apparatus that deals with networks in general, or these constraints may be introduced by neurophysiological and neuroanatomical findings which uncover certain functional or structural details in some specific cases. It is tempting, but—alas—dangerous, to translate uncritically some of the theoretical results into physiological language even in cases of some undeniable correspondences between theory and experiment. The crux of this danger lies in the fact that the overall network response (NR) is uniquely determined by the connective structure (e) of the network elements and the transfer function (TF) of these elements, but the converse is not true. In other words, we have the following inference scheme:

$$[e, TF] \rightarrow NR$$
$$[e, NR] \rightarrow Class\,[TF]$$
$$[TF, NR] \rightarrow Class\,[e]$$

Since in most cases we have either some idea of structure and function of a particular network, or some evidence about the transfer function of the neurons of a network giving certain responses, we are left either with a whole class of "neurons" or with a whole class of structures that will match the observed responses. Bad though this may sound, it represents a considerable progress in reducing the sheer combinatorial possibilities mentioned before, and it is hoped that the following account of structure and function in nervous nets will at least escape the Scylla of empty generalities and the Charybdis of doubtful specificities.

The discussion of neural networks will be presented in three different chapters. The first chapter introduces some general notions of networks, irrespective of the "agent" that is transmitted over the connective paths from element to element, and irrespective of the operations that are supposedly carried out at the nodal elements of such generalized networks. This generality has the advantage that no commitments have to be made with respect to the adoption of certain functional properties of neurons, nor with respect to certain theories as to the code in which information is passed from neuron to neuron.

Since the overall behavior of neural networks depends to a strong degree on the operations carried out by its constituents, a second chapter discusses various modalities in the operation of these elements which may respond in a variety of ways from extremely non-linear behavior to simple algebraic summation of the input signal strength. Again no claims are made as to how a neuron "really" behaves, for this—alas—has as yet not been determined. However, the attempt is made to include as much of its known properties as will be necessary to discuss some of the prominent features of networks which filter and process the information that is decisive for the survival of the organism.

The last chapter represents a series of exercises in the application of the principles of connection and operation as discussed in the earlier chapters. It is hoped that the applicability of these concepts to various concrete cases may stimulate further investigations in this fascinating complex of problems whose surface we have barely begun to scratch.

2. General Properties of Networks

In this chapter we shall make some preliminary remarks about networks in general, keeping an eye, however, on our specific needs which will arise when dealing with the physiological situation. A generalized network concept that will suit our purposes involves a set of n "elements", e_1, e_2, $\ldots, e_i, \ldots, e_n$, and the set of all ordered pairs $[e_i, e_j]$ that can be formed with these elements. The term "ordered" refers to the distinction we wish to make between a pair, say $[e_1, e_2]$ and $[e_2, e_1]$. In general:

$$[e_i, e_j] \neq [e_j, e_i].$$

This distinction is dictated by our wish to discriminate between the two cases in which an as yet undefined "agent" is transmitted either from e_i to e_j or from e_j to e_i. Furthermore, we want to incorporate the case in which an element transmits this agent to itself. Hence, the pair $[e_i, e_i]$ is also a legitimate pair. Whenever such a transmission takes place between an ordered pair $[e_i, e_j]$ we say that e_i is "actively connected with", or "acts upon", or "influences" element e_j. This may be indicated by an oriented line (arrow) leading from e_i to e_j.

With these preliminaries our generalized network can be defined as a set of n elements e_i $(i = 1 \rightarrow n)$ each ordered pair of which may or may not be actively connected by an oriented line (arrow). Hence, the connectivity of a set of elements may—in an abstract sense—be represented by a two-valued function $C(e_i, e_j)$ whose arguments are all ordered pairs $[e_i, e_j]$, and whose values are 1 or 0 for actively connected or disconnected ordered pairs respectively.

A simple example of a net consisting of five elements is given in fig. 1a. Here, for instance, the ordered pair $[3, 5]$ appears to be actively disconnected, hence $C(3, 5) = 0$, while the commuted ordered pair $[5, 3]$ shows an active connection path. The function $C(e_i, e_j)$ is, of course, an equivalent representation of any such net structure ad may best be represented in the form of a quadratic matrix with n rows and n columns (fig. 1b). The rows carry the names of the transmitting elements e_i and the columns the names of the receiving elements. Active connection is indicated by inserting a "1" into the intersection of a transmitting row with a receiving column, otherwise a "0" is inserted. Hence, the active connection between elements e_5 and e_3, indicated as an arrow leading from 5 to 3 in fig. 1a, is represented by the matrix element $C_{5,3} = 1$ in row 5 column 3.

This matrix representation permits us at once to draw a variety of conclusions. First, we may obtain an expression for the number of distinguishable networks that can be constructed with n distinguishable elements. Since a connection matrix for n elements has n^2 entries, corresponding to

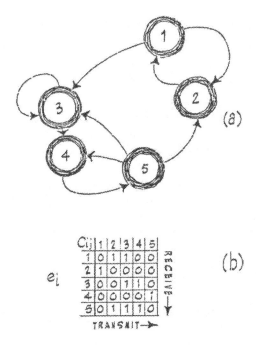

FIGURE 1. Network as directed graph (a) and its connection matrix (b).

the n^2 ordered pairs, and for each entry there are two choices, namely 0 and 1 for disconnection or active connection respectively, the number of ways in which "zeros" and "ones" can be distributed over n^2 entries is precisely

$$2^{n^2}.$$

For $n = 10$ we have $2^{100} \approx 10^{30}$ different nets and for $n = 100$ we must be prepared to deal with $2^{10000} \approx 10^{3000}$ different nets. To put the reader at ease, we promise not to explore these rich possibilities in an exhaustive manner.

We turn to another property of our connection matrix, which permits us to determine at a glance the "action field" and the "receptor field" of any particular element e_i in the whole network. We define the action field A_i of an element e_i by the set of all elements to which e_i is actively connected. These can be determined at once by going along the row e_i and noting the columns e_j which are designated by a "one". Consequently, the action field of element e_3 in fig. 1 is defined by

$$A_3 = [e_3, e_4].$$

Conversely, we define the receptor field R_i of element e_i by the set of all elements that act upon e_i. These elements can be determined at once by going down column e_i and noting the rows e_j which are designated by a "one". Consequently, the receptor field of element e_3 in fig. 1 is defined by

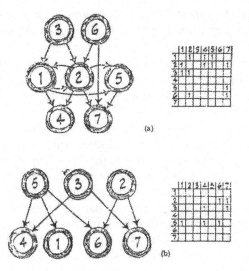

FIGURE 2. Hybrid network (a) and action network (b).

$$R_3 = [e_1, e_3, e_5].$$

Since the concepts of action field and receptor field will play an important role in the discussion of physiological nerve nets, it may be appropriate to note some special cases. Consider a network of n elements. Under certain circumstances it may be possible to divide these elements into three non-empty classes. One class, N_1, consists of all elements whose receptor field is empty; the second class, N_2, consists of all elements whose action field is empty; and the third class, C, consists of all elements for which neither the action field nor the receptor field is empty. A net for which these three classes are non-empty we shall call a "hybrid net". A net which is composed entirely of elements of the third class C we shall call an "interaction net". The net in fig. 1 represents such an interaction net. Finally we define an "action net" which does not possess an element that belongs to class C. Examples of a hybrid net and an action net are given in figs. 2a and 2b with their associated connection matrices.

In the net of fig. 2a:

$$N_1 = [e_3, e_6],$$
$$N_2 = [e_4, e_7],$$
$$C = [e_1, e_2, e_5].$$

In the net of fig. 2b:

$$N_1 = [e_2, e_3, e_5],$$
$$N_2 = [e_1, e_4, e_6, e_7],$$
$$C = [0].$$

For obvious reasons we shall call all elements belonging to class N_1 "generalized receptors" and all elements belonging to class N_2 "generalized effectors". The justification of this terminology may be derived from general usage which refers to a receptor as an element that is not stimulated by an element of its own network but rather by some outside agent. Likewise, an effector is usually thought of as an element that produces an effect outside of its own network.

This observation permits us to use hybrid networks or action networks as compound elements in networks that show some repetition in their connection scheme. An example is given in fig. 3 in which the net suggested in 3a is to be inserted into the nodes of the net indicated in 3b. The repetition of this process gives rise to the concept of periodic networks, features almost ubiquitous in the physiological situation. To expect such periodicity is not too far-fetched if one realizes for a moment that many net structures are genetically programmed. The maximum amount of information necessary to program a net of n elements is $H_n = n^2$. If this net is made up of k periods of n/k elements each, the maximum information required is only $H_{k,n} = k(n/k)^2 = n^2/k$. Consequently, periodicity—or redundancy—represents genetic economy.

Keeping this point in mind let us investigate further constraints in the structure of networks.

Consider for the moment an action net consisting of $n = 2m$ elements where the number of elements in set N_1, the generalized receptors, equals the number of elements in N_2, the generalized effectors. In this case the connection matrix has precisely half of its rows and columns empty (0), and the

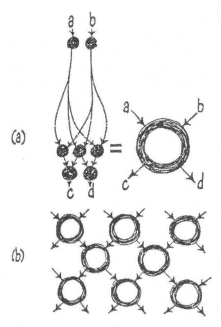

FIGURE 3. Replacement of elements by networkers. Periodic networks.

other half filled (1). This makes it possible to re-label all elements, letting the receptors as well as effectors run through labels $1 \to m$. Consequently, a new matrix can be set up, an "action matrix", which has precisely the same property as our old connection matrix, with the only difference that the elements of the effector set—which define again the columns—are labeled with the same indices as the receptor elements defining rows. Figs. 4a and 4b illustrate this transformation in a simple example.

The first advantage we may take of an action matrix is its possibility to give us an answer to the question whether or not several action nets in cascade may be replaced by a single action net; and if yes, what is the structure of this net?

The possibility of transforming a network into the form of an action-matrix has considerable advantages, because an action matrix has the properties of an algebraic square matrix, so the whole machinery of matrix manipulation that has been developed in this branch of mathematics can be applied to our network structures. Of the many possibilities that can be discussed in connection with matrix representation of networks, we shall give two examples to illustrate the power of this method.

In algebra a square matrix A_m of order m is a quadratic array of numbers arranged precisely according the pattern of our connection matrix, or our action matrix. The number found at the intersection of the ith row with the jth column is called element a_{ij}, which gives rise to another symbolism for writing a matrix:

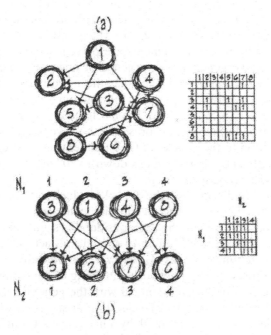

FIGURE 4. Network without feedback. Action net.

$$A_m = \|a_{ij}\|_m.$$

Addition and subtraction of two matrices of like order is simply carried out by adding or subtracting corresponding elements:

$$A_m \pm B_m = C_m = \|c_{ij}\|_m, \tag{1}$$

$$c_{ij} = a_{ij} \pm b_{ij}. \tag{2}$$

It is easy to see that matrix addition or subtraction corresponds to superposition or subposition in our networks. However, we should be prepared to obtain in some of the new entries numbers that are larger than unity or less than zero, if by change the network superimposed over an existent one has between two elements a connection that was there in the first place. Hence, the new entry will show a number "2" which is not permitted according to our old rules of representing connections in matrices (which admitted only "ones" and "zeros" as entries). We may, at our present state of insight, gracefully ignore this peculiarity by insisting that we cannot do more than connect, which gives a "one" in the matrix. Reluctantly, we may therefore adopt the rule that whenever matrix manipulation produces entries $c_{ij} > 1$, we shall substitute "1" and for entries $c_{ij} < 0$, we shall substitute "0". Nevertheless, this tour de force leaves us with an unsatisfactory aftertaste and we may look for another interpretation. Clearly, the numbers c_{ij} that will appear in the entries of the matrix indicate the numbers of parallel paths that connect element e_i with element e_j. In a situation where some "agent" is passed between these elements, this multiplicity of parallel pathways can easily be interpreted by assuming that a proportionate multiple of this agent is being passed between these elements. The present skeleton of our description of networks does not yet permit us to cope with this situation, simply because our elements are presently only symbolic blobs, indicating the convergence and divergence of lines, but incapable of any operations. However, it is significant that the mere manipulations of the concepts of our skeleton compel us to bestow our "elements" with more vitality than we were willing to grant them originally. We shall return to this point at the end of the chapter; presently, however, we shall adopt the pedestrian solution to the problem of multiple entries as suggested above, namely, by simply chopping all values down to "0" and "1" in accordance with our previous recommendations.

Having eliminated some of the scruples which otherwise may have spoiled unrestricted use of matrix calculus in dealing with our networks, we may now approach a problem that has considerable significance in the physiological case, namely, the treatment of cascades of action networks. By a cascade of two action networks A_m and B_m, symbolically represented by $Cas(AB)_m$ we simply define a network consisting of $3m$ elements, in which all general effectors of A_m are identical with the general receptors of B_m. Fig. 5a gives a simple example. The question arises as to whether or not such a cascade can be represented by an equivalent single action net.

"Equivalent" here means that a connecting pathway between a receptor in A and an effector in B should again be represented by a connection, and the same should hold for no connections.

The answer to this question is in the affirmative; the resulting action matrix C_m is the matrix product of A_m and B_m:

$$\|c_{ij}\|_m = \|a_{ij}\|_m \times \|b_{ij}\|_m, \tag{3}$$

where according to the rules of matrix multiplication the elements c_{ij} are defined by

$$c_{ij} = \sum_{k=1}^{m} a_{ik} b_{kj}. \tag{4}$$

Fig. 5b shows the transformation of the two cascaded nets into the single action net. Clearly, this process can be repeated over and over again, and we have

$$\mathrm{Cas}(A_1 A_2 A_3 A_4 \ldots A_k)_m = \prod_{1}^{k} A_{mi}. \tag{5}$$

Here we have one indication of the difficulty of establishing uniquely the receptor field of a particular element, because an observer who is aware of the presence of cascades would maintain that elements e_1 and e_3 in the second layer of fig. 5a constitute the receptor field of element e_2 in layer III, while an observer who is unaware of the intermediate layer (fig. 5b) will argue that elements e_1, e_2 and e_3 in the first layer define the receptor field of this element.

In passing, it may be pointed out that matrix multiplication preserves the multiplicity of pathways as seen in fig. 5, where in the cascaded system element e_2, bottom row, can be reached from e_3, top row, via e_1 as well as via e_3, middle row. All other connections are single-valued.

As a final example, we will apply an interesting result in matrix algebra to cascades of action networks. It can be shown that a square matrix whose rows are all alike

$$a_{ij} = a_{kj}, \tag{6}$$

and each row of which adds up to unity

$$\sum_{j=1}^{m} a_{ij} = 1 \tag{7}$$

generates the same matrix, when multiplied by itself:

$$\|a_{ij}\|_m \times \|a_{ij}\|_m = \|a_{ij}\|_m, \tag{8}$$

or

$$A_m^2 = A_m. \tag{9}$$

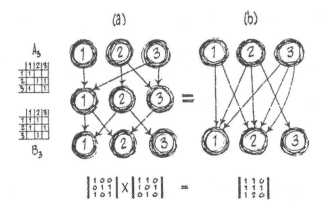

FIGURE 5. Cascade of action networks (a) and equivalent action network (b).

Translated into network language, this says that an action network with all receptors connected to the same effectors remains invariant when cascaded an arbitrary number of times. As an example, consider the action matrix

$$
A_6 = \begin{Vmatrix} 0 & 1 & 1 & 1 & 0 & 1 \\ 0 & 1 & 1 & 1 & 0 & 1 \\ 0 & 1 & 1 & 1 & 0 & 1 \\ 0 & 1 & 1 & 1 & 0 & 1 \\ 0 & 1 & 1 & 1 & 0 & 1 \\ 0 & 1 & 1 & 1 & 0 & 1 \end{Vmatrix} = 4 \times \begin{Vmatrix} 0 & \frac{1}{4} & \frac{1}{4} & \frac{1}{4} & 0 & \frac{1}{4} \\ 0 & \frac{1}{4} & \frac{1}{4} & \frac{1}{4} & 0 & \frac{1}{4} \\ 0 & \frac{1}{4} & \frac{1}{4} & \frac{1}{4} & 0 & \frac{1}{4} \\ 0 & \frac{1}{4} & \frac{1}{4} & \frac{1}{4} & 0 & \frac{1}{4} \\ 0 & \frac{1}{4} & \frac{1}{4} & \frac{1}{4} & 0 & \frac{1}{4} \\ 0 & \frac{1}{4} & \frac{1}{4} & \frac{1}{4} & 0 & \frac{1}{4} \end{Vmatrix},
$$

for which the equivalent network is represented in fig. 6. Call p the number of effectors contacted ($p = 4$ in fig. 6), and $N(A_m)$ the normalized matrix whose elements are

$$
n_{ij} = \frac{1}{p} a_{ij}. \tag{10}
$$

Clearly,

$$
\sum_{1}^{m} n_{ij} = 1,
$$

and k cascades give

$$
A_m^k = p^k N(A_m). \tag{11}
$$

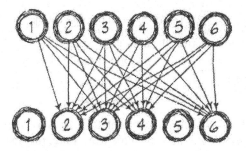

FIGURE 6. Stochastic action network.

Consequently, the multiplicity of connections will grow with p^{k-1}, while the connection scheme remains invariant.

This observation, which at this level may have the ring of triviality, will later prove to be of considerable utility when we consider variable amounts of an "agent" being passed on from element to element. This again requires a concept of what happens at the site of the elements, a question which leads us, of course, straight into the discussion of "What is a neuron"?

Before we attempt to tackle this quite difficult question—which will be approached in the next chapter—we owe our patient reader an explanation of the term "connection of two elements" for which we offered only a symbolic representation of an "oriented line", along which we occasionally passed a mysterious "agent" without even alluding to concrete entities which may be represented by these abstract concepts. We have reserved this discussion for the end of this chapter because a commitment to a particular interpretation of the term "connection" will immediately force us to make certain assumptions about some properties of our elements, and hence will lead us to the next chapter whose central theme is the discussion of precisely these properties. In our earlier remarks about networks in general we suggested that the statement "element e_i is actively connected to element e_j" may also be interpreted as "e_i acts upon e_j" or "influences e_j". This, of course, presupposes that each of our elements is capable of at least two states, otherwise even the best intentions of "influencing" may end in frustration*. Let us denote a particular state of element e_i by $S_i^{(\lambda)}$, where the superscript λ:

$$\lambda = 1, 2, 3, \ldots, s_j.$$

labels all states of element e_i, which is capable of assuming precisely s_i different states. In order to establish that element e_i may indeed have any influence on e_j we have to demand that there is at least one state of e_i that produces a state change in e_j within a prescribed interval of time, say Δt. This may be written symbolically

---

* An excellent account on finite state systems can be found in Ashby (1956).

$$\left[S_i^{(\lambda)}(t) \to S_i^{(\lambda')}(t + \Delta t)\right] = \Phi\left[S_i^{(\mu)}(t)\right]. \tag{12}$$

In this equation the function Φ relates the states $S_i(\mu)$ in e_i that produce a transition in e_j from $S_i(\lambda)$ to $S_i(\lambda')$. Consequently, Φ can be written in terms of superscripts only:

$$\lambda' = \Phi_{ij}(\lambda, \mu). \tag{13}$$

In other words Φ relates the subsequent state of e_j to its present state and to the present state of the acting element e_i. Take, for instance, $s_1 = s_2 = 3$; a hypothetical transition matrix may read as follows:

$\lambda' = \Phi(\lambda, \mu)$	μ 1 2 3
1	1 3 2
λ 2	1 2 3
3	1 1 2

The associated transition diagram is given in fig. 7, where the nodes represent the states of the reacting element, the arrows the transitions, and the labels on the arrows the states of the acting element which causes the corresponding transition. Inspection of the transition matrix of an element e_j will tell us at once whether or not another element, say e_i, is actively connected to e_j, because if there is not a single state in e_i that produces a state change in e_j we must conclude that e_i does not have any effect on e_j. Again for $s_1 = s_2 = 3$, the "transition" matrix for such an ineffective connection looks as follows

$\lambda' = \Phi(\lambda, \mu)$	μ 1 2 3
1	1 1 1
λ 2	2 2 2
3	3 3 3

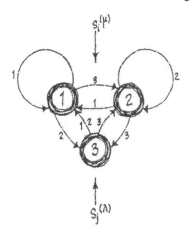

FIGURE 7. State transition diagram.

In general, if we have, for all μ:

$$\Phi_{ij}(\lambda,\mu) = \lambda,$$

we have in the connection matrix

$$c_{ij} = 0.$$

If we wish to establish whether or not an element e_i is actively connected to itself, we again set up a transition matrix, only replacing μ by λ. An example of the hypothetical transition matrix of a self-connected element, capable of three states:

$$
\begin{array}{cc|ccc}
 & & & \lambda & \\
\lambda' = \Phi\,(\lambda,\lambda) & & 1 & 2 & 3 \\
\hline\hline
 & 1 & 2 & - & - \\
\lambda & 2 & - & 1 & - \\
 & 3 & - & - & 3 \\
\end{array}
$$

This element oscillates between states 1 and 2 but stays calmly in 3 when in 3.

The state transition matrix that describes the action of, say, $(k-1)$ elements on some other element is, of course, of k dimensions. In this case the state labels for the ith element may be called λ_i. An example of such a matrix for two elements e_2, e_3, acting on e_1 is given in fig. 8.

If the states of our elements represent some physical variable which may undergo continuous changes, for instance, if these states represent the magnitudes of an electrical potential, or of a pressure, or of a pulse frequency, the symbol S_i itself may be taken to represent this magnitude and eq. (12), which described the state transitions of element e_j under the influence of element e_i, assumes now the form of a differential equation

$$\frac{dS_j}{dt} = \Phi_{ij}[S_i(t)], \tag{14}$$

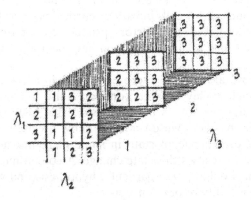

FIGURE 8. State transition matrix for three elements.

which can be solved if the time course of the activity of S_i is known. This, however, may depend on the state of elements acting upon e_i, which in turn may depend on states of elements acting upon those, etc. If we consider such a hierarchy of k levels we may eventually get:

$$\frac{dS_j}{dt} = \Phi_{ij}[\Psi_{hi}[\Psi_{gh}[\ldots S_o(t - k\Delta t)]]],$$

which is clearly a mess. Nevertheless, there are methods of solving this telescopic set of equations under certain simplifying assumptions, the most popular one being the assumption of linear dependencies.

This brief excursion into the conceptual machinery that permit us to manipulate the various states of individual elements was undertaken solely for the purpose of showing the close interdependence of the concepts of "active connection" and "elements". A crucial role in this analysis is played by the time interval Δt within which we expected some changes to take place in the reacting element as a consequence of some states of the acting element. Clearly, if we enlarge this time interval, say, to $2\Delta t$, $3\Delta t$, $4\Delta t$, . . . we shall catch more and more elements in a network which may eventually contribute to some changes of our element. This observation permits us to define "action neighbours" of the kth order, irrespective of their topographical neighborhood. Hence, in e_j we simply have an action neighbor of kth order for e_i if at least one of the states of e_j at time $t - k\Delta t$ causes a state transition in e_i at time t.

With these remarks about networks in general we are sufficiently prepared to deal with some structural properties of the networks whose operations we wish to discuss in our third chapter. In our outline of the structural skeleton of networks we kept abstract the two concepts "element" and "agent", for which we carefully avoided reference to concrete entities. However, the abstract framework of the interplay of these concepts permits us to interpret them according to our needs, taking for instance, "general receptors" for receptors proper (e.g., cones, rods, outer hair cells, Meissner's corpuscles, Krause's end-bulbs, Merkel's discs; or for intermediate relays receiving afferent information, bipolar cells, cells of the cochlear nucleus, or for cells in various cortical layers). "General effectors" may be interpreted as effectors proper (e.g., muscle fibers), and also as glia cells, which act in one way on neurons but in another way on each other.

Furthermore, we are free to interpret "agent" in a variety of ways, for instance, as a single volley on a neuron, as a pulse frequency, as a single burst of pulses, as pressure, as light intensity. This freedom is necessary, because in some instances we do not yet know precisely which physical property causes the change of state in some elements, nevertheless, we know which element causes this state change. A commitment to a particular interpretation would favor a particular hypothesis, and would thus mar the general applicability of our concepts.

Our next task is, of course, to give a description of the operational possibilities of our elements in order to put some life into the as yet dead structure of connective pathways.

3. General Properties of Network Elements

Today our globe is populated by approximately 3×10^9 people, each with his own cherished personality, his experiences and his peculiarities. The human brain is estimated to have approximately 10×10^9 neurons in operation, each with its own structural peculiarities, its scars and its metabolic and neuronal neighborhood. Each neuron, in turn, is made up of approximately 4×10^9 various building blocks—large organic compounds of about 10^6 atoms each—to which we deny individuality, either of ignorance or of necessity. When reducing neurons to a common denominator we may end up with a result that is not unlike Aristotle's reduction of man to a featherless biped. However, since it is possible to set up categories of man, say, *homo politicus, homo sapiens* and *homo faber*, categories which do not overlap but do present some human features, it might be possible to set up categories in the operation of neurons which do not overlap but do represent adequately in certain domains the activity of individual neurons. This is the method we shall employ in the following paragraphs.

We shall select some operational modalities as they have been reasonably well established to hold for single neurons under specified conditions, and shall derive from these operational modalities all that may be of significance in the subsequent discussion of neural nets.

Peculiar as it may seem, the neuron is usually associated with two operational principles that are mutually exclusive. One is known as the "All or Nothing" law, which certainly goes back to Bowditch (1871) and which states that a neuron will respond with a single pulse whose amplitude is independent of the strength of stimulus if, and only if, the stimulus equals or exceeds a certain threshold value. Clearly, this description of the behavior of a neuron attaches two states to this basic element, namely, "zero" for producing no pulse, ad "one" for producing the pulse. Since modern computer jargon has crept into neurophysiology, this neural property is usually referred to as its "digital" characteristic, for if a record of the activity of a neuron in these terms is made, the record will present itself in form of a binary number whose digits are "ones" and "zeros":

$$\ldots 01100111011110 \ldots$$

When we adopt this operational modality of a neuron as being crucial in its processing of information, we also associate with the string of "ones" and "zeros" the code in which information is transmitted in a network.

The other operational principle, which is diametrically opposed to the one just mentioned, derives its legitimacy from the observation that—at least in sensory fibers—information is coded into the lengths of time intervals between pulses. Since the length of a time interval is a continuous variable, and since under certain conditions this interval may represent monotonically a continuously varying stimulus, this behavior of a neuron is usually referred to—by again invoking computer jargon—as its "analog" characteristic. Under these conditions we may regard the behavior of a neuron as the transfer function of a more or less linear element whose input and output signal is a pulse interval code and whose function is pulse-interval modulation. A "Weber-Fechner neuron" simply has a logarithmic transfer function, a "Stevens neuron" a power-law response (Stevens, 1957) and a "Sherrington neuron" has neat, almost linear properties with threshold (Sherrington, 1906).

Although it is not at all difficult—as we shall see—to propose a single mechanism that reconciles all types of operation in neurons discussed so far (analog as well as digital), it is important to separate these operational modalities, because the overall performance of a network may change drastically if its elements move, from one operational modality to another. Consequently, we shall discuss these different modalities under two different headings: first "The Neuron as an 'All or Nothing' Element" with special attention to synchronous and a-synchronous operations, and second "The Neuron as an 'Integrating Element'". After this we shall be prepared to investigate the behavior of networks under various operation conditions of its constituents.

3.1. The Neuron as an "All or Nothing" Element

3.1.1. Synchronism

This exposé follows essentially the concepts of a "formal neuron" as proposed by McCulloch and his school (McCulloch and Pitts, 1943; McCulloch, 1962), who define this element in terms of four rules of connection and four rules of operation.

Rules of connection:
A "McCulloch formal neuron":
  i)   receives N input fibers X_i ($i = 1, 2, N$), and has precisely one output fiber *.
  ii) Each input fiber X_i may branch into n_i facilitatory (+) or inhibitory (−) synaptic junctions, but fibers may not combine with other fibers.
  iii) Through the neuron, signals may travel in one direction only.
  iv) Associated with this neuron is an integer θ ($-\infty < \theta < +\infty$), which represents a threshold.

Rules of operation:
  v)   Each input fiber X_i may be in only one of two states ($x_i = 0, 1$), being either OFF (0) or ON (1).

vi) The internal state Z of the neuron is defined by

$$Z = \sum n_i x_i - \theta. \tag{15}$$

vii) The single output Y is two-valued, either ON (1) or OFF (0) ($y = 0, 1$) and its value is determined by

$$y = \Phi(Z) = \begin{cases} 0 \text{ for } Z < -\varepsilon, \\ 1 \text{ for } Z > -\varepsilon, \end{cases}$$

where ε is a positive number smaller than unity:

$$0 < \varepsilon < 1.$$

The ith fiber will be denoted by a capital X_i, while its state by a lower case x_i.

viii) The neuron requires a time interval of Δt to complete its output.

We shall briefly elucidate these rules with the aid of fig. 9 which is a symbolic representation of this element. The neuron proper is the triangular figure, symbolizing the perikaryon, with a vertical extension upward receiving inhibitory fibers, each loop representing a single inhibitory synaptic junction. Excitatory junctions are symbolized as terminal buttons attached to the perikaryon. In fig. 9 the number of input fibers is:

$$N = 4,$$

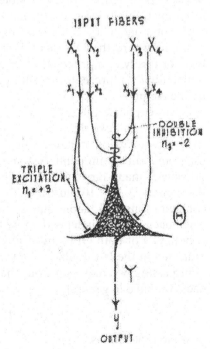

FIGURE 9. Symbolic representation of a McCulloch formal neuron.

with the following values of facilitatory and inhibitory synaptic junctions:

$$n_1 = +3, \quad n_2 = -1,$$
$$n_3 = -2, \quad n_4 = +1.$$

Consider for the moment all input fibers in the ON state and the threshold at zero:

$$x_1 = x_2 = x_3 = x_4 = 1, \quad \theta = 0. \tag{16}$$

The internal state Z is, according to eq. (16) given by

$$Z = 1 \cdot (+3) + 1 \cdot (-1) + 1 \cdot (-2) + 1 \cdot (+1) = +1 > -\varepsilon;$$

hence, according to rule (vii) eq. (16), we have:

$$y = \Phi(1) = 1,$$

so the element "fires"; its output is ON. Raising the threshold one unit, $\theta = 1$, still keeps the element in its ON state, because its internal state does not fall below zero. From this we can conclude that a completely disconnected element with zero threshold always has its output in the ON state.

If in fig. 9 the threshold is raised to +2, the simultaneous excitation of all fibers will not activate the element. Furthermore, it is easily seen that with threshold +5 this element will never fire, whatever the input configuration; with threshold −4 it will always fire.

The two-valuedness of all variables involved, as well as the possibility of negation (inhibition) and affirmation (excitation), make this element an ideal component for computing logical functions in the calculus of propositions where the ON or OFF state of each input fiber represents the truth or falsity of a proposition X_i, and where the ON or OFF state of the output fiber Y represents the truth value of the logical function Φ computed by the element.

Let us explicate this important representation with a simple example of an element with two input fibers only ($N = 2$), each attached to the element with only a single facilitatory junction ($n_1 = n_2 = +1$). We follow classical usage and call our input fibers A and B,—rather than X_1 and X_2 (which pays off only if many input fibers are involved and one runs out of letters of the alphabet). Fig. 10 illustrates the situation. First, we tabulate all input configurations—all "input states" that are possible with two input fibers when each may be independently ON or OFF. We have four cases: A and B both ON or both OFF, and A ON and B OFF, and A OFF and B ON, as indicated in the left double column in fig. 10.

In passing, we may point out that with N input fibers, two choices for each, we have in general

$$n_{\text{in}} = 2^N \tag{17}$$

possible input states.

FIGURE 10. Threshold defining the function computed by a McCulloch formal neuron.

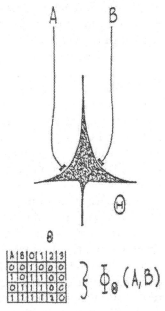

Returning to our example, we now tabulate, for a particular threshold value θ, the output $\Phi_\theta(A, B)$ which has been computed by this element for all input states, i.e., all combinations of A, B belong ON or OFF. For zero threshold (θ = 0), this element will always be in its ON state, hence, in the column θ = 0 we insert "one" only. We raise the threshold one unit (θ = 1) and observe that our element will fire only if either A or B, or both A and B are ON. We proceed in raising the threshold to higher values until a further increase in threshold will produce no changes in the output functions, all being always OFF.

In order to see that for each threshold value this element has indeed calculated a logical function on the propositions A and B, one has only to interpret the "zeros" and "ones" as "false" and "true" respectively, and the truth values in each θ-column, in conjunction with the double column representing the input states, become a table called—after Wittgenstein (1956)—the "truth table" for the particular logical function. In the example of fig. 10, the column θ = 0 represents "Tautology" because $\Phi_o(A, B)$ is always true (1), independent of whether or not A or B are true: "A or not −A, and B or not −B". For θ = 1 the logical function "A or B" is computed; it is false (0) only if both A and B are false. θ = 2 gives "A and B" which, of course, is only true if both A and B are true, etc.

Today there are numerous notations in use, all denoting these various logical functions, but based on different reasons for generating the appropriate representations, which all have their advantages or disadvantages.

The representation we have just employed is that of Wittgenstein's truth table. This representation permits us to compute at once the number of different logical functions that are possible with N propositions (arguments). Since we know that N two-valued arguments produce n_{in} different states, each of which again has two values, true or false, the total number of logical functions is, with eq. (17):

$$n_{LF} = 2^{n_{IN}} = 2^{2^N}. \tag{18}$$

For two arguments ($N = 2$) we have precisely 16 logical functions.

Another symbolism in use is that proposed by Russell and Whitehead (1925), and Carnap (1925) who employ the signs "•", "v", "→", "–" for the logical "and", "or", "implies", "non" respectively. It can be shown that all other logical functions can be represented by a combination of these functions.

Finally, we wish to mention still another form for representing logical functions, with the aid of a formalized Venn diagram. Venn, in 1881, proposed to show the relation of classes by overlapping areas whose various sections indicate joint or disjoint properties of these classes (fig. 11). McCulloch and Pitts (1943) dropped the outer contours of these areas, using only the center cross as lines of separation. Jots in the four spaces can represent all 16 logical functions. Some examples for single jots are given in fig. 11. Expressions with two or more jots have to be interpreted as the expressions with single jots connected by "or". Hence,

$$\chi = \text{"(neither } A \text{ nor } B) \text{ or } (A \text{ and } B)\text{"},$$

which, of course, represents the proposition "A is equivalent to B". The similarity of this symbol with the greek letter chi suggested the name "chiastan" symbol. The advantage of this notation is that it can be extended to accommodate logical functions of more than two arguments (Blum, 1962).

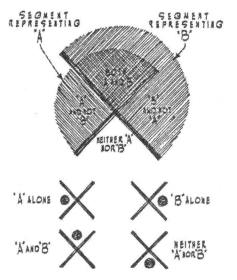

FIGURE 11. Development of the Chiastan symbol for logical functions from Venn's diagrams.

In table 1 we show that, with two exceptions all logical functions with two input lines can be computed by a single McCulloch formal neuron, if full use is made of the flexibility of this element by using various thresholds and synaptic junctions. For convenience, this table lists, in six different "languages", all logical functions for two arguments. The first column gives a digital representation of Wittgenstein's truth function (second column), taken as a binary digit number to be read downwards. The third column gives the appropriate chiastan symbol, and the fourth column shows the corresponding element with its synaptic junctions and its appropriate threshold value. The two functions which cannot be computed by a single element require a network of three elements. These are given in fig. 12 and are referred to in the appropriate entries. The fifth column shows the same

TABLE 1.

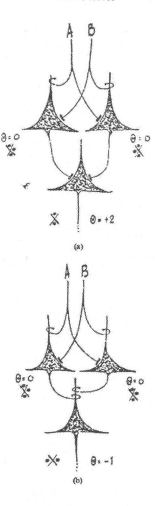

FIGURE 12. Nets of McCulloch formal neurons computing the logical functions (a) "*A* is equivalent to *B*" and (b) "either *A* or else *B*".

functions in Russell-Carnap symbols, while the sixth column translates it into English.

The seventh, and last, column lists the "logical strength" Q of each function. According to Carnap (1938) it is possible to assign to each logical function a value which expresses the intuitive feeling for its strength as a logical function. We intuitively consider a logical function to be weak if it is true in most of the cases, irrespective of whether its arguments are true or false. The tautology, which is always true, tells us nothing about the truth values of its arguments. On the other hand, a function which is true only when all its arguments are true we consider to be a strong logical function. Consequently, as a measure of strength one may take the number of ways in which the logical function is false. In other words, counting the number of zeros in Wittgenstein's truth table gives the logical strength of the function. Inspection of table 1 shows an interesting relationship between threshold and strength, because for a given synaptic distribution the logical strength increases with increasing threshold. This observation will be of importance in our discussion of adaptive nets, because by just raising the threshold to an appropriate level, the elements will be constrained to those functions which "education" accepts as "proper".

Since we have shown that all logical functions with two variables can be represented by McCulloch formal neurons, and since in drawing networks composed of elements that compute logical functions it is in many cases of no importance to refer to the detailed synaptic distributions or threshold values, we may replace the whole gamut by a single box with appropriate inputs and outputs, keeping in mind, however, that the box may contain a complex network of elements operating as McCulloch formal neurons. This box represents a universal logical element, and the function it computes may be indicated by attaching to it any one of the many available symbolic representations.

We shall make use of this simplified formalism by introducing an element that varies the functions it computes, not by manipulation of its thresholds but according to what output state was produced, say, one computational step earlier. Without specifying the particular functions this element computes, we may ask what we can expect from such an element, from an operational point of view. The mathematical formalism that represents the behavior of such an element will easily show its salient features. Let $X(t)$ be the N-tuple $(x_1, x_2, x_3, \ldots, x_N)$ representing the input state at time t for N input fibers, and $Y(t)$ the M-tuple $(y_1, y_2, \ldots, y_M)$ representing its output state at time t. Call Y' its output state at $t - \Delta t$. Hence,

$$Y = \Phi(X, Y'). \tag{19}$$

In order to solve this expression we have to know the previous output state Y' which, of course, is given by the same relation only one step earlier in time. Call X' the previous input state, then:

$$Y' = \Phi(X', Y'').$$

and so on. If we insert these expressions into eq. (19), we obtain a telescopic equation for Y in terms of its past experience $X', X'', X''', \ldots$ and Y_0, the birth state of our element:

$$Y = \Phi(X, X', X'', X''', \ldots, Y_0).$$

In other words, this element keeps track of its past and adjusts its *modus operandi* according to previous events. This is doubtless a form of "memory" (or another way of adaptation) where a particular function from a reservoir of available functions is chosen. A minimal element that is sufficient for the development of cumulatively adaptive systems has been worked out by Ashby (see Fitzhugh, 1963) (see fig. 13). It is composed of at least one, at most three, McCulloch formal neurons, depending upon the functions to be computed in the unspecified logical elements. We shall call

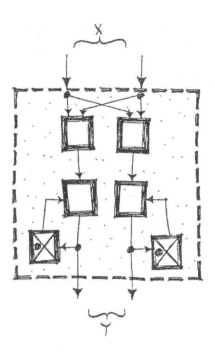

FIGURE 13. Ashby element computing recursive functions.

such a minimal element an "Ashby element". The mathematical machinery that goes along with such elements is called recursive function theory, hence elements of this general form may be called recursive elements.

We have as yet discussed elements with two inputs only. However, it is easily seen that McCulloch's concepts can be extended to neurons with many inputs as fig. 9 may remind us. However, the number of logical functions that cannot be computed by using only a single neuron increases rapidly—2 out of 16 for two inputs, and 152 out of 256 possible functions for three inputs (Verbeek, 1962)—and networks composed of several elements have to be constructed. These networks will be discussed in the following chapter.

In the preceding discussion of the operations of a McCulloch formal neuron a tacit assumption was made, namely, that the information carried on each fiber is simply its ON or OFF state. These states have to be simultaneously presented to the element, otherwise its output is meaningless with respect to these states. A term like "input strength" is alien to this calculus; a proposition is either true or false. This requires all components in these networks to operate synchronously, i.e., all volleys have not only to be fired at the same frequency, they have also to be always in phase. Although there are indications that coherency of pulse activity is favored in localized areas—otherwise an E.E.G. may show only noise—as long as we cannot propose a mechanism that synchronizes pulse activity, we have to consider synchronism as a very special case. Since this article is not the place to argue this out, we propose to investigate what happens as pulses, with various frequencies, pass over various fibers.

3.1.2. Asynchronism

Assume that along each fiber X_i travels a periodic chain of rectangular pulses with universal pulse width Δt (see fig. 14), but with time intervals τ_i varying from fiber to fiber. The probability that fiber X_i activates its synaptic junctions at an arbitrary instant of time is clearly

$$p_i = \frac{\text{duration of pulse}}{\text{duration of pulse interval}} = \frac{\Delta t}{\tau_i}, \tag{20}$$

or, replacing the periodic pulse interval on fiber X_i by the frequency f_i, we have

$$p_i = f_i \Delta t, \tag{21}$$

for the probability that X_i's synaptic junctions are activated, and the probability

$$q_i = 1 - p_i \tag{22}$$

that they are inactivated. The probability of a particular input state $X(x_1, x_2, \ldots, x_N)$ which is characterized by the distribution of "ones" and "zeros" of the input values x_i, and which may be represented by an N-digit binary number

$$0 \leq X \leq 2^N - 1, \tag{23}$$

$$X = \sum_1^N x_i 2^{i-1}, \quad x_i = 0, 1, \tag{24}$$

is given by the Bernoulli product

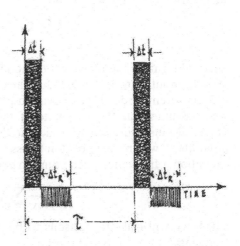

FIGURE 14. Schematic of pulse width, pulse interval and refractory period.

$$P_X = \prod_{1}^{N} p_i^{x_i} (1-p_i)^{(1-x_i)} \qquad (25)$$

with

$$\sum_{X=0}^{X=2^N-1} P_X = 1, \qquad (26)$$

which simply arises from the consideration that the simultaneous presence or absence of various events with probabilities p_i or $(1 - p_i)$ respectively is just the product of these probabilities. The presence or absence of events is governed by the exponents x_i and $(1 - x_i)$ in the Bernoulli product which are 1 and $(1 - 1) = 0$ in presence, and 0 and $(1 - 0) = 1$ in absence of the vent of interest, namely the activation of the synaptic junctions of the ith fiber.

Having established the probability of a particular input state, we have simply to find out under which conditions the element fires in order to establish its probability of firing. This, however, we know from our earlier considerations (eqs. (15), (16)) which define those input states that activate the output fiber. As we may recall, an activated output ($y = 1$) is obtained when the internal state Z equals, or exceeds, zero:

$$Z = \sum n_i x_i - \theta > -\varepsilon$$

with n_i representing the number of (positive or negative) synaptic junctions of the ith fiber and θ being, of course, the threshold. Hence, for a given threshold and a certain input state $X(x_1, x_2, \ldots, x_N)$ output state $y_{\theta,X}$ is defined by

$$y_{\theta,X} = \begin{cases} 1 \text{ for } \sum n_i x_i - \theta > -\varepsilon, \\ 0 \text{ for } \sum n_i x_i - \theta < -\varepsilon. \end{cases} \qquad (26)$$

Since whenever the output is activated y will assume a value of unity, the probability p of its activation is the sum of all probabilities of those input states that give y a value of "one":

$$p = \sum_{X=0}^{X=2^N-1} y_{\theta,X} P_X. \qquad (27)$$

Since all terms in the above expression will automatically disappear whenever an input state is present that fails to activate the output ($y = 0$), eq. (27) represents indeed the activation probability of the output fiber. If we again assume that the ON state of the output fiber conforms with the universal pulse duration Δt, which holds for all pulses traveling along the input fibers, we are in a position to associate with the probability of output excitation a frequency f according to eqs. (26), (27):

$$f = \frac{1}{\Delta t} \sum y_{\theta,X} P_X, \qquad (28)$$

which we will call—for reasons to be given in a moment—the "internal frequency".

Let us demonstrate this mathematical apparatus in the simple example of fig. 10. With letters A and B for the names of fibers X_1 and X_2 we have, of course, $A(x_1 = 0, 1)$ and $B(x_2 = 0, 1)$ and the four possible input states are again:

x_1	x_2
0	0
1	0
0	1
1	1

The four corresponding Bernoulli products are after eq. (25):

$$(1 - p_1)(1 - p_2)$$
$$p_1 \cdot (1 - p_2)$$
$$(1 - p_1) \cdot p_2$$
$$p_1 \cdot p_2$$

In order to find which of these products will contribute to our sum (eq. (27)) that defines the desired output activation probability, we simply consult the truth table for the function that is computed after a specific threshold has been selected, and add those terms to this sum for which the truth tables give a "one" ($y = 1$). The following table lists for the four values of θ as chosen in fig. 10 the resulting probabilities of output excitation according to eq. (27).

With eq. (28) we have at once what we called the internal frequency of the element when its threshold is set to compute a particular logical function. Using the notation as suggested in column 5 of table 1 for denoting logical functions as subscript, and denoting with T tautology and with C contradiction, the computer frequencies representing the various logical functions of the arguments f_1 and f_2, again represented as frequencies, are as follows.

This table indicates an interesting relationship that exists between the calculus of propositions and the calculus of probabilities (Landahl et al., 1943). Furthermore, it may be worthwhile to draw attention to the fact—which may be shown to hold in general—that low threshold values, resulting in "weak" logical functions, give this element essentially a linear characteristic; it simply adds the various stimuli f_i. This is particularly true, if the stimuli are weak and their cross-products can be neglected. For higher threshold values the element is transformed into a highly non-linear device,

TABLE 2.

θ	Sum of Bernoulli products $= p$
0	$(1 - p_1)(1 - p_2) + (1 - p_1)p_2 + p_1(1 - p_2) + p_1p_2 = 1$
1	$p_1(1 - p_2) + (1 - p_1)p_2 + p_1p_2 = p_1 + p_2 - p_1p_2$
2	$p_1p_2 = p_1p_2$
3	$- = 0$

TABLE 3.

$\theta = 0$	$f(T) = 1/\Delta t$
$\theta = 1$	$f(v) = f_1 + f_2 - tf_1f_2$
$\theta = 2$	$f(\bullet) = \Delta t f_1 f_2$
$\theta = 3$	$f(C) = 0$

taking more and more cross-products of intensities into consideration until for the strongest logical function being short of contradiction—the logical "AND"—the single cross-product of all stimuli is computed.

We have carefully avoided associating with "p" or with "f" the actual output frequency. We called f the "internal frequency". The reason will become obvious in a moment. It is well known that after the production of each pulse a physiological neuron requires a certain moment of time Δt_R—the so-called "refractory period"—to recover from its effort and to be ready for another pulse (see also fig. 14). But in table 3 the resulting frequency for zero threshold is the reciprocal of the pulse duration Δt, which, of course, is unmanageably high for a neuron whose refractory period is clearly much longer than the duration of its pulse

$$\Delta t_R > \Delta t. \tag{29}$$

It is very easy indeed to accommodate this difficulty in our calculations, if we only realize that the actual output frequency f_o of the element is the frequency f at which it "wants" to fire, reduced by the relative time span in which it cannot fire:

$$f_o = f(1 - f_o \Delta t_R). \tag{30}$$

Solving for the output frequency in terms of internal frequency and refractory period we have:

$$f_o = \frac{f}{1 + f\Delta t_R}. \tag{31}$$

From this is easily seen that the ultimate frequency at which an element can fire is asymptotically approached for $f \to 1/\Delta t$, and is given by:

$$f_{o\,max} = \frac{1}{\Delta t + \Delta t_R} < \frac{1}{\Delta t}, \tag{32}$$

which is in perfect agreement with our concept of the physiological behavior of this element. The actual values for pulse duration and refractory period may be taken from appropriate sources (Eccles, 1952; Katz, 1959).

3.2. The Neuron as an "Integrating Element"

The element discussed in the previous paragraphs is an "All or Nothing Device" *par excellence*, and in the two subtitles "Synchronism" and "Asynchronism" we investigated only how this element behaves when subjected to stimuli for which we changed our interpretation of what is a meaningful signal. While in the synchronous case a string of OFFs and ONs was the signal, in the asynchronous case pulse frequencies carried the information. However, in both cases the element always operated on a "pulse by pulse" basis, thus reflecting an important feature of a physiological neuron.

In this paragraph we wish to define an element that incorporates some of the concepts which are associated with neurons as they were first postulated by Sherrington (1952), and which are best described in the words of Eccles (1952, p. 191): "All these concepts share the important postulate that excitatory and inhibitory effects are exerted by convergence on to a common locus, the neuron, and there integrated. It is evident that such integration would be possible only if the frequency signalling of the nervous system were transmuted at such loci to graded intensities of excitation and inhibition".

In order to obtain a simple phenomenological description of this process we suggest that each pulse arriving at a facilitatory or inhibitory junction releases, or neutralizes, a certain amount q_o of a hypothetical agent, which, left alone, decays with a time constant $1/\lambda$. We further suggest that the element fires whenever a critical amount q^* of this agent has been accumulated (see fig. 15).

Again we assume N fibers X_i attached to the element, each having n_i facilitatory (+) or inhibitory (−) synaptic junction. Each fiber operates with a frequency f_i. The number of synaptic activations per unit time clearly is the algebraic sum:

$$S = \sum_1^N n_i f_i. \tag{33}$$

Hence, the differential equation that describes the rate of change in the amount of the agent q as a consequence of stimulus activity and decay is

$$\frac{dq}{dt} = Sq_o - \lambda q, \tag{34}$$

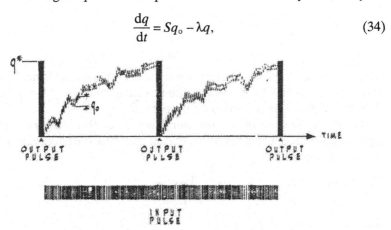

FIGURE 15. "Charging" process of an integrating element.

whose solution for q as a function of time t is:

$$q = \frac{q_o S}{\lambda}(1 - e^{-\lambda t}), \tag{35}$$

if at time $t = 0$ we also have $q = 0$.

Let Δt^* be the time required to accumulate the amount q^* at the element, and let the activity S change during many such time intervals. Clearly, the "internal frequency" of this element is

$$f = 1/\Delta t^*. \tag{36}$$

Inserting these into eq. (35):

$$q^* = \frac{q_o S}{\lambda}(1 - e^{-\lambda/f}), \tag{37}$$

we have an expression that relates the frequency f with the stimulus activity S. For convenience we introduce new variables x and y which represent normalized input and output activity respectively and which are defined by:

$$x = Sq_o/\lambda q^*, \quad y = f/\lambda. \tag{38}$$

With these, eq. (37) can be rewritten:

$$\frac{1}{x} = 1 - e^{-1/y}, \tag{39}$$

or, solved for y:

$$y = 1/[\ln x - \ln(x-1)]. \tag{40}$$

This relation shows two interesting features. First, it establishes a threshold for excitation:

$$x_o = 1$$

or

$$S_\theta = \lambda q^*/q_o.$$

Because for the logarithmic function to be real its argument must be positive, or $x \geqslant 1$. It may be noted that the threshold frequency S_θ is given only in terms of the element's intrinsic properties λ, q_o and q^*.

The second feature of the transfer function of this element is that for large values of x it becomes a linear element. This is easily seen if we use the approximations

$$\frac{1}{1-\varepsilon} \approx 1 + \varepsilon$$

and

$$\ln(1+\varepsilon) \approx \varepsilon$$

for

$$\varepsilon = 1/x \ll 1.$$

Under these conditions eq. (40) becomes simply:

$$y = x$$

or

$$f = S(q_o/q^*).$$

This "nice" feature of our element is, of course, spoiled by the considerations which were presented earlier, namely, that the activation frequency S, which is the sum-total of the impinging frequencies (see eq. (33)) might be too high to be handled by a physiological neuron. In order to adjust for the frequency limit that is expected from our element, we proceed in precisely the same way as was suggested in eq. (30): we introduce into eq. (41) a formal interaction term that reduces its potential activity x commensurate with its actual activity:

$$y_o = (1 - \mu y_o), \tag{42}$$

where the factor μ will become evident in a moment. Solving for y_o

$$y_o = \frac{x}{1 + \mu x} \tag{43}$$

and for $x \to \infty$ we have

$$y_{o\,max} = 1/\mu. \tag{44}$$

Denormalization according to (38) and comparison with (32) gives

$$f_{o\,max} = \frac{\lambda}{\mu} = \frac{1}{\Delta t + \Delta t_R}, \tag{45}$$

or

$$\mu = \lambda(\Delta t + \Delta t_R). \tag{46}$$

In other words, the parameter μ expresses all neuronic delays in units of the agent's decay constant.

In order to make the high frequency correction applicable for the whole operational range of our element, we simply replace x in eq. (40) by $x/(1 + \mu x)$ of eq. (43) which is the adjusted equivalent to the unadjusted eq. (41).

With this adjustment we have the output frequency of our element defined by

$$y_o = \frac{1}{\ln \dfrac{x}{x(1-\mu)-1}} \tag{47}$$

A graphical representation of the input–output relationship of this element according to eq. (47) is given in fig. 16 for three different values of μ (0; 0.2; 0.5). For small μ ($\mu < 0.05$) the element is linear with threshold $x_o = 1$. For large μ (≈ 0.5) the element is an almost perfect "All or Nothing" device with threshold

$$x_o = \frac{1}{1-\mu}. \tag{48}$$

For values of μ in the range $0.1 \rightarrow 0.5$ the element displays a logarithmic transfer function. Indeed, it can easily be shown that in the "vicinity" of

$$x \approx \frac{1}{\mu(1-\mu)}. \tag{49}$$

Eq. (47) can be approximated by

$$y_o = K \ln x, \tag{50}$$

$$K = \frac{1}{4} \frac{\mu}{(1-\mu)[\ln(1-\mu)]^2}. \tag{51}$$

It may be noted that this "vicinity" extends over an appreciable range as can be seen by the approximation according to (50) which in fig. 16 is super-imposed in thin line over the exact curves given in bold line.

With the choice of the parameter μ we are in a position to change the transfer function of this element considerably. We suggest the following nomenclature for the elements that arise for different values of μ:

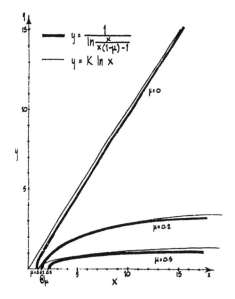

FIGURE 16. Transfer function of an integrating element under various operating conditions (μ) (bold line), and approximation by a logarithmic function (thin line).

$\mu = 0$:   "Sherrington" element (linear characteristic)
$\mu \approx 0.2$: "Weber-Fechner" element (logarithmic characteristic)
$\mu \to 1$:   "All or Nothing" element (step-function characteristic).

With these and the other elements as specified in the previous paragraph, the McCulloch element and the Ashby element, we are now prepared to discuss the behavior of networks that incorporate these elements as their basic computer components.

4. Some Properties of Computing Networks

Of the myriad networks that are not only theoretically possible but are indeed incorporated into the neural architecture of living organisms, space and ignorance will permit only a small glimpse into their vast richness. In addition, the large variety of solutions that evolution has provided in different species for their specific cognitive problems makes it difficult to present this topic from a single ordering point of view, except that here we are dealing with networks. However, in the last decades a number of general principles have been carved out from this large complex of problems and in the following an attempt will be made to do justice to some of them by briefly suggesting their conceptual framework and by giving some examples to illustrate the underlying ideas.

We shall open our discussion with two paragraphs which represent extremes in the spectrum that goes from the concrete to the abstract. The first paragraph shows the possibility of orderly behavior in a "mixed" net, the neuromuscular net in the sea urchin, where within elements of their own kind no interaction takes place, but where each kind uses the other for integrated action. The second paragraph touches briefly the McCulloch-Pitts theorem which, in a sense, ends or starts all discussions about networks.

The next paragraphs discuss the development of cognitive networks, first, in which cellular identity is recognized, while the subsequent considerations are based solely on the localizability of groups of cells, but their individuality is lost. The chapter concludes with a brief account of stability and immunology of neural networks and with some remarks on adaptive nets and how they store information.

4.1. A Neuro-muscular Net

Fulton (1943) opens his comprehensive treatise on neurophysiology with a brief account of the early evolutionary stages in the development of the nervous system. Rightly so, because the appreciation of these early stages leaves no doubt as to the ultimate purpose of this system, namely, to serve as a computer that links detection with appropriate action. Following Parker (1943) we give in fig. 17 schematically the three decisive steps which

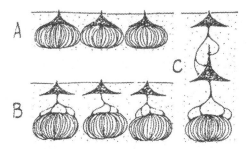

FIGURE 17. Primitive nerve nets: (a) independent effector; (b) receptor-effector system; (c) receptor-computer-effector system.

are the foundation for the emergence of neural systems with the complexity of a mammalian brain. Fig. 17a shows symbolically the "independent effector" (muscle cell in a sponge) that translates directly a general "stimulus" into action—contraction in most cases. The first step from detection to discrimination is accomplished by separating detection and action and localizing these functions in different elements (fig. 17b). This permits the development of specific sensors responsive to certain stimuli only (light, chemistry, touch, etc.). The final step in preparing the tripartite architectural organization of the nervous system—detector, computer, effector—is suggested in fig. 17c, where an intermediate ganglion cell acts as a primordial nucleus for what is to become the information processing interface between detection and action.

Although an array of such simply organized units as in fig. 17b appears not to have the properties which we would expect from a neural net, for there is no direct connection from neuron to neuron, these systems still deserve to be called interaction nets from a general point of view, because a particular state in one unit—say a contraction—may influence the state of its neighbors via the mechanical properties of the medium in which they are embedded. That such "mixed nets" are capable of highly organized behavior may be illustrated with the beautiful observations of Kinosita (1941) on the kinetics of the spines of the sea-urchin.

When a localized stimulus is given to the body surface of a sea-urchin, the spines around the stimulated spot respond so as to lean towards the stimulated spot, and this response diminishes rapidly with distance from the locus of stimulation (fig. 18). The first thought that comes to mind is to assume an anastomosing plexus of interacting nerve cells which transmit the information of this perturbation over an appropriate region to cause contraction of the muscle fibers attached to the spines (Üxküll, 1896). However, Kinosita was able to demonstrate that there are no fiber to fiber connections, only proximate fiber-muscle connections, hence each receptor pair has to operate according to local information of the deformation of its surroundings caused by deformations of the more distant regions.

FIGURE 18. Position of spines in the sea-
urchin after stimulation at the North pole.

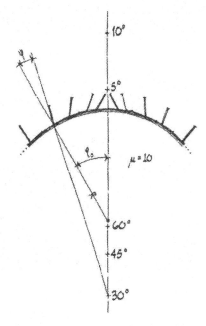

Local anatomy and geometry in the neighborhood of a spine is schematically given in fig. 19 which exaggerates certain proportions for purposes
of clarity. The spine S, centered on pivot P which is attached to a fixed shell
with radius R, can bend in all directions. Muscle fibers M contract when
stimulated by neuron N which will respond to an extension (stretch) of the
integument. If somewhere at the surface a muscle bundle contracts, it causes
the integument to follow, which produces a slight local stretch that is sensed
by the local neuron which, in turn, causes its associated muscle to contract,
and so on. Consider a spine localized at angle ϕ_o. When bent at angle ψ from
its radial rest position ($\psi = 0$) it will shift the integument surface from ϕ_o to
ϕ. Assume that a stimulus is applied at the North Pole ($\phi_o = 0$), then the shift
$\Delta\phi = \phi_o - \phi$ at angle ϕ_o is the result of the summation of differential contractions $-d(\Delta\phi)/d\phi_o$ of intermediate muscles. These, in turn, contract according
to the efferent stimulus of their associated neurons which fire in proportion
to the local perturbation, i.e., the difference between the extension of the
relaxes integument L and the stretched integument H. Hence, the differential equation that governs the *local* receptor–effector system is:

$$\frac{d(\Delta\phi)}{d\phi_o} = -k(L - H),\tag{52}$$

where the proportionality constant k represents the combined transfer
functions for neuron and muscle fiber. From inspection of fig. 19 we have
the simple geometric relations:

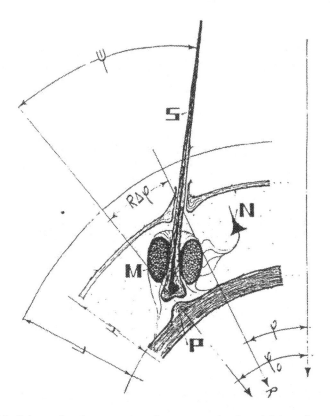

FIGURE 19. Schematic of anatomy and geometry in the vicinity of a sea-urchin spine.

$$H = L \cos \psi$$

and

$$R\Delta\phi = L \sin \psi. \tag{53}$$

Introducing a dimensionless paremeter μ which combines physiological and anatomical constants:

$$\mu = kR,$$

the differential equation (52) can be rewritten with the aid of (53) to read:

$$\frac{d \sin \psi}{d\phi_o} = -\mu(1 - \cos \psi), \tag{54}$$

which can readily be solved to yield a transcendental equation in ψ:

$$\cot\left(\frac{1}{2}\psi\right) + \psi = \mu\phi_o + \cot\left(\frac{1}{2}\psi_o\right) + \psi_o, \tag{55}$$

where ψ_o denotes the original perturbation at $\phi_o = 0$. For a chosen value of $\mu = 10$ this equation was numerically evaluated and served as a basis to conxtruct fig. 18. The entries along the axis of symmetry indicate the focal points of spines located at corresponding angles ϕ_o. The original perturbation is assumed to be strong ($\psi_o \approx 45°$).

The sole purpose of this somewhat detailed account of a relatively insignificant network was to suggest that organized behavior that seems to be governed by a central control that operates according to an "action at a distance" principle can very well arise from a localized point function that permanently links elements in an infinitesimal neighborhood. Notice how the behavior can be expressed in differential eqs. (52) or (54) which contain local properties only. The whole system swings into action whenever the "boundary value"—i.e., the stimulus—changes. The operational principle here is "action by contagion". We shall later discuss this principle in greater detail in connection with interaction networks.

4.2. The McCulloch-Pitts Theorem

A network composed of McCulloch elements we shall call a "McCulloch formal network". The central issue of the McCulloch-Pitts (1943) theorem is the synthesis of such networks, which compute any one of the 2^{2^n} logical functions that can be defined by n propositions. In other words, any behavior that can be defined at all logically, strictly, and unambiguously in a finite number of words can be realized by such a formal network. Since in my opinion this theorem not only is one of the most significant contributions to the epistemology of the 20th century, but also gives important clues as to the analysis of physiological neural nets, it is impossible in an article about nerve nets not, at least, to touch upon the basic ideas and consequences that are associated with this theorem. Its significance has best been appraised in the words of the late John Von Neumann (1951):

"It has often been claimed that the activities and functions of the human nervous system are so complicated that no ordinary mechanism could possibly perform them. It has also been attempted to name specific functions which by their nature exhibit this limitation. It has been attempted to show that such specific functions, logically completely described, are *per se* unable of mechanical neural realization. The McCulloch-Pitts result puts an end to this. It proves that anything that can be exhaustively and unambiguously described, anything that can be completely and unambiguously put into words, is *ipso facto* realizable by a suitable finite neural network".

We shall give now a brief summary of the essential points of this theorem. As already mentioned, the McCulloch-Pitts theorem shows that to any logical function of an arbitrary number of propositions (variables) a network composed of McCulloch elements can be synthesized that is equivalent to any one of these logical functions. By "equivalence" is meant that it functions so as to compute the desired logical function. This can be accomplished by singling out input fibers of some of its elements and output fibers of some other elements and then defining what original stimuli on the former are to cause what ultimate responses of the latter.

Since the basic element of the networks to be discussed—the McCulloch formal neuron—is capable of computing only some of all logical functions which can be constructed from precisely two variables, and since an arbitrary set of propositions may contain temporal relationships, we require three more steps to reach the generality claimed by the theorem. The first step involves purely logical argument and shows (a) by using substitution and the principle of induction the possibility of constructing n-variable expressions from two-variable expressions, and (b) the possibility of expressing uniquely any logical function of n-variables in a certain normal form. The second step introduces an operator S that takes care of a single synaptic delay and thus permits the representation of temporal relationships, while the third step utilizes some formal properties of this operator to obtain normal form expressions that are immediately translatable into network language.

First step:

(a) Consider a logical function of the two variables A_1 and B_1

$$[A_1, B_1]. \tag{56}$$

Let B_1 be a logical function of the two variables A_2 and B_2:

$$B_1 = [A_2, B_2],$$

and, in general,

$$B_i = [A_{i+1}, B_{i+1}]. \tag{57}$$

Iterative substitution of (57) into (56) gives:

$$[A_1, A_2, A_3, \ldots, A_n, B_n],$$

which, by induction, holds for all.

(b) It can be shown (Hilbert and Ackermann, 1928) that any logical function of n arguments can be represented by a partial conjunction ($\cdot$) of disjunctions (v) that contain each variable X_i either affirmed ($x_i = 1$) or negated ($x_i = 0$). Let each disjunction be represented by D_Z, where Z is the decimal representation of the binary number

$$0 \leqslant Z = \sum_1^n 2^{i-1} x_i \leqslant 2^n - 1, \tag{58}$$

and let K_P represent the (partial) conjunction of those disjunctions present ($D_Z = 1$, otherwise $D_Z = 0$) in the logical function, where P is the decimal representation of the binary number

$$0 \leqslant P = \sum_{0}^{2^n-1} 2^Z D_Z \leqslant 2^{2n} - 1. \tag{59}$$

The terms K_P are called "Schroeder's constituents". Consequently, there are 2^{2^n} different conjunctions possible with these constituents. Each of these conjunctions represents uniquely one logical function.

Expanding these conjunctions by virtue of the distributivity of $(\cdot)$ and (v) into a disjunction of conjunctions

$$C_1 \text{ v } C_2 \text{ v } C_3 \ldots, \tag{60}$$

one arrives, after cancellation of contradictory terms $(X_i \cdot \overline{X}_i)$, at Hilbert's disjunctive normal form which, derived this way, is again a unique representation of one of the 2^{2^n} logical functions. This form will be used in the synthesis of networks.

As an example of this procedure take the two-variable logical function expressing the equivalence of A and B. Let X_1 and X_2 be A and B respectively. The four Schroeder constituents are with Z from 0 to 3:

$$D_o = (\overline{A} \text{ v } \overline{B}),$$
$$D_1 = (A \text{ v } \overline{B}),$$
$$D_2 = (\overline{A} \text{ v } B),$$
$$D_3 = (A \text{ v } B).$$

Since

$$A \rightleftharpoons B \equiv D_1 \cdot D_2,$$

we have

$$P = 0 \cdot 2^0 + 1 \cdot 2^1 + 1 \cdot 2^2 + 0 \cdot 2^3 = 6$$

and

$$K_6 \equiv D_1 \cdot D_2 \equiv (A \text{ v } \overline{B}) \cdot (\overline{A} \text{ v } B).$$

Expanding the right hand side gives

$$(A \cdot \overline{A} \text{ v } B \cdot A) \text{ v } (\overline{A} \cdot \overline{B} \text{ v } B \cdot \overline{B}),$$

which after cancellation of contradictions yields the desired expression in Hilbert's disjunctive normal form:

$$(B \cdot A) \text{ v } (\overline{A} \cdot \overline{B}) \equiv A \rightleftharpoons B.$$

The second step considers the synaptic delay Δt at each McCulloch element. Let $N_i(t)$ denote the action performed by the ith element at time t, or for short

$$N_i \equiv N_i(t). \tag{61}$$

In order to facilitate expressions that consider n synaptic delays earlier, a recursive operator S is introduced and defined as

$$N_i(t - n\Delta t) \neq S^n N_i, \tag{62}$$

its iteration represented by its power. Clearly, this operator is applicable to propositions as well.

The third step establishes distributivity of the operator S with respect to conjunction and disjunction:

$$S(N_i \cdot N_j) \equiv SN_i \cdot SN_j,$$
$$S(N_i \cdot N_j) \equiv SN_i \vee SN_j. \tag{63}$$

Since each function of temporal propositions can be expressed in terms of Hilbert's disjunctive normal form, application of the recursive operator S permits each proposition to appear of the form $S^k X_i$ and thus can be translated into the corresponding neural expression (62), which localizes each element in the network and defines its function.

We shall illustrate this procedure with the same simple example that was chosen by the authors of this theorem. It is known as the "illusion of heat and cold".

"If a cold object is held to the skin for a moment and removed, a sensation of heat will be felt; if it is applied for a longer time, the sensation will be only of cold, with no preliminary warmth, however transient. It is known that one cutaneous receptor is affected by heat and another by cold".

We may now denote by N_1 and N_2 the propositions "heat is applied" and "cold is applied" respectively, but interchangeably we may denote by N_1 and N_2 the activity of the receptors "heat receptor active" and "cold receptor active". Similarly, we shall denote by N_3 and N_4 the propositions "heat is felt" and "cold is felt" respectively which can be translated into the activity of the elements producing the appropriate sensations *mutatis mutandis*.

The temporal propositional expression for the two observations can now be written:

Input	Output
$SN_1 \vee S^3 N_2 \cdot S\overline{N}_2$	$\equiv N_3$
$S^2 N_2 \cdot SN_2$	$\equiv N_4$

where the required persistence in the sensation of cold (N_4) is assumed to be two synaptic delays, while only one delay is required for sensation of heat (N_3).

We utilize distributivity of S

$$S(N_1 \vee S(SN_2 \cdot \overline{N}_2)) \equiv N_3,$$
$$S((SN_2) \cdot N_2) \equiv N_4,$$

and develop the whole net in individual steps of nets for two variables, working our way from inside out of the brackets. We first approach the expression for N_4 and construct a net for

$$SN_2 \equiv N_A. \qquad \text{(fig. 20.1)}$$

We complete the relation for N_4 by drawing the net (bold line):

$$N_4 \equiv S(N_A \cdot N_2), \qquad \text{(fig. 20.2)}$$

which is, of course,

$$N_4 \equiv S(SN_2 \cdot N_2).$$

We approach now the expression for N_3 and draw (bold line):

$$S(N_A \cdot \overline{N_2}) \equiv N_B, \qquad \text{(fig. 20.3)}$$

and complete the whole net by drawing (bold line):

$$N_3 \equiv S(N_1 \text{ v } N_B). \qquad \text{(fig. 20.4)}$$

Although this simple example does not do justice to the profoundness of the McCulloch-Pitts theorem, it emphasizes not only the important relationship between formal networks and formal logic but also the minimal structural necessity to accommodate functional requirements.

4.3. Interaction Networks of Discrete, Linear Elements

We now turn our attention to networks which are composed of "linear elements", i.e., of McCulloch elements operating with low thresholds ($\theta \approx 1$) and weak signals ($f_i \ll 1/(\Delta t + \Delta t_R)$) in an asynchronous network, or of Sherrington elements ($\mu = 0$) which perform algebraic summation on their inputs.

The formalism which handles the situation of an arbitrary number of interacting elements has completely been worked out by Hartline (1959) who showed in a series of brilliant experiments the mutual inhibitory action of proximate fibers in the optic stalk of the horseshoe crab by illuminating various neighbors of a particular ommatidium in the crab's compound eye.

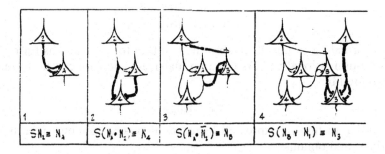

FIGURE 20. Stepwise development of a McCulloch-Pitts network that computes the "illusion of heat and cold". Bold lines represent added network elements ($\theta - 2$, everywhere).

Consider n linear elements e_i, each of which is actively connected to all others and to itself. We have a perfect connection matrix, all rows and columns being non-zero. Let $\rho(i)$ and $\sigma(i)$ represent response and external stimulus of element e_i respectively and permit a certain fraction a_{ij} of the response of element e_i to contribute to the stimulus of element e_j. The response of element e_1 is under these circumstances clearly

$$\rho(1) = \sigma(1) + a_{11}\rho(1) + a_{21}(2) + \ldots + a_{n1}\rho(n), \tag{64}$$

or in general for the jth element:

$$\rho(j) = \sigma(j) + \sum_{i=1}^{n} a_{ij}\rho(i), \tag{65}$$

where for simplicity σ and ρ are expressed in the same arbitrary units and where the coefficients a_{ij} form again a square matrix which we will call the numerical interaction matrix

$$A_n = \|a_{ij}\|_n. \tag{66}$$

In order to obtain a solution for the n unknowns $\rho(j)$ in terms of all stimuli $\sigma(1), \sigma(2), \ldots, \sigma(n)$, we first express all stimuli $\sigma(j)$ in terms of the various responses $\rho(1), \rho(2), \ldots, \rho(n)$. From (64) we have for $\sigma(1)$

$$\sigma(1) = (1 - a_{11})\rho(1) - a_{21}\rho(2) - \sigma_{31}\rho(3)\ldots$$

or in general for the jth element

$$\sigma(j) = \sum_{i=1}^{n} s_{ij}\rho(i), \tag{67}$$

where the s_{ij} form again a square matrix S_n which we will call the stimulus matrix

$$S_n = \|s_{ij}\|_n \tag{68}$$

with

$$s_{ij} = \begin{cases} -a_{ij} & \text{for } i \neq j, \\ (1 - a_{ii}) & \text{for } i = j. \end{cases} \tag{69}$$

In the formalism of matrix algebra the n values for σ as well as for ρ represent n-dimensional vectors (column matrices) and (67) can be formally represented by:

$$\sigma_n = S_n\rho_n. \tag{70}$$

In order to find ρ_n expressed in terms of σ_n one "simply" inverts the matrix S_n and obtains

$$\rho_n = S_n^{-1}\sigma_n, \tag{71}$$

which implies solving the n equations in (67) for the n unknowns $\rho(1), \rho(2)$, $\ldots, \rho(n)$. We introduce the response matrix R_n, defined by

$$R_n = \|r_{ij}\|_n = S_n^{-1}. \tag{72}$$

Let $|D|$ denote the characteristic determinant $|s_{ij}|$, and S_{ij} the product of $(-1)^{i+j}$ with the determinant obtained from $|s_{ij}|$ by striking out the ith row and jth column, then the response matrix elements r_{ij} are given by

$$r_{ij} = \frac{S_{ji}}{D}, \tag{73}$$

and we have the solution for the responses:

$$\rho_n = R_n \sigma_n. \tag{74}$$

Clearly, a solution for ρ_n can be obtained only if the characteristic determinant D does not vanish, otherwise all responses approach infinity, which implies that the system of interacting elements is unstable. It is important to note that stability is by no means guaranteed if the interactions are inhibitory, for the inhibition of an inhibition is, of course, facilitation.

The actual calculation of a response matrix, given the numerical interaction matrix, is an extremely cumbersome procedure that requires the calculation of n^2 matrices, each of which demands the calculation of $n!$ products consisting of n factors each, that is $n^3 \cdot n!$ operations all together. Under these circumstances it is clear that manual computation can be carried out for only the most simple cases, while slightly more sophisticated situations must be handled by high speed digital computers; even they prove insufficient if the number of elements goes beyond about, say, 50. However, the horseshoe crab performs these operations in a couple of milliseconds by simultaneous parallel computation in the fibers of the optic tract. The *Limulus*' eye is—so to say—made for matrix inversion.

In order to clarify procedures we give a simple example of four elements e_1, e_2, e_3, e_4 which are thought to be placed at the four corners of a square labeled clockwise with e_1 in the NW corner. We assume mutual interaction to take place between neighbors only, and all coefficients to be alike at a. The connection matrix for this configuration is

$$
\begin{array}{c|cccc}
 & \multicolumn{4}{c}{e_j} \\
 & 1 & 2 & 3 & 4 \\
\hline
1 & 0 & 1 & 0 & 1 \\
2 & 1 & 0 & 1 & 0 \\
e_i \quad 3 & 0 & 1 & 0 & 1 \\
4 & 1 & 0 & 1 & 0 \\
\end{array}
$$

and the numerical interaction matrix

$$A_4 = \begin{Vmatrix} 0 & a & 0 & a \\ a & 0 & a & 0 \\ 0 & a & 0 & a \\ a & 0 & a & 0 \end{Vmatrix}.$$

From this we obtain with eq. (69) the stimulus matrix

$$S_4 = \begin{Vmatrix} 1 & -a & 0 & -a \\ -a & 1 & -a & 0 \\ 0 & -a & 1 & -a \\ -a & 0 & -a & 1 \end{Vmatrix},$$

which inverted gives the following response matrix:

$$4 = \frac{1}{-} \begin{Vmatrix} 1-2a^2 & a & 2a^2 & a \\ a & 1-2a^2 & a & 2a^2 \\ 2a^2 & a & 1-2a^2 & a \\ a & 2a^2 & a & 1-2a^2 \end{Vmatrix}.$$

The characteristic determinant is

$$D = 1 - 4a^2$$

with the two roots $a = \pm\frac{1}{2}$. Hence, the system becomes unstable, whenever the interaction coefficient a approaches $+\frac{1}{2}$ (facilitation) or $-\frac{1}{2}$ (inhibition).

We are now in a position to write all responses $\rho(j)$ in terms of the stimuli $\sigma(i)$. For the response of the first element we have

$$D \cdot \rho(1) = (1-2a^2)\,\sigma_1 + a\sigma_2 + 2\sigma^2\sigma_3 + a\sigma_4,$$

the others are obtained by cyclic rotation of indices or directly from R_n.

For uniform stimulation of all elements, $\sigma(i) = \sigma_o$ for all i, the uniform response is

$$\rho_o = \frac{1}{1-2a}\sigma_o,$$

which clearly depends upon the sign of the interaction coefficient, giving increased or decreased responses for facilitation or inhibition respectively.

If uniform stimulation is maintained for e_2, e_3, e_4 ($\sigma_2 = \sigma_3 = \sigma_4 = \sigma_o$), but element e_1 is stimulated by a ($\pm$) superposition of σ^* ($\sigma_1 = \sigma_o + \sigma^*$), and all stimuli and responses are expressed in terms of uniform stimulus and response we have

$$\frac{\rho(i)}{\rho_o} = 1 + b_i \frac{\sigma^*}{\sigma_o}$$

$$b_1 = \frac{1-2a^2}{1+2a}, \quad b_2 = b_4 = \frac{a}{1+2a}, \quad b_3 = \frac{2a^2}{1+2a}.$$

The quadratic terms for a in the numerator of b_1 and b_3 show clearly the effect of "double negation" by "inhibition of inhibition", for these terms are independent of the sign of a.

As a final example we show the far-reaching influence of a local perturbation in a mixed net which consists of a linear array of quadrupoles as above with an inhibitory interaction coefficient of $a = -0.3$, and where each quadrupole is actively connected to its neighbor on one side, but passively connected to its neighbor on the other side (see fig. 21). Elements which actively connect quadrupoles transmit their full response to their neighbors. A unit stimulus is applied to element e_1 in the first quadrupole only. The resulting response is plotted next to each element, the length of bars representing intensity.

This example may again be taken as an instance of the principle "action by contagion" which we met earlier in a mixed interaction net (sea-urchin). In this case, however, the local perturbation spreads in the form of a decaying oscillation.

The discussion of interaction in nets composed of discrete linear elements has shown thus far two serious deficiencies. The first deficiency is clearly the insurmountable difficulty in handling efficiently even simple net configurations. We shall see later that this difficulty can be circumvented at once, if the individuality of elements is dropped and only the activity of elements associated with an infinitesimal region in space is taken into consideration. The powerful apparatus developed in the theory of integral equations will take most of the burden in establishing the response function, given a stimulus function and an interaction function.

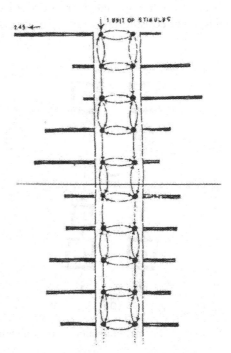

FIGURE 21. Responses in a mixed action-interaction network after a single stimulus is applied to the NW element in the first interacting quadrupole.

The second deficiency becomes obvious if we have to answer the question of where we localize the operation that transmits "a certain fraction a_{ij}" of the activity of element e_i to element e_j. Is this a property of the transmitting or of the receiving element? Clearly, our simple model of elements does not yet take care of this possibility. However, at this stage of the development it is irrelevant to decide whether we make transmitter or receiver responsible for the regulation of the amount of the transmitted agent (see fig. 22a). However, it is necessary to bestow on at least one of them the capacity to regulate the transmitted agent. We decide to make the receiver responsible for this operation, justifying this decision by the possibility of interpreting this regulatory operation—at least for neural synaptic contacts—as the number of facilitatory or inhibitory synaptic junctions of a fiber that synapses element e_i onto e_j. Hence, for our present purposes we adopt a representation of our linear elements (fig. 22b) which modifies the transmitted signal at their inputs according to the numerical value of the active connection coefficient a_{ij}. Later, however, we shall see that both, transmitter and receiver define this coefficient.

4.4. Active Networks of Discrete, Linear Elements

We draw directly from our definitions for action nets and for the various operational modalities of elements as discussed in earlier paragraphs; however, we shall introduce in this paragraph for the first time some constraints on the spatial distribution of elements. These constraints will be tightened considerably while we proceed.

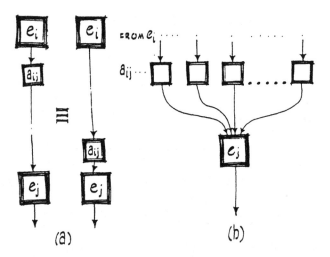

FIGURE 22. Formal equivalence of localization of operations.

4.4.1. Linear Elements

Consider a set of $n = 2m$ linear elements, half of which are general receptors and the other half general effectors. A weak geometrical constraint, which does not affect the generality of some of the following theorems but facilitates description, is to assume spatial separation of receptors and effectors. The locus of all general receptors, e_{1i}, we shall call "transmitting layer" L_1, and the locus of all general effectors, e_{2j}, $(i, j = 1, 2, 3, \ldots, m)$ the "collecting layer" L_2, regardless of the dimensionality of these loci, i.e., whether these elements are arranged in a one-dimensional array, on a two-dimensional surface, or in a specifiable volume.

We consider the fraction of activity in element e_{1i} that is passed on to an element e_{2i} as its partial stimulus a_{ij} (i). The action coefficients for the m^2 pairs define the numerical action matrix

$$A_m = \|a_{ij}\|_m. \tag{75}$$

With linear elements in the collector layer their responses are the algebraic sum of their partial stimuli:

$$\rho(j) = \sum_{i}^{m} a_{ij}\sigma(i), \tag{76}$$

or in a matrix notation

$$\rho_m = A_m \sigma_m. \tag{77}$$

Since this result is in complete analogy to interaction nets (eq. (74)) where the response matrix R_n establishes the stimulus-response relationship, we have the following theorem:

Any stable interaction network composed of m elements can be represented by a functionally equivalent action network composed of $2m$ elements (see fig. 23):

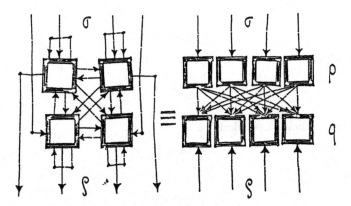

FIGURE 23. Equivalence of action network with interaction network.

$$\|r_{ij}\|_m = \|a_{ij}\|_m .$$

This result is of significance insofar as it shows that two entirely different structures have precisely the same stimulus-response characteristic. For example, Hartline's observation of inhibitory *interaction* amongst the fibers in the optic stalk of the horseshoe crab can be explained equally well by an appropriate post-ommatidial *action* net. It is only the anatomical evidence of the absence of such nets which forces us to assume that interaction processes are responsible for the observed phenomena.

The converse of the above theorem "for each action net there exists a functionally equivalent interaction net" is true only if the characteristic determinant of the inverse of A_m does not vanish.

We consider $k + 1$ cascaded layers L_i $(i = 0, 1, \ldots, k)$ with the transmitting layer L_o the locus of receptors proper, and with all elements in layer L_{i-1} acting upon all elements in layer L_i, their actions defined by an action matrix A_{mi}. The action performed by the receptors e_{oj} on the ultimate effectors e_{k1} is again (see eq. (3)) defined by the matrix product of all A_{mi}

$$\mathrm{Cas}\,(A_{m1}, A_{m2}, \ldots, A_{mk}) = \prod_1^k A_{mi}. \tag{78}$$

Hence, we have the following theorem:

Any cascaded network of a finite number of layers, each acting upon its follower with an arbitrary action matrix can be replaced by a functionally equivalent single action net with an action matrix

$$\left\| a_{ij}^{(k)} \right\|_m = \prod_{l=1}^k \left\| a_{ij}^{(l)} \right\|_m. \tag{79}$$

Again, gross structural differences may lead to indistinguishable performances.

We generalize our observation in Chapter 1, eq. (8) concerning the invariance of certain action nets to cascading. An action matrix with all rows alike

$$a_{ij}^* = a_{kj}^*,$$

and for which

$$\sum_{j=1}^m a_{ij}^* = 1$$

is invariant to being cascaded. Let

$$A_m^* = \|a_{ij}\|_m,$$

we have

$$[A_m^*]^k\,A_m^*. \tag{80}$$

4.5. Action Networks of Discrete, Localized Elements

After these general remarks we introduce more stringent geometrical constraints. We now assume that elements e_{ki} are not only traceable to a certain layer L_k, but also that each element within one layer can be localized as to its precise position in this layer (see fig. 24). We first consider only action phenomena between elements of two adjacent layers L_i and L_{i+1}, which, for simplicity, we may call L_p and L_q. We subdivide each layer into lattice elements with appropriate dimensions Δx, Δy, Δz, and $\Delta \xi$, $\Delta \eta$, $\Delta \zeta$ in L_p and L_q respectively, so as to be able to accommodate in each lattice element precisely one element e_{pi} and e_{qj} respectively. For simplicity we shall assume the corresponding lattice dimensions in both layers to be alike $\Delta x = \Delta \xi$, etc., because first, it does not infringe on the generality of the remarks we wish to make and, if there is indication to the contrary, it is not difficult to match the metric of the two layers L_p and L_q by appropriate transformations.

We are now in a position to label each element in layers L_p and L_q according to the coordinates of the lattice element in which it resides. Let p and q represent the coordinate triple that locates the lattice elements in the respective layers. We have:

$$p[x,y,z] = [x \cdot \Delta x, y \cdot \Delta y, z \cdot \Delta z],$$
$$q[u,v,w] = [u \cdot \Delta \xi, v \cdot \Delta \eta, w \cdot \Delta \zeta], \tag{81}$$

where x, y, z, u, v, w are integers 0, ±1, ±1,.... It is clear that for one-dimensional or two-dimensional layers the definitions for p and q boil down to $p[x]$, $q[u]$, and $p[x,y]$, $q[u,v]$ respectively. And it is also clear that

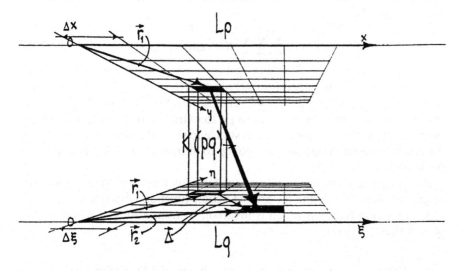

FIGURE 24. Geometry in an action network.

we may now drop the second index in the labeling of elements e_{pi} and e_{pj}, for e_p and e_q suffices to identify each element, since p and q define the locus of its position.

Again we wish to express the action exerted by element e_p to element e_q. To this end we define an "action function" $K(p,q)$ which specifies for each pair $[e_p,e_q]$ the fraction of activity in e_p that is transmitted to e_q. Of course, this action function may again be represented by an action matrix K_m. However, with our knowledge of the position of each element in both layers we may be able to associate the transmitted amount of activity with certain geometrical relationships which exist between elements of the two layers. In other words, the m^2 entries k_{pq} in the action matrix K_m may be all considered to be functions of the loci of the elements with which these entries are associated:

$$k_{pq} = k_{pq}(x, y, z; u, v, w). \tag{82}$$

The assumed dependency of the transmitted activity on geometrical relationships justifies the term action *function*. On the other hand, this is precisely the kind of relationship which is alluded to, if reference is made, say, to "cortical organization" or "organization of neural interaction". It is, to a certain extent, the genetic program that produces anatomical—read "geometrical"—constraints which prohibit, within certain limits, arbitrary developments of conceivable structures. The noteworthy feature of eq. (82) is that it links activity with geometry, in other words, function with structure.

Written out explicitly in terms of stimulus and response, the action function appears under the triple sum taken over all elements in the transmission layer L_p:

$$\rho(u,v,w) = \sum_x \sum_y \sum_z K(xyz,uvw)\,\sigma(xyz). \tag{83}$$

A discussion of the general properties of $K(p,q)$ goes beyond the scope of this article (In-selberg and Von Foerster, 1962; Von Foerster, 1962; Taylor, 1962). However, there is no need to go to extremes if the simple assumptions about prevailing geometry will suffice to show the significance of these concepts. Consequently, we are going to introduce further geometrical constraints.

Periodicity in structure is, as was suggested earlier, a ubiquitous feature in organic nets. We define a periodic action function with orders x_o, y_o, z_o to be an action function which is invariant to translations of whole multiples of these periods:

$$K(x+ix_o; y+jy_o; z+kz_o; u+ix_o; v+jx_o; w+kz_o) = K(xyz,uvw).$$
$$i, j, k = 0, \pm 1, \pm 2, \ldots . \tag{84}$$

Clearly, a network with such a periodic action function produces outputs in L_q that are invariant to any stimulus distribution which is translated with same periodicity x_o, y_o, z_o.

In order to have response invariance to stimulus translation *everywhere* alone the receptor set L_q, we must have:

$$x_o = \Delta x, \quad y_o = \Delta y, \quad z_o = \Delta z. \tag{85}$$

From this we may draw several interesting conclusions. First, the action functions so generated are independent of position, for the smallest interval over which they can be shifted is precisely the order of their period. Second, under the conditions of translatory invariance the action functions reduce to sole functions of the difference of the coordinates which localize the two connected elements in their respective layers:

$$K(p,q) = k(\Delta), \tag{86}$$

where Δ is a vector with components

$$\Delta = [(x - \xi), (y - \eta), (z - \zeta)]. \tag{87}$$

We introduce symmetric, anti-symmetric and spherically symmetric action functions which have the following properties respectively:

$$K_S(-\Delta) = K_S(\Delta), \tag{88}$$

$$K_a(-\Delta) = -K_a(\Delta), \tag{89}$$

$$K_r(\Delta) = K_r(\Delta). \tag{90}$$

It is easy to imagine the kind of abstractions these action functions perform on the set of all stimuli which are presented to the receptor set in L_p, if we assume for a moment that both layers, L_p and L_q, are planes. Clearly, in all cases the responses in the effect of set are invariant to all translations of any stimulus distribution ("pattern") in the receptor set. Moreover, K_s gives invariance to reversals of stimuli symmetric to axes $y = 0$ and $x = 0$ (e.g., 3 into $\in$, or M into W), while K_a gives invariance to reversals of stimuli symmetric to $y = \pm x$ (e.g., $\sim$ into S; and $>$ into V). Finally, action function K_r gives invariance to all stimulus rotations as well as translations (i.e., some reversals as above plus, e.g., N into Z). The planes of symmetry in three dimensions which correspond to the lines in two dimensions are clearly the three planes defined by the axes xy, yz, zx, in the first case, and, in the second case, the three planes defined by the six origin-centered diagonals that cut through the three pairs of opposite squares in the unit cube.

Although for analytic purposes the action function has desirable properties, from an experimental point of view it is by no means convenient. In order to establish in an actual case the action function of, say, element e_p^* in the receptor set, it is necessary to keep just this element stimulated while searching with a microprobe though all fibers of a higher nucleus to pick those that are activated by e_p^*. Since this is obviously an almost impossible task, the procedure is usually reversed. One enters a particular fiber e_q^* of a higher nucleus, and establishes, by stimulation of elements e_p, which one of these activates e_q^*. In this way it is the receptor field which is established, rather than the action field. However, considering the geometrical constraints so far introduced, it is easy to see that, with the exception of the anti-symmetric action function, action function $K(p,q)$ and "receptor function" $G(q,p)$ are identical:

$$G_S = K_S,$$
$$G_r = K_r,$$
$$G_a = -K_a. \tag{91}$$

For more relaxed geometrical constraints the expressions relating receptor function and action function may be more complex, but are always easy to establish.

It may have been noticed that a variety of structural properties of networks have been discussed without any reference to a particular action function. Although the actual computational labor involved to obtain stimulus-response relationships in action nets is far less than in interaction nets, the machinery is still clumsy if nets are of appreciable sophistication.

Instead of demonstrating this clumsiness in some examples, we postpone the discussion of such nets. In the next paragraph the appropriate mathematical apparatus to bypass this clumsiness will be developed. Presently, however, we will pick an extremely simple action net, and explore to a full extent the conceptual machinery so far presented with the inclusion of various examples of operation modalities of the network's constituents.

Example: Binomial Action Function

Fig. 25a represents our choice. It is a one-dimensional periodic action net with unit periodicity, its predominant feature being lateral inhibition. The universal action function and receptor function of this network are quickly found by inspection and are drawn in fig. 25b and c respectively. Obviously, these functions are symmetric, hence

$$G_1^*(q, p) = K_1^*(p, q) = K_1^*(x - u) = K_1^*(\Delta)$$
$$= (-1)^\Delta \binom{2}{1+\Delta},$$

i.e.,

$$K_1^*(\Delta) = \begin{cases} -1, \Delta = -1, \\ +2, \Delta = 0, \\ -1, \Delta = +1, \end{cases}$$

everywhere else $K_1^* = 0$.

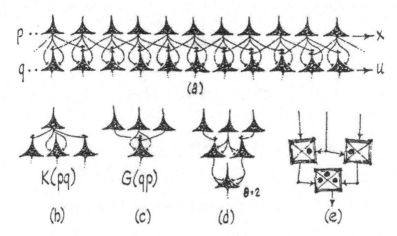

FIGURE 25. One dimensional periodic action net. Network (a); action function (b); receptor function (c); equivalent two-input element network (d); symbolic representation (e).

The equivalent net with McCulloch elements having only two inputs is given in fig. 25d. In fig. 25e its equivalent is symbolized in purely logical terms. The index 1 in G_1^* and K_1^* is adopted to discriminate this receptor and action function from a general class of such functions, K_m^*, the so-called mth order binomial action function:

$$K_m^* = (-1)^\Delta \binom{2m}{m+\Delta}, \quad \Delta = 0, \pm 1, \pm 2, \ldots, \pm m. \tag{92}$$

These arise from cascading binomial action nets m times, as suggested in fig. 26:

$$K_2^* = (-1)^\Delta \binom{4}{2+\Delta} = \begin{cases} +1, \Delta = -2, \\ -4, \Delta = -1, \\ +6, \Delta = 0, \\ -4, \Delta = +1, \\ +1, \Delta = +2. \end{cases}$$

(i) Sherrington Element

In order to obtain stimulus response relationship in the network of fig. 25 we have to specify the operational modality of the elements. First we assume a strictly linear model (Sherrington element). Let $\sigma(x)$ and $\rho(u)$ be stimulus and response at points x and u respectively. We have

$$\rho(u) = 2\sigma(x) - \sigma(x - \Delta x) - \sigma(x + \Delta x).$$

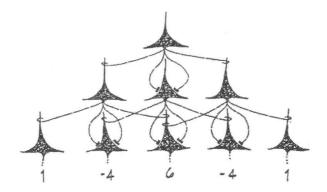

FIGURE 26. Cascading of binomial action function of mth function of $(m + 1)$th order.

Expanding σ around x we obtain:

$$\sigma(x \pm \Delta x) = \sigma(x) \pm \frac{\partial \sigma}{\partial x} \Delta x + \frac{1}{2} \frac{\partial^2 \sigma}{\partial x^2} \Delta^2 x \pm \frac{1}{6} \frac{\partial^3 \sigma}{\partial x^3} \Delta^3 x + \ldots,$$

which, inserted above, gives

$$\rho(u) = \frac{\partial^2 \sigma}{\partial x^2} \Delta^2 x + \frac{1}{12} \frac{\partial^4 \sigma}{\partial x^4} \Delta^4 x + \ldots.$$

Neglecting fourth order and higher terms, this lateral inhibition net extracts everywhere the second derivative of the stimulus distribution. It can easily be shown that the mth binomial action functions will extract the $2m$th derivative of the stimulus. In other words, for uniform stimulus, strong or weak, stationary or oscillating, these nets will not respond. However, this could have been seen by the structure of their binomial action function, since

$$\sum_{-m}^{+m} (-1)^\Delta \binom{2m}{m + \Delta} = 0.$$

(ii) McCulloch Element; Asynchronism

We change the *modus operandi* of our elements, adopt a McCulloch element with unit threshold ($\theta = 1$), and operate the net asynchronously. We ask for the output frequency of each effector element, given the stimulus distribution. For simplicity, we write our equations in terms of the ON probability p of the elements. With numbers 1, 2, 3, we label the afferent fibers in the receptor field (see fig. 25c). The truth table is easily established, giving an output ON for input states (010), (011) and (110) only. The surviving Bernoulli products (eq. (25)) are:

$$p = (1-p_1)p_2(1-p_3) + p_1p_2(1-p_3) + (1-p_1)p_2p_3,$$

which yields:

$$p = p_2 - p_1p_2p_3.$$

Uniform stimulation, i.e., $p_1 = p_2 = p_3 = p_o$, produces a "leakage" frequency

$$p_{uni} = p_o - p_o^3,$$

which disappears for strong stimulation (see fig. 27a bold line). With either element (1) or (3) OFF, i.e., with an "edge" in the stimulus field, we have

$$p_{edg} = p_o,$$

the difference between these frequencies Δp is, of course, an indication of detection sensitivity. Inspection of fig. 27a shows that this element is a poor edge detector in the dark, but does very well in bright light.

It might be worthwhile to note that a reversed action function $(1; -2; 1)$ with lateral facilitation has two operational modes, one of them with considerable sensitivity for low intensities (fig. 27b). A slight nystagmus with an amplitude of one element switches between these two modes, and thus represents an edge detector superior to the one with lateral inhibition.

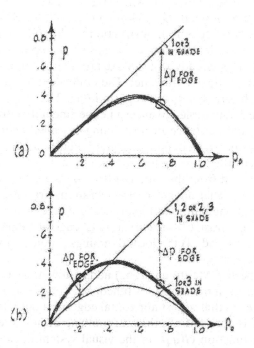

FIGURE 27. Transfer function of a McCulloch formal neuron net when operated asynchronously. (a) lateral inhibiting network. (b) lateral facilitating network.

(iii) McCulloch Element; Synchronism

We finally change the *modus operandi* of our net to synchronous operation with McCulloch element ($\theta = 1$). Clearly, for uniform stimulus distribution, strong or weak, steady or flicker, all effectors will be silent; the net shows no response. Nevertheless, an edge will be readily observed.

However, a net incorporating much simpler elements will suffice. A McCulloch element with only two inputs computing the logical function "either A or B" (see function No. 6, table 1) clearly computes an "edge". This function represents again a symmetric action function. Asymmetry may be introduced by choosing a McCulloch element that computes, say, function No. 2: "A only". A net incorporating this element in layer L_1 is given in fig. 28a. The result is, of course, the detection of an asymmetric stimulus property, the presence of a "right hand edge". Hence, in order to detect directionality in the stimulus field the net must mirror this directionality in the connectivity of its structure or in the operation of its elements. Utilizing synaptic delays that occur in layer L_1 we have attached a second layer L_2 that computes in D the function "C only". Consequently, layer L_2 detects right edges moving to the right. While C computes the presence of a right hand edge D will be silent, because the presence of a right edge implies a stimulated B which, simultaneous with an active C, gives an inactive D. Similarly, a left edge will leave D inactive; but C is inactive during the presence of a left edge. However, D will be active at once if we move the right hand edge of an obstruction to the right. Under these circumstances the synaptic delay in C will cause C to report still a right hand edge to D, while B is already without excitation. Of course, movements to the left remain unnoticed by this net. The equivalent net using the appropriate McCulloch formal neurons is given in fig. 28b.

Thanks to the remarkable advances in experimental neurophysiology, in numerous cases the existence of abstracting cascaded action networks in sensory pathways has been demonstrated. In their now classic paper "What the frog's eye tells the frog's brain" Letvvin et al. (1959) summed up their findings: "The output from the retina of the frog is a set of four distributed operations on the visual image. These operations are independent of the level of general illimunation and express the image in terms of: (1) local sharp edges and contrast; (2) the curvature of edge of a dark object; (3) the movement of edges; and (4) the local dimmings produced by movement or rapid general darkening".

To these properties Maturana (1962) adds a few more in the eye of the pigeon. Here anti-symmetric action functions produce strong directionalities. There are fibers that report horizontal edges, but only when they move. A vertical edge is detected whether it is moving or not. A careful analysis of the receptor function $G(q,p)$ in the visual system in cats and monkeys has been carried out by Hubel (1962); Mountcastle et al. (1962) explored the complicated transformations of multilayer mixed action-interaction net-

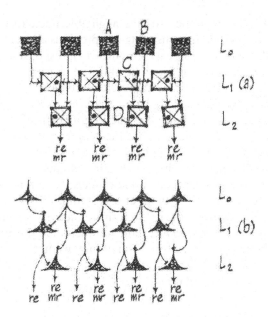

FIGURE 28. Net detecting right hand edges, and right hand edges moving to the right (a). Formal presentation. (b) Simplest equivalent neural net.

works as they occur at the thalamic relay nucleus with the somatic system as input.

3.7. Action Networks of Cell Assemblies

Sholl (1956) estimates the mean density $\bar{v}$ of neurons in the human cortex at about 10^7 neurons per cubic centimeter, although this number may vary considerably from region to region. This density implies a mean distance $\bar{l}$ from neuron to neuron.

$$\bar{l} = (\bar{v})^{-\frac{1}{3}} = 5 \times 10^{-3} \, cm$$

or approximately 50μ. This distance corresponds, of course, to the lattice constants $\Delta x, \Delta y, \Delta z$, etc. in the previous paragraph and gives the elementary lattice cell a volume of $\Delta x \Delta y \Delta z \approx 10^{-7} \, cm^3$. Even with the best equipment available today we cannot reproducibly attain a given point in the brain within this range. Consequently, our cell by cell approach describes a highly idealistic situation, and the question arises as to whether or not some of the earlier concepts can be saved if we wish to apply them to a much more realistic situation.

Let us assume optimistically that one is able to locate a certain point in the living cortex within, say 0.5 mm = 500μ. This defines a volume of uncertainty which contains approximately a thousand neurons. This number, on the other hand, is large enough to give negligible fluctuations in the total activity, so we are justified in translating our previous concepts, which apply to individual elements e_i, e_j as, e.g., stimulus $\sigma(i)$, response $\sigma(j)$, into a formalism that permits us to deal with assemblies of elements rather than with individuals. Moreover, as long as these elements connect with other elements over a distance appreciably larger than the uncertainty of its determination, and there is a considerable fraction of cortical neurons fulfilling this condition, we are still able to utilize the geometrical concepts as before. To this end we drop the cellular individuality and refer only to the activity of cell assemblies localizable within a certain volume.

In analogy to the concept of "number density" of neurons, i.e., the number of neurons per unit volume at a certain point (xyz) in the brain, we define "stimulus density" $\sigma(xyz)$ in terms of activity per unit volume as the total activity S measured in a certain volume, when this volume shrinks around the point (xyz) to "arbitrary" small dimensions:

$$\sigma(xyz) = \lim_{V \to 0} \frac{S}{V} = \frac{dS}{dV}. \tag{93}$$

Similarly, we have for the response density at (uvw):

$$\rho(uvw) = \lim_{V \to 0} \frac{R}{V} = \frac{dR}{dV}, \tag{94}$$

if R stands for the total response activity in a macroscopic region.

We wish to express the action exerted by the stimulus activity around some point in a transmitting "layer" L_1 on to a point in a receiving layer L_2. In analogy to our previous considerations we may formally introduce a "distributed action function" $K(xyz,uvw)$ which defines the incremental contribution to the response density $d\rho(uvw)$ from the stimulus activity that prevails in an incremental volume dV_1 around a point (xyz) in the transmitting layer. This activity is, with our definition of stimulus density $\sigma(xyz)$ dV_1. Consequently

$$d\rho(uvw) = K(xyz, uvw)\, \sigma(xyz)\, dV_1. \tag{95}$$

In other words, K expresses the fraction per unit volume of the activity around point (xyz) that contributes to the response at (uvw). The total response elicited at point (uvw) from all regions in layer L_1 is clearly the summation of all incremental contributions, if we assume that all cells around (uvw) are linear elements. Hence, we have

$$\rho(uvw) = \int_{V_1} K(xyz, uvw)\, \sigma(xyz)\, dV_1, \tag{96}$$

where V_1, the subscript to the integral sign, indicates that the integration has to be carried out over the whole volume V_1 representing the extension of layer L_1.

With this expression we have arrived at the desired relation that gives the response density at any point in L_2 for any stimulus density distribution in layer L_1, if the distributed action function K is specified.

In order to make any suggestions as to the form of this distributed action function, it is necessary to enliven the formalism used so far with physiologically tangible concepts. This we shall do presently. At the moment we adopt some simplifying notations. First, we may in various instances refer to cell assemblies distributed along surfaces (A) or along lines (D). In these cases we shall not change symbols for σ and ρ, although all densities refer in these cases to units of length. This may be permissible because the units will be clear from context. Second, we shall adopt for the discussion of generalities vector representation for the localization of our points of interest and introduce the point vector r. Discrimination of layers will be done by subscripts. We have the following correspondences:

transmitting "layer": $x, y, z; r_1; D_1; A_1; V_2;$

collecting "layer": $u, v, w; r_2; D_2; A_2; V_2.$

The physiological significance of the distributed action function $K(r_1,r_2)$ will become evident with the aid of figs. 29 and 30. Fig. 29a—or 29b—shows a linear array of neurons in a small interval of length dx about a point x in layer L_1. These neurons give rise to a number of axons N_x, some of which, say, $N_x(u_1)$, are destined to contact in the collecting layer L_2 with elements located in the vicinity of u_1; others, say, $N_x(u_2)$, will make contact with elements located at u_2, and so on:

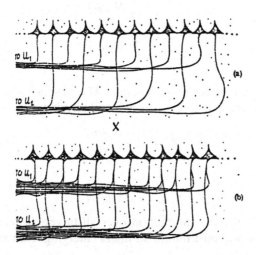

FIGURE 29. Departure of fibers in the transmitting layer of an action network of cell assemblies.

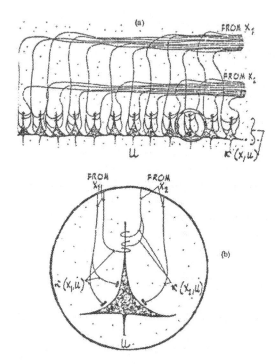

FIGURE 30. Arrival of fibers at the target layer of an action network of cell assemblies.

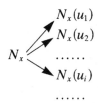

We shall define a "distribution function" $k(x, u_i)$ which is simply that fraction of all the fibers that emerge from x and terminate at u_i:

$$k(x, u_i) = \frac{N_x(u_i)}{N_x}$$

or

$$N_x(u_i) = k(x, u_i)N_x. \qquad (97)$$

Clearly, if we take the summation over all targets u_i reached by fibers emerging from x, we must obtain all fibers emerging from x:

$$N_x = \sum_{\text{all } i} N_x(u_i) = \sum_{\text{all } i} k(x, u_i)N_x = N_x \sum_{\text{all } i} k(x, u_i),$$

or, after cancellation of N_x on both sides:

$$1 = \sum_{\text{all } i} k(x, u_i).$$

If we consider now infinitesimal targets of length du, the above summation takes on the form of an integral

$$\int_{-\infty}^{+\infty} k(x, u)\, du = 1. \tag{98}$$

This suggests that $k(x,u)du$ may be interpreted as the probability for a fiber which originates at x will terminate within an interval of length du in the vicinity of u. Consequently, one interpretation of $k(x,u)$ is a "probability density function".

With this observation we may derive an expression for the contribution of region x to the fiber number density of region u. The corresponding fiber number densities are clearly defined by

$$n(x) = \frac{dN}{dx} \quad \text{and} \quad n(u) = \frac{dN}{du}.$$

Since $n(x)dx$ fibers all together emerge from an interval of length dx in the vicinity of x, their contribution to the number of fibers in the interval du at u is:

$$d^2N(u) = [k(x, u)\, du][n(x)\, dx] \tag{99}$$

with d^2 indicating that this is an infinitesimal expression of second order (an infinitesimal amount of an infinitesimal amount. Compare with eq. (97), its finite counterpart). Dividing in eq. (99) both sides by du we note that

$$\frac{d^2N_x(u)}{du} = d\left(\frac{dN}{du}\right)_x = dn_x(u) \tag{100}$$

is the contribution of x to the number density of fibers at point u. Using eqs. (99), (100), we can now express the desired relation between source and target densities by

$$dn(u) = k(x, u)n(x)dx. \tag{101}$$

From this point of view, k represents a mapping function that defines the amount of convergence or divergence of fiber bundles leaving the vicinity of point x and destined to arrive in the vicinity of point u. Clearly, k represents an important structural property of the network.

For the present discussion it is irrelevant whether certain neurons around x are the donors for elements around u, or whether we assume that after axonal bifurcation some branches are destined to contact elements around u. In both cases we obtain the same expression for the fractional contribution from x and u (compare figs. 29a and b).

If we pass over each fiber an average amount $\bar{s}$ of activity, we obtain the stimulus which is funneled from x to u by multiplying eq. (101) with this amount:

$$\bar{s}\, dn(u) = d\sigma(u) = k(x,u)\, \sigma(x)\, dx, \qquad (102)$$

because

$$\bar{s}n(x) = \sigma(x), \quad \bar{s}n(u) = \sigma(u). \qquad (103)$$

If the fractional stimulus density $d\sigma(u)$ at the target were translated directly into response density, we would have

$$\sigma(u) = \rho(u).$$

However, this is not true, for the arriving fibers will synapse with the target neurons in a variety of ways (fig. 30). Consequently, the resulting response will depend upon the kind and strength of these synaptic junctions which again may be a function of source and target points. To accommodate this observation we introduce a local transfer function $\kappa(x,u)$, that relates arriving stimulus with local response

$$d\rho(u) = \kappa(x,u)\, d\sigma(u). \qquad (104)$$

With the aid of eq. (102) we are now in a position to relate stimulus density in the source area to response density in the target area:

$$d\rho(u) = \kappa(x,u)\, k(x,u)\, \sigma(x)\, dx. \qquad (105)$$

Clearly, the product of the two functions κ and k can be combined to define one "action function"

$$K(x,u) = \kappa(x,u)\, k(x,u), \qquad (106)$$

and eq. (105) reduces simply to

$$d\rho(u) = K(x,u)\, \sigma(x)\, dx. \qquad (107)$$

Comparison of this equation with our earlier expression for the stimulus-response relationship (eq. (95)) shows an exact correspondence, eq. (107) representing the x-portion of the volume representation in eq. (95).

With this analysis we have gained the important insight that action functions—and clearly also interaction functions—are composed of two parts. A structural part $k(r_1,r_2)$ defines the geometry of connecting pathways, and a functional part $\kappa(r_1,r_2)$ defines the operational modalities of the elements involved. The possibility of subdividing the action function into two clearly separable parts introduces a welcome constraint into an otherwise unmanageable number of possibilities.

We shall demonstrate the workings of the mathematical conceptual machinery so far developed on three simple, but perhaps not trivial, examples.

(i) Ideal One-to-One Mapping

Assume two widely extending, but closely spaced, parallel surfaces, representing layers L_p and L_q. Perpendicular to L_p emerge parallel fibers which synapse with their corresponding elements in L_q without error and without deviation. In this case the mapping function k is simply Dirac's Delta Function*

$$k(r_1,r_2) = \delta^2(r_1 - r_2),$$

$$\delta(x - x_o) = \begin{cases} 0 \text{ for } x \neq x_o \\ \infty \text{ for } x = x_o \end{cases}$$

and

$$\int_{-\infty}^{+\infty} \delta(x - x_o)\, dx = 1.$$

For simplicity, let us assume that the transfer function κ is a constant a. Hence

$$K(r_1,r_2) = a\delta^2(r_1 - r_2),$$

and, after eq. (96):

$$\rho(r_2) = a \int_{L_p} \delta^2(r_1 - r_2)\, \sigma(r_1)\, dA_1 = a\sigma(r_1).$$

As was to be expected, in this simple case the response is a precise replica of the stimulus, multiplied by some proportionality constant.

(ii) Ideal Mapping with Perturbation

Assume we have the same layers as before, with the same growth program for fiber descending upon L_q, but this time the layers are thought to be much further apart. Consequently, we may expect the fibers to be affected by random perturbations, and a fiber bundle leaving at r_1 and destined for $r_2 = r_1$ will be scattered according to a normal (Gaussian) distribution. Hence, we have for the mapping function

$$k(r_1,r_2) = 1/2\pi h^2 \exp(-\Delta^2/2h^2)$$

with

$$\Delta^2 = |r_1 - r_2|^2$$

and h representing the variance of the distribution.

---

* The exponent in δ indicates the dimensionality of the manifold considered. Here it is two-dimensional, hence δ^2.

Assume furthermore that the probability distribution for facilitatory and inhibitory contacts are not alike, and let κ be a constant for either kind. The action function is now:

$$K(r_1, r_2) = K(\Delta^2)$$
$$= a_1 \exp(-\Delta^2/2h_1^2) - a_2 \exp(-\Delta^2/2h_2^2).$$

The stimulus-response relation is

$$\rho(r_2) = \int_{L_1} K(\Delta^2)\, \sigma(r_1)\, dA_1.$$

For a uniform stimulus distribution

$$\sigma = \sigma_o,$$
$$\rho(r_2) = \text{constant} = 2\pi\sigma_o (a_1 h_1^2 - a_2 h_2^2),$$

and if $$a_1 h_1^2 = a_2 h_2^2,$$

then

$$\rho = 0.$$

As one may recall, the interesting feature of giving zero-response to finite stimuli was obtained earlier for discrete action functions of the binomial form. A similar result for the Gaussian distribution should therefore not be surprising, if one realizes that the continuous Gaussian distribution emerges from a limit operation on the binomial distribution. What are the abstracting properties of this net with a random normal fiber distribution?

Fig. 31 represents the response activity for a stimulus in the form of a uniformly illuminated square. Clearly, this network operates as a computer of contours. It may be noted that the structure of this useful network arose from random perturbation. However, there was an original growth program, namely, to grow in parallel bundles. Genetically, this is not too difficult to achieve; it says: "Repeat". However, such a net is of little value, as we have just seen. It is only when noise is introduced into the program that the net acquires its useful properties. It may be mentioned that this net works best when the zero-response condition is fulfilled. This, however, requires adaptation.

(iii) Mapping into a Perendicular Plane

The previous two examples considered action functions with spherical symmetry. We shall now explore the properties of an action function of type K_S with lateral symmetry. Such an action function may arise in the following way (see fig. 32):

Assume again two surfaces layers, L_p, L_q where in L_p all neurons are aligned in parallel with their axis of symmetry perpendicular to the layer's surface, while in L_q they are also in parallel, but with their axis of symmetry lying in the surface of L_q.

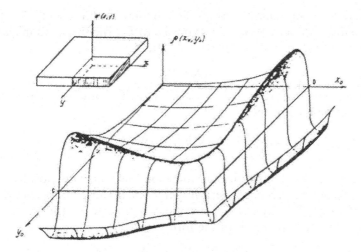

FIGURE 31. Response distribution elicited by a uniform stimulus confined to a square. Contour detection is the consequence of a distributed action function obtained by superposition of a facilitatory and inhibitory Gaussian distribution.

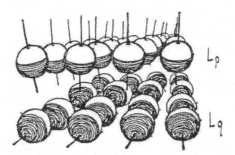

FIGURE 32. Geometrical relationship between layers of neurons with perpendicular orientation of their axes.

We consider pyramidal neurons and represent them by spheres. We let the North pole (+) coincide with basal axonal departure and the South pole (−) with the upper branchings of apical dendrites. The perikaryon is central. This spherical dipole assumes physiological significance if we associate with the neuron's structural difference when seen from north or from south different probabilities for the establishment of facilitatory or inhibitory synapses. For simplicity we assume that for an afferent fiber the probability of making facilitatory or inhibitory connection is directly proportional to the projected areas of northern southern hemisphere respectively, seen by this fiber when approaching the neuron. Hence, a fiber approaching along the equatorial plane has a 50–50 chance to either inhibit or facilitate.

A fiber descending upon the South Pole inhibits with certainty. Let there be two kinds of fibers descending from L_p to L_q. The first kind maps with a Dirac delta function the activity of L_p into L_q.

$$k_1(\Delta) = \delta(\Delta).$$

Let k_2, the mapping function of the second kind, be any spherical symmetric function that converges:

$$\int_{L_1} k_2(\Delta^2)\, dA_1 = \text{constant},$$

for instance, a normal distribution function. The associated local transfer function κ_2, however, is not spherical symmetric because of the lateral symmetry of the probability of $(\pm)$ connections. Simple geometrical considerations (fig. 33) show that κ_2 is of the form

$$\kappa_2(r_1, r_2) = a\cos\phi,$$

where ϕ is the angle between Δ and the N-S axis of elements.

The action function of the network is

$$K(\Delta) = a\delta(\Delta)\, k_2(\Delta^2)\cos\phi,$$

and the response density for a given stimulus:

$$\rho(r_2) = a\int_{L_i} \delta(\Delta)k_2(\Delta^2)\cos\phi\; \sigma(r_1)\, dA_1.$$

What does this system compute?

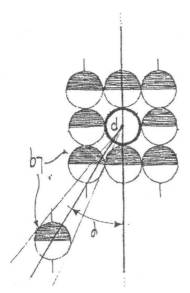

FIGURE 33. Geometrical relationship between elements of two layers. View from the top.

Its usefulness becomes obvious if we assume that the stimulus of L_q is a contour that has been computed in L_p from a preceding network say, L_o, L_p. On behalf of the δ-function this contour, and nothing else, maps from L_p into L_q. Take, for instance, a straight line with uniform intensity σ_o to be the stimulus for L_q (see fig. 34). Since two points symmetrical to any point on this line contribute

$$ak_2\sigma_o\left(\cos\phi + \cos(180 - \phi)\right)\Delta A_1 = 0,$$

because

$$\cos\phi = -\cos(180 - \phi),$$

a straight line gives no response. Curvature, however, is reported. The total net-response

$$\overline{R} = \int_{L_2} \rho(r_2)\,\mathrm{d}A_2$$

vanishes for bilateral symmetrical figures with their axis of symmetry parallel to the orientation of elements in L_q. However, when turned away from this position, $\overline{R} \neq 0$ (see fig. 35).

4.8. Interaction Networks of Cell Assemblies

We consider the case where connections between all elements in the network appear freely and a separation into layers of purely-transmitting and purely-receiving elements is impossible. When all return connections between interconnected elements are cut, the system reduces to an action network. Clearly, we are dealing here with the more general case and consequently have to be prepared for results that do not yield as easily as in the previous case. However, the methods concepts developed in the previous section are immediately applicable to our present situation.

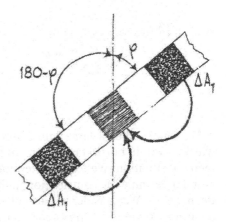

FIGURE 34. Insensitivity of antisymmetrical action net to straight lines.

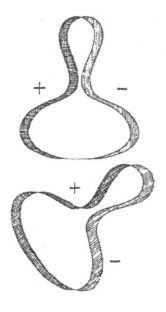

FIGURE 35. Responses of antisymmetrical action net to figures with bilateral symmetry.

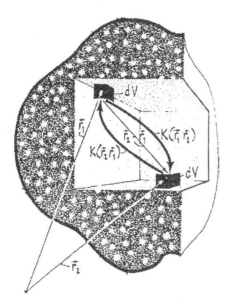

FIGURE 36. Geometry in an interaction network of cell assemblies.

Fig. 36 sketches a large interaction network, confined to volume V, in which a certain portion has been cut away in order to make clear the geometrical situation. We fix our attention on elements in the vicinity of point r_2 and determine the contribution to their stimulation from other regions of the network. We consider in particular the contribution to r_2 from the activity around point r_1. We proceed precisely as before (see eq. (95)) and write

$$d\rho(r_2) = K(r_1, r_2)\, \rho(r_1)\, dV, \tag{108}$$

where K, the distributed interaction function, defines again the fraction per unit volume of the activity prevailing around point r_1 that is transmitted to point r_2, and $\rho(r_1)dV$ is clearly the activity of elements in the vicinity of point r_1.

Of course, the same physiological interpretation that has been given to the action function is applicable to the interaction function, except that properties of symmetry refer to symmetry of exchanges of stimulation between two points. In other words, $K(r_1, r_2)$ and $K(r_2, r_1)$ describe the proportions that are transmitted from r_1 to r_2, and back from r_2 to r_1 respectively. These proportions may not necessarily be the same.

In addition to the stimuli contributed by elements of its own network, each element may, or may not, receive stimulation from fibers descending upon this network from other systems that do not receive fibers from the network under consideration. We denote the elementary stimulation so contributed to r_2 by $d\sigma(r_2)$. It is, of course, no restriction to assume that these fibers stem from another network, say V_o, that functions as action network on to our system. In this case $d\sigma(r_2)$ may be directly replaced by $d\rho(r_2)$ or eq. (95), noting, however, that the action function in this expression has to be changed into, say, $A(r_o, r_2)$, where r_o indicates positions of elements in this donor system. With $d\sigma(r_2)$ representing a stimulus from external sources to elements around r_2 of our network we have for the total elementary stimulus at r_2:

$$d\rho(r_2) = d\sigma(r_2) + K(r_1, r_2)\, \rho(r_1)\, dV, \tag{109}$$

which summed over the entire volume V gives the desired stimulus-response relationship for any point in this volume:

$$\rho(r_2) = \sigma(r_2) + \int_V K(r_1, r_2)\, \rho(r_1)\, dV. \tag{110}$$

This equation cannot be readily solved by integration, unlike the case for action networks, because here the unknown quantity ρ appears not only explicitly on the left-hand side of this equation, but also implicitly within the integral. Expressions of this type are called integral equations and (110) above belongs to the class of integral equations of the second kind. The function $K(r_1, r_2)$ is usually referred to as the "kernel", and methods of solution are known, if the kernel possesses certain properties.

It is fortunate that a general solution for eq. (110) can be obtained (Inselberg Von Foerster, 1962, p. 32) if the kernel K is a function of only the distance between points r_1 and r_2:

$$K(r_1, r_2) = K(r_1 - r_2) = K(\Delta), \tag{111}$$

where Δ stands again (see eq. (86)) for the vector expressing this distance. These kernels represent precisely the kind of interaction function we wish to consider, for it is this property that makes the computations in the network invariant to stimulus translations (see eq. (85)).

The general solution for response, given explicitly in terms of stimulus interaction function, is, for the x-component:

$$\rho(x) = 1/\sqrt{2\pi} \int_{-\infty}^{+\infty} \frac{F_S(u)}{1 - \sqrt{2\pi} F_k(u)} e^{-ixu} du, \qquad (112)$$

where F_S and F_k are the Fourier transforms of stimulus distribution σ and interaction function K respectively:

$$F_S(u) = 1/\sqrt{2\pi} \int_{-\infty}^{+\infty} \sigma(x) e^{ixu} dx, \qquad (113a)$$

$$F_k(u) = 1/\sqrt{2\pi} \int_{-\infty}^{+\infty} K(t) e^{itu} dt, \qquad (113b)$$

with t representing the x component of the distance and i the imaginary unit

$$t = \Delta_x, \quad i = \sqrt{-1}. \qquad (114)$$

The expressions are valid for the other components, *mutatis mutandis*.

With respect to these results two comments are in order. First, one should observe the analogy of eq. (110) with the result obtained in the case of interaction of individually distinguishable elements (eq. (65)).

$$\rho(j) = \sigma(j) + \sum_{i=1}^{n} a_{ij} \, \rho(i), \qquad (65)$$

where summation over the activity of individual elements and the interaction coefficients a_{ij} correspond to integration and interaction function in (110) respectively. However, the cumbersome matrix inversions as suggested in eqs. (66) to (74) disappear, because the Fourier transforms in eqs. (112) and (113) perform these inversions—so to say—in one stroke. Thus, a general study of network structures will have to proceed along the lines suggested here, otherwise sheer manipulatory efforts may attenuate the enthusiasm for exploring some worthwhile possibilities.

The other comment refers to our earlier observation of the functional equivalence of discrete action and interaction nets (see fig. 23). The question arises whether or not an action network can be found that has precisely the same stimulus-response characteristic as a given interaction network of cell assemblies. It is not insignificant that this question can be answered in the affirmative. Indeed, it can be shown that a functional equivalent action net with action function $A(t)$ can be generated from a given interaction net with interaction function $K(t)$ by the Fourier transform

$$A(t) = 1/\sqrt{2\pi} \int_{-\infty}^{+\infty} \frac{1}{1 - \sqrt{2\pi} F_k(u)} e^{itu} du. \qquad (115)$$

These transforms are extensively tabulated (Magnus and Oberhettinger, 1949) and permit one to establish quickly the desired relationships.

Since the same performance can be produced by two entirely different structural systems one may wonder what is Nature's preferred way of accomplishing these performances: by action or by interaction networks? This question can, however, be answered only from an ontogenetic point of view. Since the "easy" way to solve a particular problem is to use most of what is already available, during evolution the development of complex net structures of either kind may have arisen out of a primitive nucleus that had a slight preference for developing in one of these directions. Nevertheless, there are the two principles of "action at a distance" and "action by contagion", where the former may be employed when it comes to highly specified, localized activity, while the latter is effective for alerting a whole system and swinging it into action. The appropriate networks which easily accommodate these functions are obvious, although the equivalence principle may reverse the situation.

Examples

(i) Gaussian, Lateral Inhibition

We give as a simple example an interaction net that produces highly localized responses for not-well-defined stimuli. Consider a linear network with purely inhibitory interaction in the form of a normal distribution:

$$K(\Delta) = K(x_1 - x_2) \sim \exp\left[-p(x_1 - x_2)^2\right].$$

As a physiological example one may suggest the mutually inhibiting action in the nerve net attached to the basilar membrane which is assumed to be responsible for the sharp localization of frequencies on it.

Suppose the stimulus—in this case the displacement of the basilar membrane as a function of distance x from its basal end—is expressed in terms of a Fourier series

$$\sigma(x) = a_o + \sum a_i \sin(2\pi ix/\lambda) + \sum b_i \cos(2\pi ix/\lambda),$$
$$i = 0, 1, 2, \ldots,$$

with coefficients a_i, b_i and fundamental frequency λ. It can be shown from eq. (110) that the response will be also a periodic function which can be expressed as a Fourier series with coefficients a_i^* and b_i^*. These have the following relation to the stimulus coefficients:

$$\frac{a_i^*}{a_i} = \frac{b_i^*}{b_i} = \frac{p}{p + \exp\left[-4\pi(1/\lambda)^2\right]}.$$

Since this ratio goes up with higher mode numbers i, the higher modes are always enhanced, which shows that indeed an interaction function with inhibitory normal distribution produces considerable sharpening of the original stimulus.

(ii) Antisymmetric Interaction

As a final example of a distributed interaction function which sets the whole system into action when stimulated only by that most local stimulus, the Dirac delta function, we suggest an antisymmetric one-dimensional inter-action function

$$K(\Delta_x) = \frac{\sin^2 \Delta_x}{\Delta_x}.$$

This function inhibits to the left ($\Delta < 0$) and facilitates to the right ($\Delta_x > 0$). It is, in a sense, a close relative to the one-directional action-interaction function of the quadrupole chain (fig. 21) discussed earlier. We apply to this network at one point x a strong stimulus:

$$(x) = \delta(x).$$

The response is of the form

$$\rho(x) = \delta(x) - a \frac{\sin x}{x} \cos(x - a),$$

a and a being constants.

Before concluding this highly eclectic chapter on some properties of computing networks it is to be pointed out that a general theory of networks that compute invariances on the set of all stimuli has been developed by Pitts and McCulloch (1947). Their work has to be consulted for further expansion and deeper penetration of the cases presented here.

5. Some Properties of Network Assembles

In this approach to networks we first considered nets composed of distinguishable elements. We realized that in most practical situations the individual cell cannot be identified and we developed the notions of acting and interacting cell assemblies whose identity was associated only with geometrical concepts. The next logical step is to drop even the distinguishability of individual nets and to consider the behavior of assemblies of nets. Since talking about the behavior of such systems makes sense only if they are permitted to interact with other systems—usually called the "environment"—this topic does not properly belong to an article confined to networks and, hence, has to be studied elsewhere (Pask, 1966). Nevertheless, a few points may be made, from the network point of view, which illuminate the gross behavior of large systems of networks in general.

We shall confine ourselves to three interrelated points that bear on the question of stability of network assemblies. Stability of network structures can be understood in essentially three different ways. First, in the sense, of a constant or periodic response density within the system despite various input perturbations (Dynamic Stability); second, in terms of performance, i.e., the system's integrity of computation despite perturbations of structure or function of its constituents (Logical Stability); third, to reach stabilities in the two former senses despite permanent changes in the system's environment (Adaptation). We shall briefly touch upon these points.

5.1. Dynamic Stability

Beurle (1962) in England and Farley and Clark (1962) at MIT were probably the first to consider seriously the behavior of nets of randomly connected elements with transfer functions comparable to eq. (47). Both investigated the behavior of about a thousand elements in a planar topology and a neighborhood connection scheme. Beurle used his network to study computation with distributed memory. To this end, elements were constructed in such a way that each activation caused a slight threshold reduction at the site of activity and made the element more prone to fire for subsequent stimuli. The system as a whole shows remarkable tendencies to stabilize itself, and it develops dynamic "engrams" in the form of pulsating patterns. Farley and Clark's work is carried out by network simulation on the Lincoln Laboratory's TX-s computer; and the dynamic behavior resulting from defined stimuli applied to selected elements is recorded with a motion picture camera. Since elements light up when activated, and the calculation of the next state in the network takes TX-2 about 0.5 seconds, the film can be presented at normal speed and one can get a "feeling" for the remarkable variety of patterns that are caused by variations of the parameters in the network. However, these "feelings" are at the moment our best clues to determine our next steps in the approach to these complicated structures.

Networks composed of approximately one thousand Ashby elements (see fig. 13) were studied by Fitzhugh (1963) who made the significant observation that slowly adding connections to the element defines with reproducible accuracy, a "connectedness" by which the system swings from almost zero activity to full operation, with a relatively small region of intermediate activity. This is an important corollary to an observation made by Ashby et al. (1962), who showed that networks composed of randomly connected McCulloch elements with facilitatory inputs only, but controlled by a fixed threshold, show no stability for intermediate activity, only fit or coma, unless threshold is regulated by the activity of elements.

In all these examples, the transfer function of the elements is varied in some way or another in order to stabilize the behavior of the system. This, however, implies that in order to maintain dynamic stability one has to sacrifice logical stability, for as we have seen in numerous examples (e.g., fig. 10) variation in threshold changes the function computed by the element. Hence, to achieve both dynamic and logical stability it is necessary to consider logically stable networks that are immune to threshold variation.

5.2. Logical Stability

McCulloch (1958) and later Blum (1962), Verbeek (1962) and Cowan (1962) were probably the first to consider the distinction between proper computation based on erroneous arguments (calculus of probability) and erroneous calculation based on correct arguments (probabilistic logic). It is precisely the latter situation one encounters if threshold in a system is subjected to variations. On the other hand, it is known that living organisms are reasonably immune to considerable variations of threshold changes produced by, say, chemical agents. Clearly, this can only be accomplished by incorporating into the neural network structure nets that possess logical stability.

The theory developed by the authors mentioned above permits the construction of reliable networks from unreliable components by arranging the various components so that, if threshold changes occur in the net, an erroneous computation at one point will be compensated by the computation at another point. However, the theory is further developed for independent changes of threshold everywhere, and nets can be developed for which the probability of malfunctioning can be kept below an arbitrarily small value, if sufficient components are added to the system. This, of course, increases its redundancy. However, this method requires substantially fewer elements to achieve a certain degree of reliability than usual "multiplexing" requires in order to operate with equal reliability. Due to a multivaluedness in the structure of these nets, an additional bonus offered by this theory is immunity against perturbation of connections.

All these comments seem to imply that variability of function tends to increase the stability of systems to such a degree that they will become too rigid when secular environmental changes demand flexibility. On the contrary, their very complexity adds a store of potentially available functions that enables these networks to adapt.

5.3. Adaptation

This ability may be demonstrated on an extraordinarily simple network composed of McCulloch elements operating in synchrony (fig. 37). It consists of two arrays of elements, black and white denoting sensory and computer elements respectively. Each computer element possesses two inputs proper that originate in two neighboring sensory elements A, B. The output of each computer element leads to a nucleus Σ that takes the sum of the outputs of all computer elements. The logical function computed by these is not specified. Instead it is proposed that each element is capable—in principle—of computing all 16 logical functions. These functions are to change from functions of lowest logical strength Q (see table 1 or fig. 38) to functions of higher logical strength in response to a command given by an "improper input", whose activity is defined by Σ that operates on the functions—the inner structure—of these elements and not, in a direct sense, on their outputs.

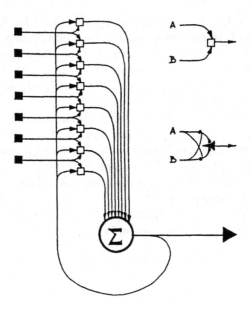

FIGURE 37. Adaptive network.

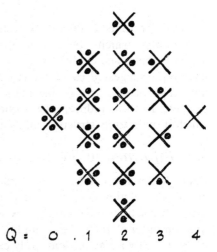

FIGURE 38. Logical functions ordered according to increasing logical strength Q.

Consider this net exposed to a variety of stimuli which consist of shadows of one-dimensional objects that are in the "visual field" of the sensors. With all computing elements operating on functions with low Q the feedback loop from Σ is highly active and shifts all functions to higher Q's. It is not difficult to see that this process will go on with decreasing activity in the loop until function No. 6 or even functions No. 4, or 2 of table 1 are reached, at which instant the net is ready to compute the presence of edges in the stimulus field—as has been shown in fig. 25—and the loop activity is reduced to the small amount that remains when objects happen to be in the visual field. Since it is clear that the output of the whole system represents the number of edges present at any moment, it represents at the same time a count of the number of objects, regardless of their size and position, and independent of the strength of illumination.

Consider this system as being in contact with an environment which has the peculiar property of being populated by a fixed number of objects that freely move about so that the limited visual field consisting of, say, N receptors perceives only a fraction of the number of these objects. In the long run, our system will count objects normally distributed around a mean value with a standard deviation of, say, σ.

The amount of information (entropy) H_{OUT} as reported by the system about its environment is (Shannon and Weaver, 1949):

$$H_{\text{OUT}} = \ln \sqrt{2\pi e}$$

with

$$e = 2.71828.\ldots$$

On the other hand, with N binary receptors its input information is

$$H_{\text{IN}} = N \gg H_{\text{OUT}}.$$

This represents a drastic reduction in information—or reduction in uncertainty—which is performed by this network and one may wonder why and how this is accomplished. That such a reduction is to be expected may have been suggested by the system's indifference to a variety of environmental particulars as, e.g., size and location of objects, strength of illumination, etc. But this indifference is due to the network's abstracting powers, which it owes to its structure and the functioning of its constituents.

5.4. Information Storage in Network Structures

Let us make a rough estimate of the information stored by the choice of a particular function computed by the network elements. As we have seen, these abstractions are computed by sets of neighbor elements that act upon one computer element. Let n_s and n_h be the number of neighbors of the kth order in a two-dimensional body-centered square lattice and hexagonal lattice respectively (see fig. 39):

$$n_s = (2k)^2, \quad n_h = 3k(k+1). \tag{116}$$

The number of logical functions with n inputs is (eq. (18))

$$N = 2^{2^n},$$

and the amount of information necessary to define a particular function is:

$$H_F = \log_2 N = 2^n. \tag{117}$$

FIGURE 39. First and second order neighbors in a body centered cubic lattice (two dimensional "cube").

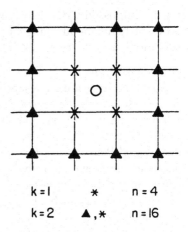

$$k = 1 \qquad * \qquad n = 4$$
$$k = 2 \qquad \blacktriangle, * \qquad n = 16$$

On the other hand, the input information on a sensory organ with N binary receptors is

$$H_{\text{IN}} = N.$$

A sensory network is properly matched to its structure if

$$H_{\text{IN}} = H_{\text{F}}, \tag{119}$$

or, in other words, if its input information corresponds to its computation capacity stored in its structure. We have with eqs. (116), (117), (119):

$$N = \begin{cases} 2^{4k^2} & \text{square lattice} \\ 2^{3k(k+1)} & \text{hexagonal lattice} \end{cases}$$

The following table relates the size of the sensory organ that is properly matched to its computing network which utilizes kth order neighbors, constituting a receptor field of n elements:

N	k_{s}	k_{h}	$n_{\text{s}} = n_{\text{h}}$
10^2	1.29	1.07	6.65
10^4	1.82	1.68	13.2
10^5	2.04	1.91	16.7
10^6	2.24	2.12	20.0
10^7	2.41	2.31	23.2
10^8	2.58	2.52	26.5

This table indicates that in an eye of, say, 10^6 receptor elements a receptor field of more than 20 elements is very unlikely to occur.

In conclusion it may be pointed out that the evolution of abstracting network structures as a consequence of interactions with an environment gives rise to new concepts of "memory" which do not require the faithful recording of data. In fact, it can be shown that a theory of memory that is based on mechanisms that store events is not only uneconomical bordering on the impossible, but also is incapable of explaining the most primitive types of behavior in living organisms that show one or another form of retention (Von Foerster, 1965; Von Foerster et al., 1966).

Acknowledgements. I am indebted to numerous friends and colleagues who contributed their ideas presented in this article, and who helped in shaping their presentation. I owe to Alfred Inselberg the development of the elegant mathematics in networks of cell assemblies, and I am grateful to George W. Zopf for critical comments and advice in matters of physiology, to W. Ross Ashby and Paul Weston for reading and helping to polish the manuscript, and—last not least—to Lebbius B. Woods III for his congenial drawings. If errors and faults of exposition remain, it is not these friends and colleagues who gave so generously of their time and advice, but I alone who must bear the blame.

The work reported in this paper was made possible jointly by the U.S. Air Force Office of Scientific Research under Grant AF-OSR 7-63, the U.S. Department of Public Health under Grant GM 10718-01, the U.S. Air Force Engineering Group Contract AF 33(615)-3890, and the U.S. Air Force Office of Scientific Research under Contract AF 49(638)-1680.

References

Ashby, W. R., 1956, *An introduction to cybernetics* (Chapman and Hall, London).

Ashby, W. R., H. Von Foerster and C. C. Walker, 1962, Instability of pulse activity in a net with threshold, Nature 196, 561.

Bar-Hillel, Y., 1955, Semantic information and its measure, in: *Cybernetics*, eds. H. Von Foerster, M. Mead and H. L. Teuber (Josiah Macy Jr. Foundation, New York) p. 33.

Beurle, R. L., 1962, Functional organizations in random networks, in: *Principles of self-organization*, eds. H. Von Foerster and G. W. Zopf Jr. (Pergamon Press, New York) p. 291.

Blum, M., 1962, Properties of a neuron with many inputs, in: *Principles of self-organization*, eds. H. Von Foerster and G. W. Zopf Jr. (Pergamon Press, New York) p. 98.

Carnap, R., 1938, *The logical syntax of language* (Harcourt, Brace and Co., New York).

Cowan, J., 1962, Many-valued logics and reliable automata, in: *Principles of self-organization*, eds. H. Von Foerster and G. W. Zopf Jr. (Pergamon Press, New York) p. 135.

Eccles, J. C., 1952, *The neurophysiological basis of mind* (Clarendon Press, Oxford).

Farley, B., 1962, Some similarities between the behavior of a neural network model and electrophysiological experiments, in: *Self-organizing systems*, eds. M. C. Yovits et al. (Spartan Books, Washington, D. C., 1962) p. 207.

Fitzhugh II, H. S., 1963, Some considerations of polystable systems, *Bionics Symposium*, Dayton, Ohio, USAF Aeronautical Systems Division, Aerospace Medical Division, 19–21 March 1963.

Fulton, J. F., 1943, *Physiology of the nervous system* (Oxford University Press, New York).

Hartline, H. K., 1959, Receptor mechanisms and the integration of sensory information in the eye, *Biophysical science*, ed. J. L. Oncley (John Wiley and Sons, New York) p. 515.

Hilbert, D. and W. Ackermann, 1928, *Grundzüge der theoretischen Logik* (Springer, Berlin).

Hubel, D. H., 1962, Transformation of information in the cat's visual system, in: *Information processing in the nervous system*, eds. R. W. Gerard and J. W. Duyff (Excerpta Medica Foundation, Amsterdam) p. 160.

Inselberg, A. and H. Von Foerster, 1962, *Property extraction in linear networks*, NSF Grant 17414, Tech. Rep. No. 2, (Electrical Engineering Research Laboratory, Engineering Experiment Station, University of Illinois, Urbana).

Katz, B., 1959, Mechanisms of synaptic transmission, in: *Biophysical science*, ed. J. L. Oncley (John Wiley Sons, New York) p. 524.

Kinosita, H., 1941, Conduction of impulses in superficial nervous systems of sea urchin, Jap. J. Zool. 9, 221.

Landahl, H. D., W. S. McCulloch and W. Pitts, 1943, A statistical consequence of the logical calculus of nervous nets, Bull. Math. Biophys. 5, 135.

Lettvin, J. Y., H. R. Maturana, W. S. McCulloch and W. Pitts, What the frog's eye tells the frog's brain, Proc. Inst. Radio Engrs. 47, 1940.

Magnus, W. and F. Oberhettinger, 1949, *Special functions of mathematical physics* (Chelsea, New York) p. 115.

Maturana, H. R., 1962, Functional organization of the pigeon retina, in: *Information processing in the nervous system*, eds. R. W. Gerard and J. W. Duyff (Excerpta Medica Foundation, Amsterdam) p. 170.

McCulloch, W. S., 1958, The stability of biological systems, in: *Homeostatic mechanisms*, ed. H. Quastler (Brookhaven Natl. Lab. Upton) p. 207.

McCulloch, W. S., 1962, Symbolic representation of the neuron as an unreliable function, in: *Principles of self-organization*, eds. H. Von Foerster and G. W. Zopf Jr. (Pergamon Press, New York) p. 91.

McCulloch, W. S. and W. Pitts, A logical calculus of the ideas immanent in nervous activity, Bull. Math. Biophys, 5, 115.

Mountcastle, V. B., G. F. Poggio and G. Werner, 1962, The neural transformation of the sensory stimulus at the cortical input level of the somatic afferent system, in: *Information processing in the nervous system*, eds. R. W. Gerard and J. W. Duyff (Excerpta Medica Foundation, Amsterdam) p. 196.

Parker, G. H., 1943, *The elementary nervous system* (Lippincott, Philadelphia).

Pask, G., 1966, Comments on the cybernetics of ethical, sociological and psychological systems, in: *Progress in biocybernetics*, vol. 3, eds. N. Wiener and J. P. Schadé (Elsevier, Amsterdam) p. 158.

Pitts, W. and W. S. McCulloch, 1947, How we know universals, Bull. Math. Biophys. 9, 127.

Russell, B. and A. N. Whitehead, 1925, *Principia mathematica* (Cambridge University Press, Cambridge).

Shannon, C. E. and W. Weaver, 1949, *The mathematical theory of communication* (University of Illinois Press, Urbana) p. 56.

Sherrington, C. S., 1906, *Integrative action of the nervous system* (Yale University Press, New Haven).

Sherrington, C. S., 1952, Remarks on some aspects of reflex inhibition, Proc. Roy. Soc. B 95, 519.

Sholl, D. A., 1956, *The organization of the cerebral cortex* (Methuen and Co., London).

Stevens, S. S., 1957, Psych. Rev. 64, 153.

Taylor, D. H., 1962, *Discrete two-dimensional property detectors*, Tech. Rep. No. 5, NSF G-17414, (Electrical Engineering Research Laboratory, Engineering Experiment Station, University of Illinois, Urbana).

Üxküll, J. V., 1896. Über Reflexe bei den Seeigeln, Z. Biol. 34, 298.

Verbeek, L., 1962. On error minimizing neural nets, in: *Principles of self-organization*, eds. H. Von Foerster and G. W. Zopf Jr. (Pergamon Press, New York) p. 122.

Von Foerster, H., 1962, Circuitry of clues to platonic ideation, in: *Artificial intelligence*, ed. C. A. Muses (Plenum Press, New York) p. 43.

Von Foerster, H., 1965, Memory without record, in: *The anatomy of memory*, ed. D. P. Kimble (Science and Behavior Books Inc., Palo Alto) p. 388.

Von Foerster, H., A. Inselberg and P. Weston, 1966, Memory and inductive inference, in: *Bionics Symposium* 1965, ed. H. Oestreicher (John Wiley and Sons, New York) in press.

Von Neumann, J., 1951, The general and logical theory of automata, in: *Cerebral mechanisms in behavior*, ed. L. A. Jeffres (John Wiley and Sons, New York) p. 1.

Wittgenstein, L., 1956, *Tractatus logico-philosophicus* (Humanities Pupl., New York).

3
What Is Memory that It May Have Hindsight and Foresight as well?*

HEINZ VON FOERSTER

"What is Time?" According to Legend, Augustine's reply to this question was: "If no one asks me, I know: but if I wish to explain it to one that asketh, I know not." Memory has a similar quality, for if not asked, we all know what memory is, but when asked, we have to call for an International Conference on the Future of Brain Sciences. However, with a minimal change of the question, we could have made it much easier for Augustine. If asked "What's the time?" he may have observed the position of the sun and replied: "Since it grazes the horizon in the west, it is about the sixth hour after noon."

A theory of memory that is worth its name must not only be able to account for Augustine's or anybody else's intelligent conduct in response to these questions, moreover, it also must be able to account for the recognition of the subtle but fundamental difference in meaning of the two questions regarding time or memory of before, a distinction that is achieved by merely inserting a syntactic "operator"—the definite article "the"—at a strategic point in the otherwise unchanged string of symbols. At first glance it seems that to aim at a theory of memory which accounts for such subtle distinctions is overambitious and preposterous. On second thought, however, we shall see that models of mentation that ignore such aims and merely account for a hypothetical mapping of sensations into indelible representations on higher levels within the neural fabric of the brain or—slightly less naively—account for habituation, adaptation and conditioning by replacing "indelibility" by "plasticity", do not only fall pitiably short of explaining anything that may go on at the semantic level or, to put it differently, that is associated with "information" in the dictionary sense, i.e., "*knowledge* acquired in any manner",[1] but also appear to inhibit the development of notions that will eventually account for these so-called "higher functions" of cerebral activity.

* This article was originally published in *The Future of the Brain Sciences*, Proceedings of a Conference held at the New York Academy of Medicine, S. Bogoch (ed.), Plenum Press, New York, pp. 19–64 (1969). Reprinted with permission of Kluwer Academic/Plenum Publishers.

Since an approach that attempts to integrate the enigmatic faculty of memory into the even more enigmatic processes of cognition veers off under a considerable angle from well established modes of thinking about this problem, it may be profitable to develop the argument carefully step by step, first exposing and circumventing some of the semantic traps that have become visible in the course of this study, and then showing that even at the possible risk of losing track of some operational details a conceptual frame work is gained which, hopefully, allows the various bits and pieces to fall smoothly into place.

At this moment it appears to me that this objective may be best achieved by delivering the argument in four short "chapters". I shall open the discussion with an attempt to clarify some of the most frequently used terms in discussing memory and related mental functions. In the second chapter I shall state my thesis which is central to the whole argument, and I shall develop this thesis in details that are commensurate with the scope of this paper in Chapter III. Finally, I shall venture to present a conjecture regarding the possibility of computing recursive function on the molecular level.

Throughout this paper I shall be using examples and metaphors as explanatory tools, rather than the frightful machinery of mathematical and logical calculi. I am aware of the dangers of misrepresentation and misunderstanding that are inherent in these explanatory devices, and I shall try to be as unambiguous as my descriptive powers permit me to be. For those who wish to become acquainted with a more rigorous treatment of this subject matter I must refer to the widely scattered technical literature as much—or as little—as there exists such literature today.[2–10]

I. Clarification of Terminology

There are two pairs of terms that occur and re-occur with considerable frequency in discussions of memory and related topics. They are (i) "storage and retrieval" and (ii) "recognition and recall". Unfortunately—in my opinion—they are used freely and interchangeably as if they were to refer to the s\ame processes. Permit me, therefore to restore their distinctive features:

(i) Storage and Retrieval

I wish to associate with these terms a certain *invariance of quality* of that which is stored at one time and then retrieved at a later time.

Example: Consider Mrs. X who wishes to store her mink coat during the hot months in summer, takes this coat to her furrier for storage in his vault in spring and returns in the fall for retrieving it in time for the opening night at the opera.

Please note that Mrs. X is counting on getting precisely her mink coat back and not any other coat, not to speak of a token of this coat. It is up to everybody's imagination to predict what would happen if in the fall her furrier would tell her "Here is your mink coat" by handing over to her a slip on which is printed "HERE IS YOUR MINK COAT".

At this point I don't believe that anybody may disagree with my insistence on invariance of quality of entities when stored and retrieved and with my choice of example illustrating this invariance. Consequently, one may be tempted to use this concept for somewhat more esoteric entities than mink coats as, for instance, "information". Indeed, it may be argued that there exist reasonable well functioning and huge information storage and retrieval systems in the form of some advanced library search and retrieval systems, the nationwide Educational Resources Information Center (ERIC), etc., etc., which may well serve as appropriate models or analogies for the functional organization of physiological memory.

Unfortunately, there is one crucial flaw in this analogy inasmuch as these systems store books, tapes, micro-fiches or other forms of documents, because, of course, they can't store "information". And it is again these books, tapes, micro-fiches or other documents that are retrieved which only when looked upon by a human mind, may yield the desired "information". By confusing *vehicles* for potential information with *information*, one puts the problem of cognition nicely into one's blind spot of intellectual vision, and the problem conveniently disappears. If indeed the brain were seriously compared with one of these document storage and retrieval systems, distinct from these only by its quantity of storage rather than by quality of process, such theory would require a little demon, bestowed with cognitive powers, who zooms through this huge storage system in order to extract the necessary information for the owner of this brain to be a viable organism.

It is the aim of this paper to explore the brain of this demon in terms of the little I know of neurophysiology so that we may ultimately dismiss the demon and put his brain right there where ours is.

If there should be any doubt left as to the distinction between vehicles of potential information and information proper, I suggest experimenting with existent so-called "information storage and retrieval systems" by actually requesting answers to some queries. He who did not yet have the chance to work with these systems may be amused or shocked—depending on his view of such systems—by the sheer amount of pounds or tens of pounds, of documents that may arrive in response to a harmless query, some of which—if he is lucky—may indeed carry the information he requested in the first place.

I shall now turn to the second pair of terms I promised to discuss, namely "Recognition and Recall".

(ii) Recognition and Recall

I wish to associate with these terms the overt manifestations of *results* of certain operations, and I wish not to confuse the results of these operations with either the *operations themselves* or the *mechanisms* that implement these operations.

Example: After arrival from a flight I am asked about the food served by this airline. My answer:

"FILET MIGNON
WITH FRENCH FRIES AND SOME SALAD,
AND AN UNDEFINABLE DESSERT."

My behavior in response to this question—I believe—appears reasonable and proper. Please note that nobody expects me to produce in response to this question a real

filet mignon
with french fries and some salad
and an undefinable dessert.

I hope that after my previous discussion of storage and retrieval systems it is clear that my verbal response cannot be accounted for by any such system. For in order that the suspicion may arise that I am nothing but a storage and retrieval system first the *sentence*:

"FILET MIGNON
WITH FRENCH FRIES AND SOME SALAD,
AND AN UNDEFINABLE DESSERT."

had to be "read in" into my system where it is stored until a querier pushes the appropriate retrieval button (the query) whereupon I reproduce with admirable invariance of quality (high fidelity) the sentence:

"FILET MIGNON
WITH FRENCH FRIES AND SOME SALAD,
AND AN UNDEFINABLE DESSERT."

However, I must ask the generous reader to take my word for it that nobody ever *told* me what the courses of my menu were, I just *ate* them.

Clearly something fundamentally different from storage and retrieval is going on in this example in which my verbal behavior is the result of a set of complex processes or operations which transform my experiences into utterances, i.e., symbolic representations of these experiences.

The neural mechanisms that perform the operations which permit me to identify experiences and to classify these with other earlier experiences determine my faculty to recognize (Re-cognition). Those mechanisms and operations which allow me to make symbolic representations of these experiences, say, in the form of utterances, determine my faculty to recall (Re-call).

The hierarchy of mechanisms, transformational operations and processes that lead from sensation over perception of particulars to the manipulation of generalized internal representations of the perceived, as well as the inverse transformations that lead from general commands to specific actions, or from general concepts to specific utterances I shall call "Cognitive Processes". In the analysis of these processes we should be prepared to find that terms like "recall" and "recognition"—as convenient as they may be for referring quickly to certain aspects of cognition—are useless as descriptors of actual processes and mechanisms that can be identified in the functional organization of nervous tissue.

It could be that already at this point of my exposé the crucial significance of cognitive processes may have become visible, namely, to supply an organism with the operations that "lift"—so to say—the information from its carriers, the signals, may they be sensations of external or internal events, signs or symbols,[11] and to provide the organism with mechanisms that allow it to compute inferences from the information so obtained.

To use more colloquial terms, cognition may well be identified with all the processes that establish "meaning" from experience. I may mention that a somewhat generalized interpretation of "meaning" as "all that which can be inferred from a signal" leads to a semantic rationale of considerable analytic power, independent of whether the signal is a sign or a symbol. Moreover, I may add in passing that this interpretation allows not only for qualitative distinctions of "meaning" depending on the mode of inference that is operative—i.e., the deductive, inductive or abductive mode[12]— but also for quantitative estimates of the "amount of meaning"—straightforwardly at least in the deductive mode[13]—that is carried by a given signal for a given recipient.

I hope to make these points more transparent later on in the development of my thesis which, after these preliminary remarks, we are ready to hear now.

II. Thesis

In the stream of cognitive processes one can conceptually isolate certain components, for instance

     (i) the faculty to perceive,
     (ii) the faculty to remember,
and (iii) the faculty to infer.

But if one wishes to isolate these faculties functionally or locally, one is doomed to fail. Consequently, if the mechanisms that are responsible for any of these faculties are to be discovered, then the totality of cognitive processes must be considered.

Before going on with a detailed defense of this thesis by developing a model of an "integrated functional circuit" for cognition, let me briefly suggest the inseparability of these faculties on two simple examples.

First Example: If only one of the three faculties mentioned above is omitted, the system is devoid of cognition:

(i) Omit perception: the system is incapable of representing internally environmental regularities.

(ii) Omit memory: the system has only throughput.

(iii) Omit prediction, i.e., the faculty of drawing inferences: perception degenerates to sensation, and memory to recording.

Second Example: If the conceptual linkages of memory with the other two faculties are removed one by one, *nolens, volens* "memory" degenerates first to a storage and retrieval system and, ultimately, to an inaccessible storage bin that is void of any content.

After these *reductiones ad absurdum* I shall now turn to a more constructive enterprise, namely, to the development of a crude and—alas—as yet incomplete skeleton of cognitive processes.

III. Cognitive Elements and Complexes

I shall now develop my thesis in several steps of ascending complexity of quality, rather than of quantity, beginning with the most elementary case of apparent functional isolation of memory but of zero inferential powers, concluding with the most elementary case of functionally unidentifiable memory, but of considerable inferential powers. Throughout this discussion I shall use examples of minimal structural complexity for the sake of clarity in presenting the argument. I am well aware of many of the fascinating results that can be derived from a rigorous extension of these minimal cases, but in this context I feel that these findings may divert us from the central issue of my thesis.

My first case deals with the computation of concomitance. The detection of concomitances in the outside world is of considerable economic significance for an organism immersed in this world, for the larger an equivalence class of events becomes the fewer specific response patterns have to be developed by the organism. The power of inductive inference rests on the ability to detect concomitance of properties, and—as was believed until not long ago—the efficacy of the conditioned reflex rests on the ability to detect concomitance of events.

The principle of inductive inference is essentially a principle of generalization. It says that, since all things examined that exhibited property P_1 also exhibited property P_2, all as yet unexamined things that have property P_1 will likewise exhibit property P_2. In other words, inductive inference generalizes the concomitance of properties P_1 and P_2. In its naive formulation the "conditioned reflex" can be put into a similar logical schema which I will call "Elementary Conditioned Reflex" (ECR) in order to establish a clear distinction between this model and the complex processes that regulate conditioned reflexive behavior in mammals and other higher vertebrates. However, I shall return to these in a moment.

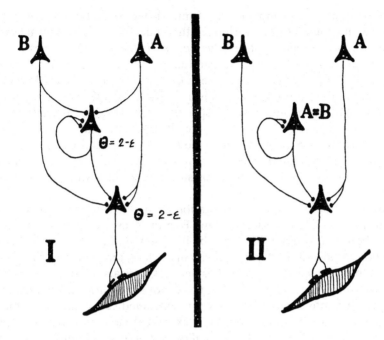

FIGURE 1. Minimal network for computing an elementary conditioned reflex (ECR). (I) Before conditioning; (II) After conditioning.

Part 1 of Figure 1 shows the minimal net capable of computing an ECR. Neurons A, B transmit the conditioning and conditioned stimulus respectively to the motoneuron with threshold $\theta = 2 - \varepsilon$, where $0 < \varepsilon \ll 1$, which always fires when A fires, since the double excitation on its two synapses override its threshold of less than two units (A single synapse represented one unit of excitation). The motoneuron cannot be triggered by B only, for one synapse is insufficient to override its threshold.

However, when first concomitance of A and B occurs, the internuncial is activated and provides sufficient facilitation for B to initiate the reflex. The internuncial's recurrent collaterals secure its permanent excited state and, henceforth, B only is sufficient to elicit a response.

In spite of the structural simplicity of this four-element network, it exhibits some features that are instructive in this context. First it should be noticed that it alters its function as a consequence of the occurrence of certain stimulus configurations: before concomitance of A with B, the net is impervious to B, while afterwards it is responsive to B as it is and was to A. Unfortunately, some authors seem to associate with this simple alteration higher mental functions by calling this "learning through experience". Whether this misrepresentation is caused by underestimating the complexity of the processes that establish algorithms for solving certain classes of problems—i.e., "learning" in its proper sense—or by overestimating the sophistication of this simple circuit, I must leave to anybody's judgement. However, it should not be overlooked that indeed a specific external event caused this net to change its *modus operandi*, and that the occurrence of this event is recorded in the reverberating loop of the self-excitatory inter-nuncial. This is particularly clearly seen in Part II of Fig. 1 which despite its degenerated internuncial afferents is a functional equivalent of the net in Part I *after* it was modified by the specific event. The internuncial "holds" the equivalence relation A = B, and one may be tempted to associate with this store some form of elementary "memory" that indeed, can be localized and functionally isolated. Alas, this is not so, as I shall show in a moment, for nothing can be inferred from this store except its own truth: "A = B is the case".

However, note that the representation of the concomitance of A and B is in form of a *relation* between A and B in the sense of establishing equiv-alence between A and B. This may be interpreted as an elementary repre-sentation of "meaning", for the activity in the loop represents nothing more nor less but "B means A", hence this network appears to be an elementary inductive inference computer. Alas, this is not so. In order for induction to be operative, inference about "as yet unexamined cases" has to be made. But this network reiterates only the examined case, and more complex structures have to be considered to allow for inductive inference. Can we get some leads as to these structures from the examples of this net? Perhaps, yes.

The answer may be gleaned from the fact that if equivalence of stimuli can be computed by the net, the general notion of "equivalence" must be somewhere "stored" in this net. Indeed it is, but not in a single element as one may be prone to believe, but in the whole functional and structural organization of the network *before* the event took place that caused it to become a record of the specificity of this event. We may conclude from this that an inductive net must keep its "equivalence structure" intact in order to be ready for every new case of "class B" to be classified as being also cases of "class A", and if this should prove to be false, to either drop this hypothesis or else switch to another one.

I would like to conclude the discussion of this simple net by citing a keen observation by Susan Langer[14] who considers the ontogenesis of mentation as being initiated by an ECR. In a passage devoted to the clarification of the distinction between symbol and sign she writes:

"There is a profound difference between using symbols and signs. The use of signs is the very first manifestation of mind. It arises as early in bio-logical history as the famous 'conditioned reflex', by which a concomitant of a stimulus takes over the stimulus-function. The concomitant becomes a sign of the condition to which the reaction is really appropriate. This is the real beginning of mentality, for here is the birthplace of *error*, and herewith of *truth*."

As much as can be said about some features of this elementary four-element network, I wish to stress again the utter inadequacy of this net to account for even the most straight-forward cases of conditioned reflexive behavior in higher animals. The belief harbored perhaps by early reflexologists, that ultimately such behavior can be reduced to a logical or neural schema of the sort shown in Fig. 1 has—to my knowledge—been completely destroyed by the superb work of Jerzy Konorski[15] who showed that, for instance, in dogs at the first application of the positive conditioned stimulus it elicits a quite distinct "orientation reaction", i.e., pricking up the ears, turning the head, etc., while salivation as response is negligible. He goes on to demonstrate that in almost all experimental set-ups conditioned stimuli ". . . do not usually possess a single modality, but they supply a number of cues . . ." which the animal utilizes and evaluates as to theis significance in determining future action. Konorski reaches the conclusion that essentially two principles govern the acquisition of various types of conditioned reflexes, one, a principle of selection, the other one a principle of inseparability of information from its utilization. Since I consider these principles of considerable importance in my argument, I shall state them more explicitly in Konorski's own words:

(i) Selection

"In solving a given conditioning problem the animal does not utilize *all* the information supplied by the conditioned stimuli, but it *definitely selects* certain cues, neglecting the other ones."

(ii) Inseparability

". . . it is not so as we would be inclined to think according to our introspection, that receipt of information and its utilization are two separate processes which can be combined one with the other in any way." Hence: "Information and its utilization are inseparable constituting, as a matter of fact, one single process."

If I may translate these observations into my terminology of before, the principle of selection becomes a "search for meaning" in the sense that animal selects those cues—i.e., that information—from which it can optimally draw inferences; while the principle of inseparability becomes a "recourse to self-reference" in the sense that the animal evaluates the inferences drawn from that information always with regard to its utilization favorable of its own self.

In search for a minimal network that would exhibit these two principles of selection and of inseparability of information from its utilization—or of "search for meaning" and "self-reference"—I came across J.Z. Young's drawing of a network representing a single memory unit or a "mnemon" as he calls it.[16] Although Eccles' "The Cerebellum as a Neural Machine"[17] abounds with examples of such networks, they exhibit many more functions than needed at this moment and thus I cannot consider them to be "minimal" in this context.

Figure 2 reproduces Young's drawing of the organization of such a single memory unit. He describes its general features as follows:

". . . each unit consists of a classifying neuron that responds to the occurrence of some particular type of external event that is likely to be relevant to the life of the species. The resulting impulse may initially activate either two or more channels by branching of the axon. More than one line of conduct is therefore possible. The mnemon includes other cells whose metabolism is so triggered as to alter the probable future use of the channels on receipt of signals indicating the consequences of the actions that were taken after the classifying cell had first been stimulated."

From this it is easily seen that Young's mnemon indeed incorporates minimally the two principles mentioned earlier. The principle of "selection" or of "search for meaning" of a particular stimulus is incorporated by the choice of pathways that lead to different actions. What that stimulus "means" becomes clear to the animal of course, only *after* a test. "Attack" may under certain stimulus conditions mean "Pain", under others "Pleasure". Note here the important point that neither pain nor pleasure are objective states of the external universe. They are states that are generated purely within the animal, they are "self-states" or—to use terminology of physics—"Eigen-States" of the organism which permit it to refer each incoming signal to its own self, i.e., to establish self-reference with respect to the outside world.

With this observation, the second principle of inseparability of information and its utilization falls smoothly into place, for this system checks the incoming information as to its usefulness by comparison with its eigenstates where upon it initiates the appropriate actions.

With regard to the functional organization of this memory element I wish to make two points that will later become important in the synthesis of a cognitive element. For this purpose, I have redrawn Young's anatomical schema in order to let the relations of the various functions become more transparent, rather than the anatomical ones. Figure 3 represents an information flow diagram that is functionally equivalent to mnemon of Figure 2. Again a classifying cell (cl.c.) allows for two alternate actions that are initiated in the memory and motor cell complexes (A) or (B). Young's collaterals pick up information of the action state from the thick axon of the motoneurons (A) (B) and feed it to comparators (A^+) (A^-) or (B^+) (B^-) which evaluate the action states by comparing them with the information of resulting eigen-states, either desired (+) or else undesired (−).

The two points I wished to make earlier are now as follows:

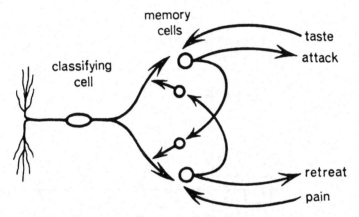

FIGURE 2. The components of a mnemon. "The classifying cell records the occurrence of a particular event. It has two outputs, producing alternative possible motor actions. The system is biased to one of these (say 'attack'). Following this action signals instigating its result arrive and either reinforce what was done or produce the opposite action. Collaterals of the higher motor cells then activate the small cells, which produce inhibitory transmitter and close the unused pathway. These may be called "memory cells" because their synapses can be changed." (Reproduced with kind permission from J.Z. Young[16]).

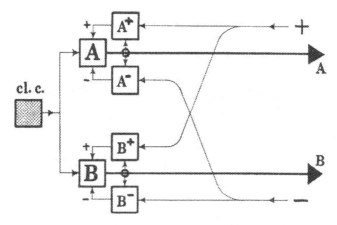

FIGURE 3. Information flow diagram of a mnemon. (cl.c.) classifying cell; (A) (B) motoneurons and memory cell complex; (+) (−) information of eigen-states "good", "bad", or positive and negative internal reinforcement signals; (A+) (A−) (B+) (B−) comparators of action states with eigen-states.

(i) Self-Reference

Self-reference enters the system through two channels: one, via *a priori* established "good" or "bad" signals (+) (−) that report the *consequences* of an action; the other one via the loop (A) → (A+) → A or, *mutatis mutandis*, via corresponding other loops that report the states of its own actions.

(ii) Experience

Experience enters the systems through two operations: one, which modifies the synapses on the memory cells in cell complexes (A) (B) so as to inhibit undesired or facilitate desired actions; the other one, which compares past actions with its present consequences in comparators (A^+) (A^-) (B^+) (B^-) and transmits the results (+) (−) to complexes (A) (B) for appropriate modifications.

I shall now show with respect to point (i) that self-reference is an ubiquitous feature and is computed over and over again in neural organizations, mostly by a resolution of paradoxes in representation, and not necessarily by reference of *a priori* signals; and with respect to point (ii) that experience is gathered in a much more powerful and economic way by modifying the function of the recursive loop $A(t - \Delta) \to A^+(t) \to A(t^+\Delta)$—t indicating time and Δ cumulative synaptic delays—than by storing the outcome of each particular action in a corresponding synaptic modification of a memory cell.

Let me now develop these comments in somewhat more detail, by using again minimal examples. First on self-reference:

Figure 4 shows two objects (a) white, and (b) black, whose images are focused on the retinas of the two eyes of a binocular animal. Amongst many other operations[18] that may be applied to these images by post-retinal networks or at higher nuclei, I assume that there is one that computes a relation which indicates that one thing is to the left of another thing. I symbolize this relation by L(x,y), read "x is to the left of y". The existence of such anisotropic nets has been demonstrated, for instance in pigeons,[19] and their functional and structural organization is well-established.[20]

Since with respect to the animal object (a) is behind (b), the left eye Left-computer reports L(b,a) while the right eye Left-computer reports an opposite state of affairs, namely, L(a,b). This apparent paradox can be resolved by a computer $B(L_1, L_r)$ which realizes that the information L(b,a) is supplied by the left eye, L_1 (subscript l = left), while L(a,b) is supplied by the right eye L_r (subscript r = right). With this observation the paradox disappears, for the two apparently contradictory results are in fact obtained from two distinct and locally separated sensory systems which by no necessity should deliver the same picture of the outside world. However, it is significant that a consistent picture of the outside world can be computed by generating a new space, "depth", in which the relation B(a,b)—read "a is

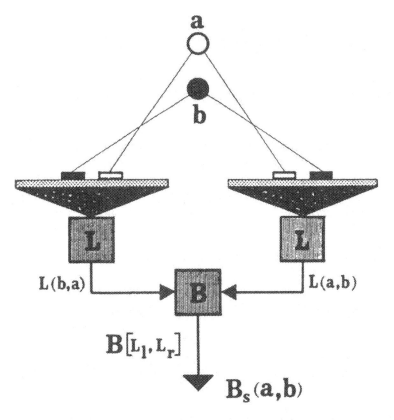

FIGURE 4. Computation of "depth" by resolving a sensory paradox in binocular vision. (A) Networks computing the relation "x is left of y"; (B) Networks computing the relation "x is behind y".

behind b" is now established. Note, however, that this resolution could never have been obtained without reference to the animal's own left and right eye. Consequently, the relation B(a,b) too must carry a subscript B_s, to indicate reference to "self" or to indicate the system's geometrical relation to objects of the outside world. This becomes particularly transparent if one lets the binocular system encircle the two fixed objects. In this case the arguments in the relation B(a,b) begin to rotate

$$B_s(b,a)$$
$$B_s{}^{(b)} \quad {}^{(a)}_{}\!{a \atop b}{}^{(b)} \quad B_s{}^{(a)}$$
$$B_s(a,b)$$

mirroring the relativity of the relation "Behind". (Absence of a detectable difference in the two L computers I have symbolized by writing the arguments of B in a vertical column ${}^{(a)}_{(b)}$ or ${}^{(b)}_{(a)}$.

Of course, I could have used other examples, as, for instance, the generation of a "color space" by the resolution of a triple-paradox which is produced by the divergent reports of the three types of cones with different pigmentation regarding the appearance of one and the same spot in the external world. However, this case and other cases are not minimal.

I shall now enlarge on my earlier brief comment regarding the use of recursive functions as a more powerful tool in accounting for past experience than simple storage of the outcomes of individual acts. This comment was prompted by Young's observation of recursive loops that report back to a central station via some synaptic delays. Turning again to Figure 3 and following the arrows leading from (A) to (A⁺) back to (A), we realize that an action A that took place in the past, say, one cumulative synaptic delay Δ ago, i.e., $A(t - \Delta)$, is evaluated by (A⁺) at time t, i.e., $A^+(t)$, which in turn modifies the cellular aggregate in (A) that will at best respond with new action after a cumulative synaptic delay Δ, i.e., with $A(t + \Delta)$.

I propose now to make changes neither in the structure nor in the function of this subsystem, but only in the *interpretation* of the modifications that are supposed to take place. Instead of interpreting the synaptic modifications in the cellular complex A as stores of the outcomes of various individual actions, I propose that these modifications should be interpreted as a modification of the transfer function of the whole subsystem (A, A^+). Let me demonstrate this idea again with a minimal example, this time of recursive functions.

First, I have to point out that the term "recursive function" is a misnomer, for these functions are like any other function, and it is only that they are not as usual defined explicitly but are defined recursively. By this is meant that a function which relates a dependent variable y to an independent variable, say time t, is not explicitly given in terms of this independent variable, say $y = t^2$, y-sinωt, or in general $y = f(t)$, but is given in terms of its own values at earlier instances $y(t) = F(y\{t - \Delta\})$, where Δ expresses the interval between the earlier instance and the instance of reference t. A typical example of a recursive definition of a function is, for instance, the description of growth of a bacterial colony:

"The number of bacteria in a bacterial colony at any time is twice the number it was one generation ago."

If it takes on the average the time Δ for a bacterium to divide—i.e., one generation extends over a time interval Δ—then the recursive description of the size y of this colony is

$$y(t) = 2 \cdot y(t - \Delta)$$

I shall not discuss the mathematical machinery that "solves" these expressions, i.e., transforms them into explicit statements with respect to the independent variable t only. For instance, in the above case the "solution" is, of course,

$$y(t) = y(0) \cdot e^{\frac{t}{\Delta} \ln 2}$$

where $y(0)$ is the initial size of the colony, i.e., its size at time $t = 0$. The methods of solution are of no concern to us here, the point I wanted to make is only to assure you that a recursive definition of a function is as good as any other and, in some cases, may be even more powerful than an explicit expression (e.g., compare the terse recursive definition of above with the cumbersome explicit expression).

If we now go back to our original problem of finding an appropriate description for a system that acts according to the outcome of previous actions, then it seems—at least to me—that the conceptual tool of recursive function theory is just tailor-made for this purpose.

I shall now discuss the minimal case that corresponds to the "mnemonic" part of Young's mnemon. Figure 5 shows a three-element system (F,T,D) whose functional correspondence with the mnemonic features of the subsystem (A,A$^+$) of Figure 3 will emerge in a moment.

Box F stands for the mechanism that computes the function* $Y = F(X,Y')$ on its two arguments X and Y'. The argument X is an explicit function of time $X(t)$, and is called the "input proper". The argument Y' is a representation of the "output" Y of mechanism F at an earlier time, say $t - \Delta$, and is called the "recursive input". In order that F can be informed about its previous output—or action—the intensity of this action has to be measured by an element, T, which translates this intensity into a signal that is accepted ("understood") by F, and feeds this information with a delay D back to F.

The functional correspondences of these elements with some of the physiological features in Young's mnemon seem to be clear. D corresponds to a cumulative synaptic delay that "holds" the whole picture of this system's output activity for a while, Δ, before it informs the cell aggregate in A of this activity. T represents the motoneuron's collaterals or terminations of sensory afferents that generate the information of A's activity. F is, of course, the aggregate (A,A$^+$), as yet without input of an eigen-state (+) (–), but with an input proper, X, which represents the signal from the classifying cell (cl.c.).

Let us now watch this three-element system in operation. Foremost, we wish to know its output $Y(t)$ at time t for a given input $X(t)$ at that time. Since F is given, we have

$$Y(t) = F\{X(t), Y(t - \Delta)\},$$

---

* According to standard notation, capital letters X,Y represent a set of components $(x_1, x_2 \ldots x_n)$, $(Y_1, y_2 \ldots y_m)$, the value of each component representing, say, the stimulus or response intensity along a corresponding fiber. In other words, X and Y represent the activity along whole fiber bundles and not necessarily that of single fibers.

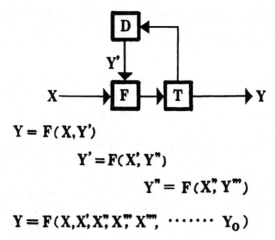

$$Y = F(X, Y')$$

$$Y' = F(X', Y'')$$

$$Y'' = F(X'', Y''')$$

$$Y = F(X, X', X'', X''', X'''', \cdots \cdots Y_0)$$

FIGURE 5. Circuitry and basic components of a recursive function computer. F computing element; X input proper; Y', recursive input of a previous output Y' D, delay; T, translates action Y into a representation of Y acceptable to F.

or, more conveniently, if we drop the reference to t by merely writing Y and X, and for the previous state Y' and X':

$$Y = F(X, Y').$$

However, this does not yet tell us the actual output of the system, because we do not know the value of one of its inputs, namely Y', i.e., its previous output. But this can be determined, using the recursive definition of Y:

$$Y' = F(X', Y'')$$

or in words: the previous output is a function of the previous input and its previous-previous output. Hence, by inserting the expression for Y' into the earlier equation for Y, the present output becomes:

$$Y = F(X, X', Y'').$$

Again we may ask for the value of the previous-previous output, and by applying again the recursion, we will arrive at an expression which leads us three steps back in time, and so on, until we arrive at the "birth-date state" Y_0 of the system:

$$Y = F(X, X', X'', X''', \ldots, Y_0).$$

The remarkable feature of this expression is that it clearly shows the dependence of the present output of this system on the *history* of the previous inputs, rather than on just its present input or to put this into more poetic terms, this system's present actions depend on its past experiences.

Two features should be noted here. First, no storage of representations of past events—save for those traveling through the delay loop—take place here. Reference to the past is completely taken care of by the specific function F that is operative. F is, so to say, the "hypothesis" that predicts from previous cases future actions. Physiologically, F is determined by the functional organization of the cellular aggregate (A, A^+). Second, an external observer who wishes to predict the behavior of this system in terms of its input-output or stimulus-response pattern

$$Y = f(X),$$

and who has no access to its internal structure may soon find to his dismay that he is unable to determine the elusive function "f", for after each experimental session the system behaves differently unless—by lucky circumstances—he finds a repeated sequence of inputs that will given him—by the very nature of that particular F—repeatedly a corresponding sequence of outputs. In the former case, this experimenter will in disgust turn away with the remark "unpredictable!"; in the latter case he will say in delight: "I taught it something!" and may turn around to develop a theory of memory.

Although such recursive function elements exhibit some interesting properties, in the restricted and isolated way in which I have discussed them, they are as yet incapable of responding to OK-signals (+), HANDS-OFF-signals (–), and any other signals that report eigen-states of affairs, or, in general, to self-referential information. Assume such information were available. The question is now to which of the elements in the three-element recursive function computer must this information go in order to modify its *modus operandi* in accordance with a desired eigen-state configuration? It seems to me that the question already carries its own answer: if such change is necessary at all, then the only effective way to modify the general properties of this computer is to change its "hypothesis" by which it computes future states from past experience, i.e., the recursive function F_1 which was operative until this moment must be altered to become, say, F_2, and perhaps later F_3, F_4 and so on, in order to achieve the properties that are commensurate with the system's eigen-state configurations. In other words, F itself has to be treated as a variable, as an element in a range of functions $\Phi(F)$, whose particular value F_i is determined by the eigen-states. Physiologically this means that the recurrent fibers that carry self-referential information are synapsing with cells in the (A, A^+) aggregate so as to change it from a computer that calculates F_i to one that calculates F_j. Mechanisms that achieve such modifications are well-known as, e.g., long range inhibitions and facilitations. However, I doubt that it will ever be possible to establish a detailed account of the relations between individual synaptic changes and the computational properties of the whole aggregate, the main reason being that this is a problem that does not have a unique solution, on the contrary, it can be shown that with just a few cells making up this computer, the number of different solutions is, for all practical purposes, infinite. On the other hand, I do not believe that such detailed knowledge is of importance, as long as the principles are understood that make such modifications possible.

Let me now briefly summarize some of the essential results of this discussion. Most of the neural machinery is functionally organized to establish from sensory information—whether about states of the outside world or about internal states—relations between observed entities with respect to the observing organism. This relational information modifies the *modus operandi* of a computer system that computes new actions recursively on the basis of the outcome of previous actions and, hence, on the basis of the history of the stream of external and internal information. Figure 6 is a graphical representation of this summary in form of a block diagram. I shall call this whole system a "Cognitive Element", for it represents a minimal case of a cognitive process, or a "Cognitive Tile", for it may be used in conjunction with other such tiles to form whole mosaics—or "tessellations"— which, as a whole, permit the high flexibility in representing relational structures not only of what has been perceived but also of the symbols— the "linguistic operators"—that ultimately are to convey in natural language all that which can be inferred from what has been perceived.

The various components of this cognitive tile are quickly explained. X stands for (external) sensory input, and Y for the output of the system as seen by an outside observer. Hence, this elementary component is a "through-put" system, as suggested by the small inset, lower right. However, because of its internal organization, this element is quite a different animal from a simple stimulus-response mechanism with fixed transfer function.

First, sensory information, X, is operated on to yield relations $R_s(X)$ between observed activities with respect to "self" (note subscript s), and is then used as input proper for the recursive function computer which may be operative at this moment with any one of the functions F belonging to range Φ. Its output is fed back over two channels, one being the recursive loop with delay D to allow F to assess its earlier actions, the other carrying all the relational information of the system's own actions $R_s(Y)$ as they refer to "self", and operates on $\Phi(F)$ in order to set the recursive function computer straight as to this tile's internal goals and desires.

This element incorporates all those faculties which I considered earlier to be necessary components of cognitive processes: to perceive, to remember and to infer. However, in this element none of these faculties can be

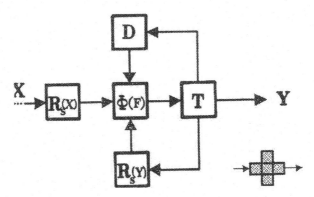

FIGURE 6. Circuitry and basic components of a cognitive tile. $\Phi(F)$ general purpose computer for a range Φ of computable functions F; X input, Y output; $R_S(X)$, $R_S(Y)$ compute relations in the spatio-temporal configurations of input and output respectively, with reference to intrinsic properties of this particular tile; D a delay element; T translates Y into a representation of Y that is acceptable to this and other tiles.

isolated functionally: it is the interaction of all the processes here involved that "life" the information from the input signal and translate it into action meaningful for this tile.

Nevertheless, if forced to interpret some of this tile's functional components in terms of those conceptual components I would reluctantly give the following breakdown: (i) Perception is accomplished by the elements that establish self-referential relations in the spatio-temporal configurations of stimuli and responses; (ii) Memory is represented by the particular *modus operandi* of the central computer whose gross functional organization is determined and redetermined by evaluation of eigen-states or relations; (iii) Inference in this tile appears on three levels, depending on the type of functions that are in range Φ and on the type of processes one wishes to focus on. Adductive inference is operative in the cumulative absorption of comparisons of past external and internal experiences that give rise to the functional organization of the central computer. Inductive or deductive inferences are computed by the central system concurrently with any new signal, the inferential mode being solely dependent on strings of earlier failures or successes *and* of some of this tile's internal dispositions to "disregard" false inductions or to take them "seriously" by converting to more stringent logical deductions.

I shall now conclude my thesis with only a brief report on some properties as they may be relevant to this topic of aggregates of such tiles or "tessellations" as they are usually referred to in the literature. John von Neumann was the first to realize the high computational potential of these structures in his studies of self-reproducing automata,[21] and later Löfgren applied similar principles to the problem of self-repair.[22] We use these, however, in connection with problems of self-reference and self-representation.

Two features of cognitive tiles permit them to mate with other tiles: one is its inconspicuous element T which translates into a universal "internal language" whatever the "output language may be; the other one is its essential character as a "through-put" element. Consequently, one may assemble these tiles into a tessellation as suggested in Figure 7, each cross, white or black, corresponding to a single tile, while each square in a cross represents the corresponding functional element as suggested in the earlier Figure 6. Information exchange between tiles can take place on all interfaces, however, under observance of transmission rules implicit in the flow diagram of Figure 6. For instance, one tile may incorporate into its own delay loop preprocessed information from an adjacent tile, but eigen-state information of one tile cannot retroactively modify the operations of a "left" tile, although it can—via its own output—modify that of a "right" tile, and so on.

When in operation, this system shifts kaleido-scopically from one particular configuration of cooperating sets of adjacent tiles to other configurations, in an ever changing dynamic mode, giving the impression of

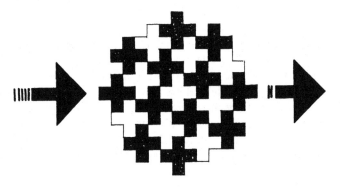

FIGURE 7. Example of a tessellation of cognitive tiles.

"clouds" of activity shifting, disappearing and reforming as the task may demand.

We have studied such systems as yet only in the "representative mode", i.e., in which these tiles correspond to "linguistic operators" with their multiple ramifications into various depth of meaning. These systems are now simulated by complex computer programs, one being a particularly interesting three-dimensional extension of the two-dimensional scheme presented here and being called "Cylinders" by its inventor Paul Weston.[10,23] These novel program structures represent at this stage prototypes of systems which permit symbolic discourse between man and machine in the form of natural language. We do not foresee fundamental difficulties when switching to the "perceptive mode", i.e., in which the inputs to some specified "sensory tiles" are not symbols but signals from some restricted, but relevant, environment.

We hope to provide with these studies the foundation for a new architecture of future computers that may well serve as models for a cognitive memory that has hindsight and foresight as well.

IV. A Conjecture

Eccles' *grand oeuvre The Cerebellum as a Neuronal Machine*[17] is extremely encouraging to look for small, highly organized, cell assemblies that could be represented by the operational unit I have developed above, namely, by a cognitive tile. I have convinced—at least—myself that there are numerous examples of networks whose actions can be described by individual tiles or by smaller or larger tessellations.[24] The question however, which plagues a theoretician is as to the *minimal* physiological unit that could be described by the corresponding minimal operational unit, i.e., a single cognitive tile. Learning about the tremendous complexity of a single Purkinje cell, with its wide range of response activity and with a convergence of inputs up to about 200,000 synapses, I believe most of my tile's functional properties can be found in a single specimen of these cells, were it not for one operational feature of the tile, its computation of recursive functions, that would require the cooperation of at least one other cell to form a single cognitive tile. However, I believe there is a way out of this dilemma by following the ideas proposed, for instance, by Holger Hyden,[25,26] which suggest to look *into* the cell, i.e., into modifications of the cell's molecular constituents, in order to account for some mnemonic properties of a single cell.

The most pedestrian way to look at the potentialities of a complex molecule is to look at it as a storage and retrieval device.[27,28] This possibility offers itself readily by the large number of excitable states that go hand in hand with the large number of atoms that constitute such molecules. Consequently the chances are enhanced for the occurrence of metastable states which owe their existence to quantum mechanically

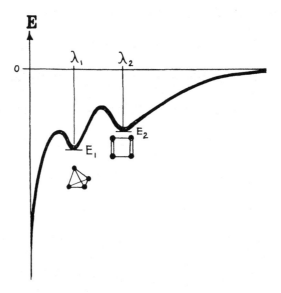

FIGURE 8. Association of energy levels with the two different configurations (isomeres) of a molecule composed of four atoms, each having three valences (n = 4;
V = 3). λ_1 and λ_2 represent the eigen-values in the solution of the Schrödinger wave
equation.

"forbidden" transitions.[28] Since being in such a state is the result of a particular energy transaction, selective "read-out" that triggers the transition
to the groundstate—as in the optical maser—permits retrieval of the information stored in the excited states.

There is, however, another way to allow for information storage in macro-
molecules where the "read-out" is defined by structural matching (templet).
It is obvious that, m, the number of ways (isomeres) in which n atoms with
V valences can form a molecule Z_n will increase with the number of atoms
that constitute this molecule. Each of these configurations is associated with
two characteristic energy levels (quantum states), one that gives the potential energy of this configuration, the other one which is the next higher level
at which this configuration becomes unstable. Figure 8 sketches this situa-

tion for the two isomeric states of an hypothetical molecule Z_4 composed of four 3-valence atoms Z. Simple considerations show that the tetrahedral configuration is more stable than the quadratic form, hence some energy must be supplied to the tetrahedron to transform it into the square. However, it will not stay indefinitely in this configuration because of the quantum mechanical "tunnel effect" which gives each state a "life-span" of

$$\tau = \tau_0 e^{\frac{\Delta E}{kT}}$$

where ΔE is the height of the energy "trough" which keeps the configuration stable, k is Boltzmann's constant, T is the absolute temperature surrounding this molecule and τ_0 is an intrinsic oscillatory time constant associated with orbital or lattice vibrations.

It is these spontaneous transitions from one configuration into another one which tempt me to consider such a molecule as a basic computer element, particularly if one contemplates the large number of configurations which such macro-molecules can assume. Estimates of the lower and upper bounds of the number of isomeres are[29]

$$\underline{m} \approx \frac{5}{8} \cdot n$$

and

$$\overline{m} \approx \left(\frac{nV}{2p(V)} \right)^{p(V)}$$

where p(N) is the number of unrestricted partitions of the positive integer N, and V and n are again the number of valences and the number of atoms respectively.

Since each different configuration of the same chemical compound Z_n is associated with a different potential energy, the fine-structure of this molecule may not only represent a single energy transaction that has taken place in the past but may represent a segment of the *history* of events during which this particular configuration evolved. This consideration brings me right to my conjecture; namely to interpret the responses of such macro-molecules to some energy transactions as those of a recursive function computer element.

The idea to look upon various structural transformations which many of the macro-molecules perpetually undergo as being outcomes of computations is not at all new. Pattee, for instance, has demonstrated in a delightful paper[30] the isomorphism between the growth of some helical macro-molecules with the operation of a finite, binary autonomous shift-register. In his example the recursive relation is only between a present state Y and an earlier one Y':

$$Y = F(Y').$$

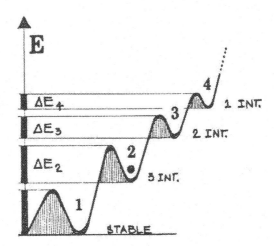

FIGURE 9. Four of the lowest energy levels corresponding to some molecular configurations together with the threshold energies that keep these configurations stable within some definite intervals of time.

We, however, need to account of an "input proper" X, in order to be able to interact with this system, i.e., to allow for a "read-in" and "read-out" operation:

$$Y = F(X, Y').$$

Figure 9 sketches the four lowest energy states numbered 1,2,3,4 of a molecule together with the three energy thresholds, ΔE_2, ΔE_3 and ΔE_4 which keep the corresponding configurations stable at least during the "life-span" of these states. For simplicity, I assume these life-spans to be multiples of the shortest life span τ^*, i.e., under normal temperature conditions ΔE_4 will give state #4 a lifespan of τ^*, and the others as suggested by Table 1.

Assume now that at a particular instant, t_0, this molecule is in state 2 (black dot suggests this position), and within three intervals no energy is supplied to kick it into a higher state. As a consequence, it will flip back into state 1, giving off the stored energy difference between state 2 and 1.

TABLE 1.

State	Threshold	Life Span
# 1	large	∞
# 2	ΔE_2	$3\tau^*$
# 3	ΔE_3	$2\tau^*$
# 4	ΔE_4	$1\tau^*$

TABLE 2.*

			t_0		
t_2	t_1	1	2	3	4
0	0	1	2	2	3
0	1	2	3	4	4
1	0	2	3	3	4
1	1	3	4	4	4

* The initial states are assumed to have been acquired within one earlier interval. A more elaborate table is needed to indicate "aged" states.

I shall now consider the general situation in which two events follow each other at times t_1 and t_2, each spaced approximately at intervals corresponding to τ^*, and each event either supplying (1) or else not supplying (0) the energy to lift the molecule into its next higher state.

Table 2 gives the result of these operations, indicating on the left whether or not the events at times t_1 and t_2 carried the required energies, and giving at the head of columns under t_0 the initial state of the molecule.

Clearly, for each of the different initial conditions this molecule "computes according to the four possible input-configurations (00) (01) (10) (11) a different set of outcomes, in other words, this computer changes its operations depending on its initial state which is, of course, nothing else but the result of previous operations.

It is easy to see how this idea can be extended to accommodate an arbitrary number of sequential events $t_1, t_2, t_3 \ldots t_s$ and an arbitrary number of molecular states 1, 2, 3, 4, 5, ... m, and thus gives rise to the possibility to interpret the various induced and spontaneous states of a macro-molecule as those of a recursive function computer of considerable flexibility and latitude of range.

However, there remains the question still open as to the external mechanisms that will induce these changes. To this end we have to evaluate numerically the equation that was given earlier and which relates the various quantities here involved, namely, the threshold energies ΔE, the average life span of a state τ, and the other two quantities, τ_0, an intrinsic time constant, and the temperature T of the system. If we assume a constant body temperature of 36.6°C, then T = 309.8° Kelvin and with the known value of Boltzmann's constant there remain only the three variables

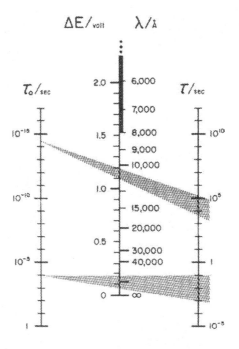

FIGURE 10. Nomogram for τ, τ_0, ΔE, representing the $\tau = \tau_0 e^{\frac{\Delta E}{kT}}$ for a fixed temperature of T = 310°K.

τ_0, τ and ΔE to be related. This is most clearly done in form of a nomogram as given in Figure 10. Values along the three scales that are read off at points which are connected by a straight line always represent a solution of our equation that relates these quantities. The three scales represent the values of τ_0, the period of intrinsic oscillations in seconds, of ΔE the energy threshold in electron volts, and τ and life span of a state in seconds. Since a delivery of an energy quantum of size ΔE is always associated with an electromagnetic radiation of wave length λ, this quantity is given along the middle scale in Angstrom units, the visible spectrum being represented by the heavy bar (4000 A to 8000 A).

The numerical evaluation is now particularly simple since there are essentially only two values with small spread for τ_0, the intrinsic oscillation period, to be considered. One is of the order of 3.10^{-15} seconds[31] and is associated with electron orbits within the crystal. Life spans that are controlled by this time constant are those of configurational change. The energy amount necessary to accomplish configurational change one can calculate from the amount of kinetic energy per mole that molecules must acquire before they can react. This amount is well-established for proteins and enzymes—it is the μ-value of the Arhenius equation of reactions—and is found to be in the vicinity of 28,000 calories.[32] Changing these thermal units into electrical units we obtain a ΔE of about 1.1 and 1.2 electron volts. Drawing the straight lines that connect the appropriate values on the τ_0-scale and the ΔE-scale, we find life spans of configurational changes on the τ-scale between 10^4 and 10^5 seconds, i.e., between about three hours and one day.

Apparently, these life spans are, on the one hand, too long to make an effective recursive element, on the other hand, they seem to be too short to account for long term memory traces. However, if one admits chemical processes to participate in these operations, these might be just the proper intervals to compute recursively over an arbitrary long stretch of time those configurations that give a neuron certain operational properties. Be that as it may, the significance of these slow configurational changes will become obvious if we turn now to the other value of τ_0, which is associated with the intrinsic oscillations of the lattice structure of these macro-molecules and is of the order of 10^{-4} seconds.[33] The amounts of energy ΔE that have to be supplied to these quantum states in order that they may jump from one state to another at intervals that correspond approximately to various intervals in which fiber volleys arrive at a neuron, say, between one and one hundred milli-seconds, are again found by the intersections of straight lines that connect these points with the ΔE-scale. The corresponding ΔE values are between 50 milli-volts and 180 milli-volts, i.e., just in the proper range to have an action potential of about 80 milli-volts to excite the lattice vibrational states. In other words, in this mode a macro-molecule may well operate as a recursive element, responding directly to the frequency of neural activity. Moreover, as can be read off from the nomogram if a train of more than about 15 volleys of 80 milli-volts each and each volley following the other at intervals not longer than about 3 milli-seconds act on the molecule, then it will not have the time to go into lower energy states and will be "pumped" up into an energy level of about 1.2 volts which corresponds to levels in which configurational changes take place.

Now the game of recursion can be played including configurational changes whose relatively long life spans allow us to make an almost unlimited number of working hypotheses where only our imagination seems to be the limit.

I have presented this conjecture of molecular computation only to suggest that there are avenues open that point to a participation of molecules in the grand spectacle of mentation in which they play a dynamic rather than a static role.

Acknowledgements. Some of the ideas and results presented in this paper grew out of work jointly sponsored by the Office of Education under Grant OEC-1-7-071213-4557, by the Air Force Office of Scientific Research under Grant AF-OSR 7-68 and by the Air Force Systems Engineering Group under Contract AF 33(615)-3890.

References

1. Webster's New World Dictionary of the American Language.
2. Von Foerster, H., "Memory with Record" in: The Anatomy of Memory, D. P. Kimble (ed.), Science and Behavior Books, Palo Alto, pp. 388–433, 1965.
3. Von Foerster, H., "Computation in Neural Nets", Currents Mod. Biol., 1, 47–93, 1967.
4. Von Foerster, H., "Time and Memory" in: Interdisciplinary Perspectives of Time, R. Fisher (ed.), New York Academy of Sciences, New York, pp. 866–873, 1967.
5. Newell, A. and H. A. Simon, "The Logic Theory Machine" in: I.R.E. Transaction on Information Theory IT-2, pp. 61–79, 1956.
6. Minsky, M., "Steps toward Artificial Intelligence" in Proc. I.R.E. 49, pp. 8–30, 1961.
7. Lindsay, R. K., Inferential Memory as the Basis of Machines which Understand Natural Language. In: Computers and Thought, E. Feigenbaum and J. Feldman (eds.), McGraw-Hill, New York, 217–233, 1963.
8. Raphael, R., A Computer Program which Understands. In: Proc. AFIPS, F.J.C.C., 577–589, 1964.
9. Von Foerster, H. and R. T. Chien, Cognitive Memory, Coordinated Science Laboratory, University of Illinois, Urbana, 1967.
10. Weston, P., Cylinders: A Data Structure Concept Based on a Novel Use of Rings. In: Accomplishment Summary 1968, BCL Report 68.2, Biological Computer Laboratory, Department of Electrical Engineering, University of Illinois, Urbana, 42–61, 1968.
11. Langer, S. K., Philosophy in a New Key, New American Library, New York, 1951.
12. McCulloch, W. S., Embodiments of Mind, M.I.T. Press, Cambridge, 1965.
13. Bar-Hillel, Y., Semantic Information and Its Measures. In: Cybernetics: Transactions of the Tenth Conference, H. Von Foerster, Margaret Mead, and H. L. Teuber (eds.), Josia Macy, Jr. Foundation, New York, 33–48, 1955.
14. Langer, S. K., op. cit., 30.
15. Konorski, J., The Role of Central Factors in Differentiation. In: Information Processing in the Nervous System, R. W. Gerard and J. W. Duyff (eds.), Excerpta Medica Foundation, Amsterdam, 3, 318–329, 1962.
16. Young, J. Z., The Organization of a Memory System. In: Proceedings of the Royal Society, B, 163, The Groonian Lecture, 285–320, 1965.

17. Eccles, J. C., M. Ito, and J. Szentagothai, The Cerebellum as a Neuronal Machine, Springer-Verlag, New York, 1967.
18. Lettvin, J. Y., H. R. Maturana, W. S. McCulloch, and W. Pitts, What the Frog's Eye tells the Frog's Brain. In: Proc. I.R.E., 47, 1940–1951, 1959.
19. Maturana, H. R., Functional Organization of the Pigeon Retina In: Information Processing in the Nervous System, R. W. Gerard and J. W. Duyff (eds.), Excerpta Medica Foundation, Amsterdam, 3, 170–178, 1962.
20. Von Foerster, H., Structural Models of Functional Interactions in Information Processing in the Nervous System, R. W. Gerard and J. W. Duyff (eds.), Excerpta Medica Foundation, Amsterdam, 3, 370–383, 1962.
21. von Neumann, J., The Theory of Automata: Construction, Reproduction and Homogeneity In: John von Neumann's Collected Works, A. Burks (ed.), University of Illinois Press, Urbana, 1964.22.
22. Löfgren, L., Kinematic and Tessellation Models of Self-Repair, TR 8, Contract NONR 1834(21) Electrical Engineering Research Laboratory, Engineering Experiment Station, University of Illinois, Urbana, 61, 1961.
23. Weston, P. and H. Tuttle, Data Structures for Computations within Networks of Relations. In: Accomplishment Summary 1967, BCL Report 67.2 Biological Computer Laboratory, Department of Electrical Engineering, University of Illinois, Urbana, 35–37, 1967.
24. See Reference 17, Figures 114, 115 and p. 311ff.
25. Hydén, H., Activation of Nuclear RNA of Neurons and Glia in Learning. In: The Anatomy of Memory, D. P. Kimble (ed.), Science and Behavior Books, Inc., Palo Alto, 1, 178–239, 1965.
26. Hydén, H., Studies on Learning and Memory. In: this volume.
27. Von Foerster, H., Das Gedachtnis; Eine Quanten-mechanische Untersuchung, F. Deuticke, Vienna, 40, 1948.
28. Von Foerster, H., "Quantum Mechanical Theory of Memory. In: Cybernetics: Transactions of the Sixth Conference, H. Von Foerster (ed.), Josiah Macy Jr. Foundation, New York, 112–145, 1949.
29. Von Foerster, H., Molecular Bionics. In: 1963 Bionics Symposium: Information Processing by Living Organisms and Machines, H. L. Oestreicher (ed.), Aerospace Medical Division, Wright-Patterson AFB, Ohio, 161–190, 1964.
30. Pattee, H. H., On the Origin of Macro-molecular Sequences, Biophys. J., 1, 683–710, 1961.
31. Schrodinger, E., What Is Life?, University Press, Cambridge, 1945.
32. Hoagland, H., Consciousness and the Chemistry of Time. In: Problems of Consciousness, H. A. Abramson (ed.), Josiah Macy Jr. Foundation, New York, 164–200, 1951.
33. Landau, L. D. and E. M. Lifshitz, Statistical Physics, Pergamon Press, London, 1958.

4
Molecular Ethology, An Immodest Proposal for Semantic Clarification*

HEINZ VON FOERSTER

Departments of Biophysics and Electrical Engineering, University of Illinois, Urbana, Illinois

I. Introduction

Molecular genetics is one example of a successful bridge that links a phenomenology of macroscopic things experienced directly (a taxonomy of species; intraspecies variations; etc.) with the structure and function of a few microscopic elementary units (in this case a specific set of organic macromolecules) whose properties are derived from other, independent observations. An important step in building this bridge is the recognition that these elementary units are not necessarily the sole constituents of the macroscopic properties observable in things, but are determiners for the synthesis of units that constitute the macroscopic entities. Equally helpful is the metaphor which considers these units as a "program," and the synthesized constituents in their macroscopic manifestation as the result of a "computation," controlled and initiated by the appropriate program. The genes for determining blue eyes are not blue eyes, but in blue eyes one will find replicas of genes that determine the development of blue eyes.

Stimulated by the success of molecular genetics, one is tempted to search again for a bridge that links another set of macroscopic phenomena, namely the behavior of living things, with the structure and function of a few microscopic elementary units, most likely the same ones that are responsible for shape and organization of the living organism. However, "molecular ethology" has so far not yet been blessed by success, and it may be worthwhile to investigate the causes.

One of these appears to be man's superior cognitive powers in discriminating and identifying forms and shapes as compared to those powers which allow him to discriminate and identify change and movement. Indeed, there is a distinction between these two cognitive processes, and this distinction is reflected by a difference in semantic structure of the linguistic elements

* This article was originally published in *Molecular Mechanisms in Memory and Learning*, Georges Ungar (ed.), Plenum Press, New York, pp. 213–248 (1970). Reprinted with permission of Kluwer Academic/Plenum Publishers.

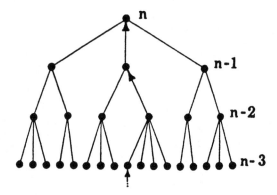

FIGURE 1. Ascending hierarchical definition structure for nouns. (Nouns are at nodes; arrow heads: definiens; arrow tails: definiendum.)

which represent the two kinds of apparitions, namely different nouns for things distinct in form and shape, and verbs for change and motion.

The structural distinction between nouns (cl_i^k) and verbs (v_i) becomes apparent when lexical definitions of these are established. Essentially, a noun signifies a class (cl^1) of objects. When defined, it is shown to be a member of a more inclusive class (cl^2), denoted also by a noun which, in turn, when defined is shown to be a member of a more inclusive class (cl^3), etc., [pheasant → bird → animal → organism → thing]. We have the following scheme for representing the definition paradigm for nouns:

$$cl^n = \left\{ cl_{i_{n-1}}^{n-1} \left\{ cl_{i_{n-2}}^{n-2} \left\{ \ldots \left\{ cl_{i_m}^{m} \right\} \right\} \right\} \right\} \tag{1}$$

where the notation $\{\varepsilon_i\}$ stands for a class of elements ε_i $(i = 1, 2, \ldots, p)$, and subscripted subscripts are used to associate these subscripts with the appropriate superscripts. The highest order n in this hierarchy of classes is always represented by a single undefined term "thing," "entity," "act," etc., which appeals to basic notions of being able to perceive at all. A graphic representation of the hierarchical order of nouns is given in Fig. 1 and a more detailed discussion of the properties of these (inverted) "noun-chain-trees" can be found elsewhere (Weston, 1964; Von Foerster, 1967a).

Essentially, a verb (v_i) signifies an action, and when defined is given by a set of synonyms $\{v_j\}$, by the union or by the intersection of the meaning of verbs denoting similar actions. [hit → {strike, blow, knock} → {(hit, blow, ...) (stir, move air, sound, soothe, lay eggs, ..., boast) (strike, blow, bump, collide ...)} → etc.]

$$v_i = \{v_j\} \vee \sum v_k \vee \prod v_e \tag{2}$$

A graphic representation of this basically closed heterarchical structure is given in Fig. 2, and its corresponding representation in form of finite matrices is discussed elsewhere (Von Foerster, 1966).

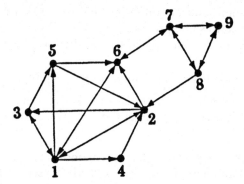

FIGURE 2. Closed heterarchical definition structure for verbs. (Verbs are at nodes; arrow heads: definiens; arrow tails: definiendum.)

The essential difference in the cognitive processes that allow for identification of forms and those of change of forms is not only reflected in the entirely different formalisms needed for representing the different definition structures of nouns [Eq. (1)] and of verbs [Eq. (2)], but also by the fact that the set of invariants that identify shape under various transformations can be computed by a single *deductive* algorithm (Pitts and McCulloch, 1947), while identification of even elementary notions of behavior requires *inductive* algorithms that can only be computed by perpetual comparison of present states with earlier states of the system under consideration (Von Foerster *et al.*, 1968).

These cognitive handicaps put the ethologist at a considerable disadvantage in developing a phenomenology for his subject matter when compared to his colleague the geneticist. Not only are the tools of expressing his phenomena devoid of the beautiful isomorphism which prevails between the hierarchical structures of all taxonomies and the definition of nouns that describe them, but, he may fall victim to a semantic trap which tempts him to associate with a conceptually isolable function a corresponding isolable mechanism that generates this function. This temptation seems to be particularly strong when our vocabulary suggests a variety of conceptually separable higher mental faculties as, for instance "to learn," "to remember," "to perceive," "to recall," "to predict," etc., and the attempt is made to identify and localize within the various parts of our brain the mechanisms that learn, remember, perceive, recall, predict, etc. The hopelessness of a search for mechanisms that represent these functions in isolation does not have a physiological basis as, for instance, "the great complexity of the brain," "the difficulty of measurement," etc. This hopelessness has a purely semantic basis. Memory, for instance, contemplated in isolation is reduced to "recording," learning to "change," perception to "input," and so on. In other words, in separating these functions from the totality of cognitive processes, one has abandoned the original problem and is now searching for mechanisms that implement entirely different functions which may or may not have any semblance to some processes that are subservient to the maintenance of the integrity of the organism as a functioning unit (Maturana, 1969).

Consider the two conceivable definitions for memory:

(a) An organism's potential awareness of past experiences.
(b) An observed change of an organism's response to like sequences of events.

While definition A postulates a faculty (memory$_A$) in an organism whose inner experience cannot be shared by an outside observer, definition B postulates the same faculty (memory$_A$) to be operative in the observer only—otherwise he could not have developed the concept of "change"—but ignores this faculty in the organism under observation, for an observer cannot "in principle" share the organism's inner experience. From this follows definition B.

It is definition B which is generally believed to be the one which obeys the ground rules of "the scientific method," as if it were impossible to cope scientifically with self-reference, self-description, and self-explanation, i.e., closed logical systems that include the referee in the reference, the descriptor in the description, and the axioms in the explanation.

This belief is unfounded. Not only are such logical systems extensively studied (e.g., Gunther, 1967; Löfgren, 1968), but also neurophysiologists (Maturana *et al.*, 1968), experimental psychologists (Konorski, 1962), and others (Pask, 1968; Von Foerster, 1969) have penetrated to such notions.

These preliminaries suggest that the explorer of mechanisms of mentation has to resolve two kinds of problems, only one of which belongs to physiology or, as it were, to physics; the other one is that of semantics. Consequently, it is proposed to reexamine some present notions of learning and memory as to the category to which they belong, and to sketch a conceptual framework in which these notions may find their proper place.

The next section, "Theory," reviews and defines concepts associated with learning and memory in the framework of a unifying mathematical formalism. In the Section III various models of interaction of molecules with functional units of higher organization are discussed.

II. Theory

A. General Remarks

Since we have as yet no comprehensive theory of behavior, we have no theory of learning and, consequently, no theory of memory. Nevertheless, there exists today a whole spectrum of conceptual frameworks ranging from the most naive interpretations of learning to the most sophisticated approaches to this phenomenon. On the naive side, "learning" is interpreted as a change of ratios of the occurrence of an organism's actions which are predetermined by an experimenter's ability to discriminate such actions and his value system, which classifies these actions into "hits" and "misses." Changes are induced by manipulating the organism through electric shocks, presentations of food, etc., or more drastically by mutilating, or even removing, some of the organism's organs. "Teaching" in this frame of mind is the administration of such "reinforcements" which induce the changes observed on other occasions.

On the sophisticated side, learning is seen as a process of evolving algorithms for solving categories of problems of ever-increasing complexity (Pask, 1968), or of evolving domains of relations between the organism and the outside world, of relations between these domains, etc. (Maturana, 1969). Teaching in this frame of mind is the facilitation of these evolutionary processes.

Almost directly related to the level of conceptual sophistication of these approaches is their mathematical naiveté, with the conceptually primitive theories obscuring their simplicity by a smoke screen of mathematical proficiency, and the sophisticated ones failing to communicate their depth by the lack of a rigorous formalism. Among the many causes for this unhappy state of affairs one seems to be most prominent, namely, the extraordinary difficulties that are quickly encountered when attempts are made to develop mathematical models that are commensurate with our epistemological insight. It may require the universal mind of a John von Neumann to give us the appropriate tools. In their absence, however, we may just browse around in the mathematical tool shop, and see what is available and what fits best for a particular purpose.

In this paper the theory of "finite state machines" has been chosen as a vehicle for demonstrating potentialities and limitations of some concepts in theories of memory, learning, and behavior mainly for two reasons. One is that it provides the most direct approach to linking a system's external variables as, e.g., stimulus, response, input, output, cause, effect, etc., to states and operations that are internal to the system. Since the central issue of a book on "molecular mechanisms in memory and learning" must be the development of a link which connects these internal mechanisms with their manifestations in overt behavior, the "finite state machine" appears to be a useful model for this task.

The other reason for this choice is that the interpretations of its formalism are left completely open, and may as well be applied to the animal as a whole; to cell assemblies within the animal; to single cells and their operational modalities, for instance, to the single neuron; to subcellular constituents; and, finally, to the molecular building blocks of these constituents.

With due apologies to the reader who is used to a more extensive and rigorous treatment of this topic, the essential features of this theory will be briefly sketched to save those who may be unfamiliar with this formalism from having to consult other sources (Ashby, 1956; Ashby, 1962; Gill, 1962).

B. Finite State Machines

1. Deterministic Machines

Essentially, the theory of finite state machines is that of computation. It postulates two finite sets of external states called "input states" and "output states," one finite set of "internal states," and two explicitly defined operations (computations) which determine the instantaneous and temporal relations between these states.*

Let X_i ($i = 1, 2, \ldots, n_x$) be the n_x receptacles for inputs x_i each of which can assume a finite number, $v_i > 0$, of different values. The number of distinguishable input states is then

$$X = \prod_{i=1}^{n_x} v_i \tag{3}$$

A particular input state $x(t)$ at time t (or x for short) is then the identification of the values x_i on all n_x input receptacles X_i at that "moment":

$$x(t) \equiv x = \{x_i\}_t \tag{4}$$

Similarly, let y_j ($j = 1, 2, \ldots, n_y$) be the n_y outlets for outputs y_j, each of which can assume a finite number, $v_j > 0$, of different values. The number of distinguishable output states is then

$$Y = \prod_{j=1}^{n_y} v_j \tag{5}$$

A particular output state $y(t)$ at time t (or y for short) is then the identification of the values y_i on all n_y outlets y_j at that "moment":

$$y(t) \equiv y = \{y_i\} \tag{6}$$

Finally, let Z be the number of internal states z which, for this discussion (unless specified otherwise), may be considered as being not further ana-

---

* Although the interpretation of states and operations with regard to observables is left completely open, some caution is advisable at this point if these are to serve as mathematical models, say, for the behavior of a living organism. A specific physical spatiotemporal configuration which is identifiable by the experimenter who wishes that this configuration be appreciated by the organism as a "stimulus" cannot *sui modo* be taken as "input state" for the machine. Such a stimulus may be a stimulant for the experimenter, but be ignored by the organism. An input state, on the other hand, cannot be ignored by the machine, except when explicitly instructed to do so. More appropriately, the distribution of the activity of the afferent fibers has to be taken as an input, and similarly, the distribution of activity of efferent fibers may be taken as the output of the system.

lyzable. Consequently, the values of z may just be taken to be the natural numbers from 1 to Z, and a particular output state $z(t)$ at time t (or z for short) is the identification of z's value at that "moment":

$$z(t) \equiv z \tag{7}$$

Each of these "moments" is to last a finite interval of time, Δ, during which the values of all variables x, y, z are identifiable. After this period, i.e., at time $t + \Delta$, they assume values $x(t + \Delta)$, $y(t + \Delta)$, $z(t + \Delta)$ (or x', y', z' for short), while during the previous period $t - \Delta$ they had values $x(t - \Delta)$, $y(t - \Delta)$, $z(t - \Delta)$ (or x^*, y^*, z^* for short).

After having defined the variables that will be operative in the machine we are now prepared to define the operations on these variables. These are two kinds and may be specified in a variety of ways. The most popular procedure is first to define a "driving function" which determines at each instant the output state, given the input state and the internal state at that instant:

$$y = f_y(x, z) \tag{8}$$

Although the driving function f_y may be known and the time course of input states x may be controlled by the experimenter, the output states y as time goes on are unpredictable as long as the values of z, the internal states of the machine, are not yet specified. A large variety of choices are open to specify the time course of z as depending on x, on y, or on other newly to be defined internal or external variables. The most profitable specification for the purposes at hand is to define z recursively as being dependent on previous states of affairs. Consequently, we define the "state function" f_z of the machine to be:

$$z = f_z(x^*, z^*) \tag{9a}$$

or alternately and equivalently

$$z' = f_z(x, z) \tag{9b}$$

that is, the present internal state of the machine is a function of its previous internal state and its previous input state; or alternately and equivalently, the next internal machine state is a function of both its present internal and input states.

With the three sets of states $\{x\}$, $\{y\}$, $\{z\}$ and the two functions f_y and f_z, the behavior of the machine, i.e., its output sequence, is completely determined if the input sequence is given.

Such a machine is called a sequential, state-determined, "nontrivial" machine and in Fig. 3a the relations of its various parts are schematically indicated.

Such a nontrivial machine reduces to a "trivial" machine if it is insensitive to changes of internal states, or if the internal states do not change (Fig. 3b):

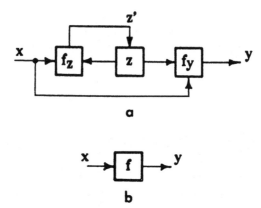

FIGURE 3. Signal flow in a finite state machine (a); input–output relation in a trivial machine (b).

$$z' = z = z_0 = \text{constant} \tag{10a}$$

$$y = f_y(x, \text{constant}) = f(x) \tag{10b}$$

In other words, a trivial machine is one which couples deterministically a particular input state with a specific output state or, in the language of naive reflexologists, a particular stimulus with a specific response.

Since the concept of "internal states" is crucial in appreciating the difference between a trivial and a nontrivial machine, we shall now give various formal interpretations of these states to lift them from the limbo of "being not further analyzable."

First, it may appear that by an artifice one can get rid of these mysterious states by defining the driving function f_y in a recursive form. However, as we shall see shortly, these states reappear in just another form.

Consider the driving function [Eq. (8)] at time t and one step later $(t + \Delta)$:

$$y = f_y(x, z)$$
$$y' = f_y(x', z') \tag{8'}$$

and assume there exists an "inverse function" to f_y:

$$z = \phi_y(x, y) \tag{11}$$

We now enter the state function [Eq. (9b)] for z' into Eq. (8') and replace z by Eq. (11):

$$y' = f_y(x', f_z(x, \phi_y(x, y))) = F_y^{(1)}(x', x, y) \tag{12}$$

or alternately and equivalently

$$y = F_y^{(1)}(x, x^*, y^*) \tag{13}$$

However, y^* is given recursively through Eq. (13)

$$y^* = F_y^{(1)}(x^*, x^{**}, y^{**}) \tag{13*}$$

and inserting this into Eq. (13) we have

$$y = F_y^{(2)}(x, x^*, x^{**}, y^{**})$$

and for n recursive steps

$$y = F_y^{(n)}(x, x^*, x^{**}, x^{***} \ldots x^{(n)*}, y^{(n)*}) \tag{14}$$

This expression suggests that in a nontrivial machine the output is not merely a function of its present input, but may be dependent on the particular sequence of inputs reaching into the remote past, and an output state at this remote past. While this is only to a certain extent true—the "remoteness" is carried only over Z recursive steps and, moreover, Eq. (14) does not uniquely determine the properties of the machine—this dependence of the machine's behavior on its past history should not tempt one to project into this system a capacity for memory, for at best it may look upon its present internal state which may well serve as *token* for the past, but without the powers to recapture for the system all that which has gone by.

This may be most easily seen when Eq. (13) is rewritten in its full recursive form for a linear machine (with x and y now real numbers)

$$y(t + \Delta) - ay(t) = bx(t) \tag{15a}$$

or in its differential analog expanding $y(t + \Delta) = y(t) + \Delta dy/dt$:

$$\frac{dy}{dt} - \alpha y = x(t) \tag{15b}$$

with the corresponding solutions

$$y(n\Delta) = a^n \left[y(0) + b \sum_{i=0}^{n} a^{-i} x(i\Delta) \right] \tag{16a}$$

and

$$y(t) = e^{\alpha t} \left[y(0) + \int_0^t e^{-\alpha \tau} x(\tau) d\tau \right] \tag{16b}$$

From these expressions it is clear that the course of events represented by $x(i\Delta)$ (or $x(\tau)$) is "integrated out," and is manifest only in an additive term which, nevertheless, changes as time goes on.

However, the failure of this simple machine to account for memory should not discourage one from contemplating it as a possible useful element in a system that remembers.

While in these examples the internal states z provided the machine with an appreciation—however small—of its past history, we shall now give an

TABLE 1. Computing Z logical function $F_z(x)$ on inputs x

z	1	1	1	...1	2	2	2	...2	...	Z	Z	Z	...Z
x	a	b	c	...X	a	b	c	...X	...	a	b	c	...X
y	γ	α	β	...δ	α	γ	β	...∈	...	β	∈	γ	...δ

interpretation of the internal states z as being a selector for a specific function in a set of multivalued logical functions. This is most easily seen when writing the driving function f_y in form of a table.

Let $a, b, c \ldots X$ be the input values x; $\alpha, \beta, \gamma \ldots Y$ be the output values y; and $1, 2, 3 \ldots Z$ be the values of the internal states. A particular driving function f_y is defined if to all pairs $\{xz\}$ an appropriate value of y is associated. This is suggested in Table 1.

Clearly, under $z = 1$ a particular logical function, $y = F_1(x)$, relating y with x is defined; under $z = 2$ another logical function, $y = F_2(x)$, is defined; and, in general, under each z a certain logical function $y = F_z(x)$ is defined.

Hence, the driving function f_y can be rewritten to read

$$y = F_z(x), \tag{17}$$

which means that this machine computes another logical function $F_{z'}$ on its inputs x, whenever its internal state z changes according to the state function $z' = f_z(x, z)$.

Or, in other words, whenever z changes, the machine becomes a *different* trivial machine.

While this observation may be significant in grasping the fundamental difference between nontrivial and trivial machines, and in appreciating the significance of this difference in a theory of behavior, it permits us to calculate the number of internal states that can be effective in changing the *modus operandi* of this machine.

Following the paradigm of calculating the number n of logical functions as the number of states of the dependent variable raised to the power of the number of states of the independent variables

$$n = (\text{no. of states of dep. variables})^{(\text{no. of states of indep. variables})} \tag{18}$$

we have for the number of possible trivial machines which connect y with x

$$n_T = Y^X \tag{19}$$

This, however, is the largest number of internal states which can effectively produce a change in the function $F_z(x)$, for any additional state has to be paired up with a function to which a state has been already assigned, hence such additional internal states are redundant or at least indistinguishable. Consequently

$$Z \le Y^X$$

TABLE 2. The number of effective internal states Z, the number of possible driving functions n_D, and the number of effective state functions n_S for machines with one two-valued output and with from one to four two-valued inputs

n	Z	n_D	n_S
1	4	256	65536
2	16	2.10^{19}	6.10^{76}
3	256	10^{600}	$300.10^{4.10^3}$
4	65536	$300.10^{4.10^3}$	$1600.10^{7.10^4}$

Since the total number of driving functions $f_y(x, z)$ is

$$n_D = Y^{XZ}, \tag{20}$$

its largest value is:

$$\bar{n}_D = Y^{XY^X} \tag{21}$$

Similarly, for the number of state functions $f_z(z, x)$ we have

$$n_S = Z^{Y \cdot Z} \tag{22}$$

whose largest effective value is

$$\bar{n}_s = Y^{X \cdot XY^X} = [\bar{n}_D]^X \tag{23}$$

These numbers grow very quickly into meta-astronomical magnitudes even for machines with most modest aspirations.

Let a machine have only one two-valued output ($n_y = 1$; $v_y = 2$; $y = \{0; 1\}$; $Y = 2$) and n two-valued inputs ($n_x = n$; $v_x = 2$; $x = \{0; 1\}$; $X = 2^n$). Table 2 gives the number of effective internal states, the number of possible driving functions, and the number of effective state functions for machines with from one to four "afferents" according to the equations

$$Z = 2^{2^n}$$

$$n_D = 2^{2^{2n+n}}$$

$$n_S = 2^{2^{2n+2n}}$$

These fast-rising numbers suggest that already on the molecular level without much ado a computational variety can be met which defies imagination. Apparently, the large variety of results of genetic computation, as manifest in the variety of living forms even within a single species, suggests such possibilities. However, the discussion of these possibilities will be reserved for the next section.

2. *Interacting Machines*

We shall now discuss the more general case in which two or more such machines interact with each other. If some aspects of the behavior of an organism can be modeled by a finite state machine, then the interaction of the organism with its environment may be such a case in question, if the environment is likewise representable by a finite state machine. In fact, such two-machine interactions constitute a popular paradigm for interpreting the behavior of animals in experimental learning situations, with the usual relaxation of the general complexity of the situation, by chosing for the experimental environment a trivial machine. "Criterion" in these learning experiments is then said to have been reached by the animal when the experimenter succeeded in transforming the animal from a nontrivial machine into a trivial machine, the result of these experiments being the interaction of just two trivial machines.

We shall denote quantities pertaining to the environment (E) by Roman letters, and those to the organism (Ω) by the corresponding Greek letters. As long as E and Ω are independent, six equations determine their destiny. The four "machine equations," two for each system

$$E: \quad y = f_y(x, z) \tag{24a}$$

$$z' = f_z(x, z) \tag{24b}$$

$$\Omega: \quad \eta = f_\eta(\xi, \zeta) \tag{25a}$$

$$\zeta' = f_\zeta(\xi, \zeta) \tag{25b}$$

and the two equations that describe the course of events at the "receptacles" of the two systems

$$x = x(t); \quad \xi = \xi(t) \tag{26a, b}$$

We now let these two systems interact with each other by connecting the (one step delayed) output of each machine with the input of the other. The delay is to represent a "reaction time" (time of computation) of each system to a given input (stimulus, cause) (see Fig. 4). With these connections the following relations between the external variables of the two systems are now established:

$$x' = \eta = u'; \quad \xi' = y = v' \tag{27a, b}$$

where the new variables u, v represent the "messages" transmitted from $\Omega \to E$ and $E \to \Omega$ respectively. Replacing x, y, η, ξ, in Eqs. (24) (25) by u, v according to Eq. (27) we have

$$v' = f_y(u, z); \quad u' = f_\eta(v, \zeta)$$

$$z' = f_z(u, z); \quad \zeta' = f_\zeta(v, \zeta) \tag{28}$$

These are four recursive equations for the four variables u, v, z, ζ, and if the four functions $f_y, f_z, f_\eta, f_\zeta$ are given, the problem of 'solving" for $u(t)$,

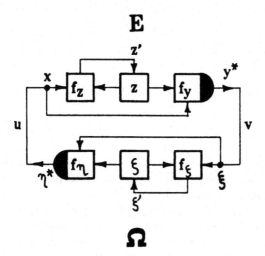

FIGURE 4. Two finite state machines (E) (Ω) connected via delays (black semicircles).

$v(t)$, $z(t)$, $\zeta(t)$, i.e., expressing these variables explicitly as functions of time, is purely mathematical. In other words, the "meta-system" $(E\Omega)$ composed of the subsystems E and Ω, is physically as well as mathematically "closed," and its behavior is completely determined for all times. Moreover, if at a particular time, say $t = 0$ (initial condition), the values of all variables $u(0)$, $v(0)$, $z(0)$, $\zeta(0)$ are known, it is also completely predictable. Since this meta-system is without input, it churns away according to its own rules, coming ultimately to a static or dynamic equilibrium, depending on the rules and the initial conditions.

In the general case the behavior of such systems has been extensively studied by computer simulation (Walker, 1965; Ashby and Walker, 1966; Fitzhugh, 1963), while in the linear case the solutions for Eqs. (28) can be obtained in straight-forward manner, particularly if the recursions can be assumed to extend over infinitesimally small steps:

$$w' = w(t + \Delta) = w(t) + \Delta \frac{dw}{dt} \tag{29}$$

Under these conditions the four Eqs. (28) become

$$\dot{w}_i - \sum_{j=1}^{4} \alpha_{ij} w_j \tag{30}$$

where the w_i $(i = 1, 2, 3, 4)$ are now the real numbers and replace the four variables in question, $\dot{w}$ represents the first derivative with respect to time, and the 16 coefficients a_{ij} $(i, j = 1, 2, 3, 4)$ define the four linear functions under consideration. This system of simultaneous, first-order, linear differential equations is solved by

$$w_i(t) = \sum_{j=1}^{4} A_{ij} e^{\lambda_{ji} t} \qquad (31)$$

in which λ_j are the roots of the determinant

$$|a_{ij} - \delta_{ij}\lambda| = 0 \qquad (32)$$

$$\delta_{ij} = \begin{cases} 1 \dots i = j \\ 0 \dots i \neq j \end{cases}$$

and the A_{ij} depend on the initial conditions. Depending on whether the λ_j turn out to be complex, real negative or real positive, the system will ultimately oscillate, die out, or explode.*

While a discussion of the various modes of behavior of such systems goes beyond this summary, it should be noted that a common behavioral feature in all cases is an initial transitory phase that may move over a very large number of states until one is reached that initiates a stable cyclic trajectory, the dynamic equilibrium. Form and length of both the transitory and final equilibrial phases are dependent on the initial conditions, a fact which led Ashby (1956) to call such systems "multistable." Since usually a large set of initial conditions maps into a single equilibrium, this equilibrium may be taken as a *dynamic representation* of a set of events, and in a multistable system each cycle as an "abstract" for these events.

With these notions let us see what can be inferred from a typical learning experiment (e.g., John *et al.*, 1969) in which an experimental animal in a Y-maze is given a choice ($\xi_0 \equiv C$, for "choice") between two actions ($\eta_1 \equiv L$, for "left turn"; $\eta_2 \equiv R$, for "right turn"). To these the environment E, a trivial machine, responds with new inputs to the animal ($\eta_1 = x_1' \to y_1' = \xi_1'' \equiv S$, for "shock"; or $\eta_2 = x_2' \to y_2' = \xi_2'' \equiv F$, for "food"), which, in turn, elicit in the animal a pain ($\eta_3 \equiv$ "−") or pleasure ($\eta_4 \equiv$ "+") response. These responses cause E to return the animal to the original choice situation ($\xi_0 \equiv C$).

Consider the simple survival strategy built into the animal by which under neutral and pleasant conditions it maintains its internal state [$\zeta' = \zeta$, for ($C\zeta$) and ($F\zeta$)], while under painful conditions it changes it [$\zeta' \neq \zeta$, for ($S\zeta$)]. We shall assume eight internal states ($\zeta = i; i = 1, 2, 3, \dots, 8$).

With these rules the whole system (ΩE) is specified and its behavior completely determined. For convenience, the three functions, $f_y = f$ for the trivial machine E, f_η and f_ζ for Ω are tabulated in Tables 3a, b, c.

With the aid of these tables the eight behavioral trajectories for the (ΩE) system, corresponding to the eight initial conditions, can be written. This

---

* This result is, of course, impossible in a finite state machine. It is obtained here only because of the replacement of the discrete and finite variables u, v, z, ζ, by w_i which are continuous and unlimited quantities.

TABLE 3a.

$$y = f(x)$$

$x\ (= \eta^*)$	$y\ (= \xi')$
L	S
R	F
−	C
+	C

TABLE 3b.

$$\eta = f_\eta(\xi, \zeta)$$

$\eta\ (=x')$		ζ							
		1	2	3	4	5	6	7	8
$\xi\ (= y^*)$	C	L	L	L	L	R	R	R	R
	S	−	−	−	−	−	−	−	−
	F	+	+	+	+	+	+	+	+

TABLE 3c.

$$\zeta' = f_\zeta(\xi, \zeta)$$

ζ'		ζ							
		1	2	3	4	5	6	7	8
$\xi\ (= y^*)$	C	1	2	3	4	5	6	7	8
	S	2	3	4	5	6	7	8	1
	F	1	2	3	4	5	6	7	8

has been done below, indicating only the values of the pairs $\xi\zeta$ as they follow each other as consequences of the organism's responses and the environment's reactions.

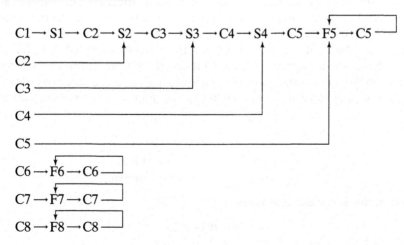

These trajectories show indeed the behavior as suggested before, initial transients depending in length on the initial conditions, and ultimately a dynamic equilibrium flipping back and forth between two external states without internal state change. The whole system, and its parts, have become trivial machines. Since, even with maximum semantic tolerance, one cannot say a trivial machine has memory, one wonders what is intended to be measured when at this stage it is opened and the internal workings are examined. Does one wish to inspect its present workings? Or, to see how much it has changed since earlier examinations? At best, these are tests of the experimenter's memory, but whether the machine can appreciate any changes cannot, in principle, be inferred from experiments whose concep-tual design eliminates the quality which they intend to measure.

3. Probabilistic Machines

This dilemma can be seen in still another light if we adopt for the moment the position of statistical learning theory (Skinner, 1959; Estes, 1959; Logan, 1959). Here either the concept of internal states is rejected or the existence of internal states is ignored. But whenever the laws which connect causes with effects are ignored, either through ignorance or else by choice, the theory becomes that of probabilities.

If we are ignorant of the initial state in the previous example, the chances are 50/50 that the animal will turn left or right on its first trial. After one run the chances are 5/8 for turning right, and so on, until the animal has turned from a "probabilistic (nontrivial) machine" to a "deterministic (trivial) machine," and henceforth always turns right. While a statistical learning theoretician will elevate the changing probabilities in each of the subsequent trials to a "first principle," for the finite state machinist this is an obvious consequence of the effect of certain inputs on the internal states of his machine: they become inaccessible when paired with "painful inputs." Indeed, the whole mathematical machinery of statistical learning theory can be reduced to the paradigm of drawing balls of different color from an urn while observing certain non-replacement rules.

Let there be an urn with balls of m different colors labeled $0, 1, 2, \ldots,$ $(m-1)$. As yet unspecified rules permit or prohibit the return of a certain colored ball when drawn. Consider the outcomes of a sequence of n draw-ings, an "n-sequence," as being an n digit m-ary number (e.g., $m = 10; n = 12$):

$$v = 1\ 5\ 7\ 3\ 0\ 2\ 1\ 8\ 6\ 2\ 1\ 4$$

↑	↑
Last	First
drawn	drawn

From this it is clear that there are

$$\mathfrak{n}(n, m) = m^n$$

different n-sequences. A *particular* n-sequence will be called a v-number, i.e.:

$$0 \le v(m, n) = \sum_{i=1}^{n} j(i) m^{(i-1)} \le m_{-1}^{n} \tag{33}$$

where $0 \le j(i) \le (m - 1)$ represents the outcome labeled j at the ith trial.

The probability of a *particular* n-sequence (represented by a v-number) is then

$$p_n(v) = \prod_{i=1}^{n} p_i[j(i)] \tag{34}$$

where $p_i[j(i)]$ gives the probability of the color labeled j to occur at the ith trial in accordance with the specific v-number as defined in Eq. (33).

Since after each trial with a 'don't return" outcome all probabilities are changed, the probability of an event at the nth trial is said to depend on the "path," i.e., on the past history of events, that led to this event. Since there are m^{n-1} possible paths that may precede the drawing of j at the nth trial, we have for the probability of this event:

$$p_n(j) = \sum_{v=0}^{m^{n-2}-1} p_n(j \cdot m^{n-1} + v(n-1, m))$$

where $j \cdot m^{n-1} + v(n - 1, m)$ represent a $v(n, m)$-number which begins with j.

From this a useful recursion can be derived. Let j^* be the colors of balls which when drawn are *not* replaced, and j the others. Let n_{j^*} and n_j be the number of preceding trials on which j^* and j came up respectively ($\Sigma n_{j^*} + \Sigma n_j = n - 1$), then the probability for drawing j (or j^*) at the nth trial with a path of Σn_{j^*} withdrawals is

$$p_n(j) = \frac{N_j}{N - \sum n_{j^*}} \cdot p_{n-1}\left(\sum n_{j^*}\right) \tag{35a}$$

and

$$p_n(j^*) = \frac{N_{j^*} - n_{j^*}}{N - \sum n_{j^*}} \cdot p_{n-1}\left(\sum n_{j^*}\right) \tag{35b}$$

where $N = \Sigma N_j + \Sigma N_{j^*}$ is the initial number of balls, and N_j and N_{j^*} the initial number of balls with colors j and j^* respectively.

Let there be N balls to begin with in an urn, N_w of which are white, and $(N - N_w)$ are black. When a white ball is drawn, it is returned; a black ball, however, is removed. With "white" $\equiv 0$, and "black" $\equiv 1$, a particular n-sequence ($n = 3$) may be

$$v(3, 2) = 1 \ 0 \ 1$$

and its probability is:

$$p_3(1\,0\,1) = \frac{N-N_w-1}{N-1} \cdot \frac{N_w-1}{N-1} \cdot \frac{N-N_w}{N}$$

The probability of drawing a black ball at the third trial is them:

$$p_3(1) = p_3(1\,0\,0) + p_3(1\,0\,1) + p_3(1\,1\,0) + p_3(1\,1\,1)$$

We wish to know the probability of drawing a white ball at the nth trial. We shall denote this probability now by $p(n)$, and that of drawing a black ball $q(n) = 1 - p(n)$.

By iteratively approximating [through Eq. (35)] trial tails of length in as being path independent $[p_i(j) = p_1(j)]$ one obtains a first-order approximation for a recursion in $p(n)$:

$$p(n) = p(n-m) + \frac{m}{N} q(n-m) \tag{36}$$

or for $m = n - 1$ (good for $p(1) \approx 1$, and $n/N \ll 1$):

$$p(n) = p(1) + \frac{n-1}{N} q(1) \tag{37}$$

and for $m = 1$ (good for $p(1) \approx 1$):

$$p(n) = p(n-1) + \frac{1}{N} q(n-1) \tag{38}$$

A second approximation changes the above expression to

$$p(n) = p(n-1) + \theta q(n-1) \tag{39}$$

where $\theta = \theta(N, N_w)$ is a constant for all trials. With this we have

$$p(n) - p(n-1) = \Delta p = \theta(1-p) \tag{40}$$

which, in the limit for

$$\lim_{\Delta n \to 0} \frac{\Delta p}{\Delta n} = \frac{dp}{dn}$$

gives

$$\frac{dp}{dn} = \theta(1 - p(n))$$

with the solution

$$p(n) = 1 - (1 - p_0)e^{-\theta n} \tag{41}$$

This, in turn, is an approximation for $p \approx 1$ of

$$p(n) = \frac{p_0}{p_0 + (1-p_0)e^{-\theta n}} \tag{42}$$

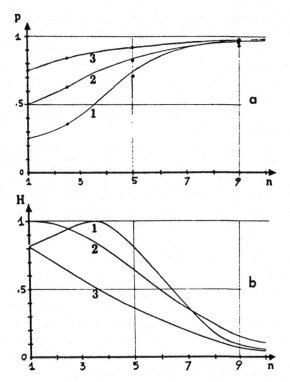

FIGURE 5. Probability for drawing a white ball at the nth trial from an urn having initially four balls of which 1, 2, or 3 are white, the others black. White balls are replaced, black are not (a). Entropy at the nth trial (b).

which is the solution of

$$\frac{dp}{dn} = \theta p(1-p) \tag{43}$$

or, recursively expressed, of

$$p(n) = p(n-1) + \theta p(n-1) \cdot q(n-1) \tag{44}$$

Figure 5a compares the probabilities $p(n)$ for drawing a white ball at the nth trial, as calculated through approximation [Eq. (42)] (solid curves), with the exact values computed by an IBM 360/50 system with a program kindly supplied by Mr. Atwood for an urn with initially four balls ($N = 4$) and for the three cases in which one, two, or three of these are white ($N_w = 1$, $N_w = 2$, $N_w = 3$). The entropy* $H(n)$ in bits per trial corresponding to these cases is shown in Fig. 5b, and one may note that while for some cases [$p(1) \leq 0.5$] it reaches a maximum in the course of this game,

---

* Or the "amount of uncertainty", or the "amount of information" received by the outcome of each trial, defined by $-H(n) = p(n)\log_2 p(n) + q(n)\log_2 q(n)$.

it vanishes in all cases when certainty of the outcome is approached $[p(n) \rightarrow 1]$.

Although the sketch on probabilities dealt exclusively with urns, balls, and draws, students of statistical learning theory will have recognized in Eqs. (39), (41), and (42) the basic axioms of this theory [Estes, 1959; Eqs. (5), (6), and (9)], and there is today no doubt that under the given experimental conditions animals will indeed trace out the learning curves derived for these conditions.

Since the formalism that applies to the behavior of these experimental animals applies as well to our urn, the question now arises: can we say an urn learns? If the answer is "yes," then apparently there is no need for *memory* in learning, for there is no trace of black balls left in our urn when it finally "responds" correctly with white balls when "stimulated" by each draw; if the answer is "no," then by analogy we must conclude it is not *learning* that is observed in these animal experiments.

To escape this dilemma it is only necessary to recall that an urn is just an urn, and it is animals that learn. Indeed, in these experiments learning takes place on two levels. First, the experimental animals learned to behave "urnlike," or better, to behave in a way which allows the experimenter to apply urnlike criteria. Second, the experimenter learned something about the animals by turning them from nontrivial (probabilistic) machines into trivial (deterministic) machines. Hence, it is from studying the experimenter whence we get the clues for memory and learning.

C. Finite Function Machines

1. Deterministic Machines

With this observation the question of where to look for memory and learning is turned into the opposite direction. Instead of searching for mechanisms in the environment that turn organisms into trivial machines, we have to find the mechanisms within the organisms that enable them to turn their environment into a trivial machine.

In this formulation of the problem it seems to be clear that in order to manipulate its environment an organism has to construct—somehow—an internal representation of whatever environmental regularities it can get hold of. Neurophysiologists have long since been aware of these abstracting computations performed by neural nets from right at the receptor level up to higher nuclei (Lettvin *et al.*, 1959; Maturana *et al.*, 1968; Eccles *et al.*, 1967). In other words, the question here is how to compute functions rather than states, or how to build a machine that computes programs rather than numerical results. This means that we have to look for a formalism that handles "finite *function* machines." Such a formalism is, of course, one level higher up than the one discussed before, but by maintaining some pertinent analogies its essential features may become apparent.

Our variables are now functions, and since relations between functions are usually referred to as "functionals," the essential features of a calculus of recursive functionals will be briefly sketched.

Consider a system like the one suggested in Fig. 3a, with the only difference that it operates on a finite set of functions of two kinds, $\{f_{yi}\}$ and $\{f_{zj}\}$. These functions, in turn, operate on their appropriate set of states $\{y_i\}$ and $\{z_j\}$. The rules of operation for such a finite function machine are modeled exactly according to the rules of finite state machines. Hence:

$$f_y = F_y[x, f_z] \tag{45a}$$

$$f_z' = F_z[x, f_z] \tag{45b}$$

where F_y and F_z are the functionals which generate the driving functions f_y and the subsequent internal function f_z' from the present internal function f_z and an input x. One should note, however, that the input here is still a state. This indicates an important feature of this formalism, namely, the provision of a link between the domain of states with the entirely different domain of functions. In other words, this formalism takes notice of the distinction between entities and their representations and establishes a relation between these two domains.

Following a procedure similar to that carried out in Eqs. (10) through (14), the functions of type f_z can be eliminated by expressing the present driving function as result of earlier states of affairs. However, due to some properties that distinguish functionals from functions, these earlier states of affairs include both input states as well as output functions. We have for n recursive steps:

$$f_y = \Phi_y^{(n)}\left[x, x^*, x^{**}, x^{***}, \ldots, x^{(n)*}; f_y^*, f_y^{**}, \ldots, f_y^{(n)*}\right] \tag{46}$$

Comparing this expression with its analog for finite state machines [Eq. (14)], it is clear that here the reference to past events is not only to those events that were the system's history of inputs $\{x^{(i)*}\}$, but also to its history of potential actions $\{f_y^{(i)*}\}$. Moreover, when this recursive functional is solved explicitly for time $(t = k\Delta; k = 0, 1, 2, 3, \ldots ;)$ [compare with Eq. (16)], it is again the history of inputs that is "integrated out"; however, the history of potential actions remains intact, because of a set of n "eigenfunctions" which satisfy Eq. (46). We have explicitly for $k\Delta$), and for the ith eigenfunction:

$$f_y^i(k\Delta) = K_i(k\Delta) \cdot [\pi_i(f_y^{(i)*}) + G_i(x, x^*, x^{**}, \ldots, x^{(n)*})] \tag{47}$$

$$i = 1, 2, 3, \ldots, n$$

with K_i and G_i being functions of $(k\Delta)$, the latter one giving a value that depends on a tail of values in $x^{(i)*}$ which is n steps long. π_i is again a functional, representing the output function f_y of i steps in the past in terms of another function.

Although this formalism does not specify any mechanism capable of performing the required computations, it provides us, at least, with an adequate description of the functional organization of memory. Access to "past experience" is given here by the availability of the system's own *modus operandi* at earlier occasions, and it is comfortable to see from expression (47) that the subtle distinction between an experience in the past ($f_y^{(i)}*$), and the present experience of an experience in the past $[\pi_i(f_y^{(i)}*)]$—i.e., the distinction between "experience" and "memory"—is indeed properly taken care of in this formalism. Moreover, by the system's access to its earlier states of functioning, rather than to a recorded collection of accidental pairs $\{x_i, y_i\}$ that manifest this functioning, it can compute a stream of "data" which are consistent with the system's past experience. These data, however, may or may not contain the output values $\{y_i\}$ of those accidental pairs. This is the price one has to pay for switching domains, from states to functions and back again to states. But this is a small price indeed for the gain of an infinitely more powerful "storage system" which computes the answer to a question, rather than stores all answers together with all possible questions in order to respond with the answer when it can find the question (Von Foerster, 1965).

These examples may suffice to interpret without difficulty another property of the finite function machine that is in strict analogy to the finite state machine. As with the finite state machine, a finite function machine will, when interacting with another system, go through initial transients depending on initial conditions and settle in a dynamic equilibrium. Again, if there is no internal function change ($f_z' = f_z = f_0$) we have a "trivial finite function machine" with its "goal function" f_0. It is easy to see that a trivial finite function machine is equivalent to a nontrivial finite state machine.*

Instead of citing further properties of the functional organization of finite function machines, it may be profitable to have a glance at various possibilities of their structural organization. Clearly, here we have to deal with aggregates of large numbers of finite state machines, and a more efficient system of notation is required to keep track of the operations that are performed by such aggregates.

2. Tesselations

Although a finite state machine consists of three distinct parts, the two computers, f_y and f_z, and the store for z, (see Fig. 3a), we shall represent the entire machine by a single square (or rectangle); its input region denoted white, the output region black (Fig. 6). We shall now treat this unit as an elementary computer—a "computational tile," T_i—which, when combined with other tiles, T_j, may form a mosaic of tiles—a "computational tessela-

---

* In the case of several equilibria $\{f_{oi}\}$, we have, of course, a *set* of nontrivial finite state machines that are the outcomes of various initial conditions.

FIGURE 6. Symbolization of a finite state machine by a computational tile. Input region white; output region black.

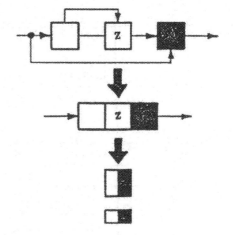

tion," 3. The operations performed by the ith tile shall be those of a finite state machine, but different letters, rather than subscripts, will be used to distinguish the two characteristic functions. Subscripts shall refer to tiles.

$$y_i = f_i(x_i, z_i)$$

$$z_i = g_i(x_i, z_i) \tag{48}$$

Figure 7/I sketches the eight possible ways (four each for the parallel and the antiparallel case) in which two tiles can be connected. This results in three classes of elementary tesselations whose structures are suggested in Fig. 7/II. Cases I/1 and I/3, and I/2 and I/4 are equivalent in the parallel case, and are represented in II/1 ("chain") and II/2 ("stack") respectively. In the antiparallel case the two configurations I/1 and I/3 are ineffective, for outputs cannot act on outputs, nor inputs on inputs; cases I/2 and I/4 produce two autonomous elementary tesselations $A = [a^+, a^-]$, distinct only by the sense of rotation in which the signals are processed.

Iterations of the same concatenations result in tesselations with the following functional properties (for n iterations):

1. *Stack*

$$nT: \quad y = \sum_{1}^{n} f_i(x_i, z_i) \tag{49}$$

2. *Chain*

$$T^n: \quad y = f_n\left(f_{n-1}\left(f_{n-2} \ldots (x^{(n)*}, z^{(n)*}) \ldots z_{n-2}^{**}\right)z_{n-1}^{*}\right)z_n \tag{50}$$

3. $A = \{a^+, a^-\}$

$$\left.\begin{matrix} a^+a^- \\ a^-a^+ \end{matrix}\right\} = 0 \qquad \left.\begin{matrix} a^+a^+ \\ a^-a^- \end{matrix}\right\} \neq 0$$

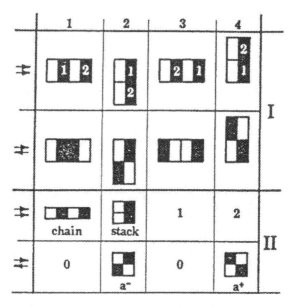

FIGURE 7. Elementary tesselations.

(i) *Stack* nA^n (51)

(ii) *Chain* A^n (52)

Introducing a fourth elementary tesselation by connecting horizontally $T \rightarrow A \rightarrow T$, or *TAT*, we have

4. *TAT*

(i) *Stack* $n(TA^nT)$ (53)

(ii) *Chain* $(TAT)^n$ (54)

Figure 8 suggests further compositions of elementary tesselations. All of these contain autonomous elements, for it is the presence of at least two such elements as, e.g., in $(TAT)^2$, which constitute a finite function machine. If none of these elements happens to be "dead"—i.e., are locked into a single state static equilibrium—they will by their interaction force each other from one dynamic equilibrium into another one. In other words, under certain circumstances they will turn each other from one trivial finite function machine into another one, but this is exactly the criterion for being a nontrivial finite function machine.

It should be pointed out that this concept of formal mathematical entities interacting with each other is not new. John von Neumann (1966) developed this concept for self-reproducing "automata" which have many properties in common with our tiles. Lars Löfgren (1962) expanded this

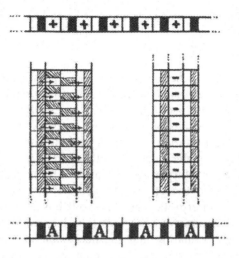

FIGURE 8. Some examples of simple tesselations.

concept to include self-repair of certain computational elements which are either stationary or freely moving in their tesselations, and Gordon Pask (1962) developed similar ideas for discussing the social self-organization of aggregates of such automata.

It may be noted that in all these studies ensembles of elements are contemplated in order to achieve logical closure in discussing the proprietory concept and autonomous property regarding the elements in question as, e.g., *self*-replication, *self*-repair, *self*-organization, *self*-explanation, etc. This is no accident, as Löfgren (1968) observed, for the prefix "self-" can be replaced by the term to which it is a prefix to generate a second-order concept, a concept of a concept. Self-explanation is the explanation of an explanation; self-organization is the organization of an organization (Selfridge, 1962), etc. Since cognition is essentially a self-referring process (Von Foerster, 1969), it is to be expected that in discussing its underlying mechanisms we have to contemplate function of functions and structure of structures.

Since with the build-up of these structures their functional complexity grows rapidly, a detailed discussion of their properties would go beyond the scope of this article. However, one feature of these computational tesselations can be easily recognized, and this is that their operational modalities are closely linked to their structural organization. Here function and structure go hand in hand, and one should not overlook that perhaps the lion's share of computing has been already achieved when the system's topology is established (Werner, 1969). In organisms this is, of course, done mainly by genetic computations.

This observation leads us directly to the physiology and physics of organic tesselations.

III. Biophysics

A. General Remarks

The question now arises whether or not one can identify structural or functional units in living organisms which can be interpreted in terms of the purely mathematical objects mentioned previously, the "tiles," the "automata," the "finite function machines," etc. This method of approach, first making an interpretation and then looking for confirming entities, seems to run counter to "the scientific method" in which the "facts" are supposed to precede their interpretation. However, what is reported as "fact" has gone through the observer's cognitive system which provides him, so to say, with a priori interpretations. Since our business here is to identify the mechanisms that observe observers (i.e., becoming "self-observers"), we are justified in postulating first the necessary functional structure of these mechanisms. Moreover, this is indeed a popular approach, as seen by the frequent use of terms like "trace," "engram," "store," "read-in," "read-out," etc., when mechanisms of memory are discussed. Clearly, here too the metaphor precedes the observations. But metaphors have in common with interpretations the quality of being neither true nor false; they are only useful, useless, or misleading.

When a functional unit is conceptually isolated—an *animal*, a *brain*, the *cerebellum*, *neural nuclei*, a *single neuron*, a *synapse*, a *cell*, the *organelles*, the *genomes*, and other molecular building blocks—in its abstract sense the concept of "machine" applied to these units is useful, if it were only to discipline the user of this concept to identify properly the structural and functional components of his "machine." Indeed, the notions of the finite state machine, or all its methodological relatives, have contributed—explicitly or implicitly—much to the understanding of a large variety of such functional units. For instance, the utility of the concepts "transcript," "en-coding," "de-coding," "computation," etc., in molecular genetics cannot be denied.

Let the n-sequence of the four bases ($b = 4$) of a particular DNA molecule be represented by a v-number $v(n, b)$ [see Eq. (33)]; let $Tr(v) = \bar{v}$ be an operation which transforms the symbols $(0, 1, 2, 3) \to (3, 2, 1, \emptyset)$, in that order, with $0 \equiv$ thymine, $1 \equiv$ cytosine, $2 \equiv$ guanine, $3 \equiv$ adenine, and $\emptyset \equiv$ uracil, and I be the identity operation $I(v) = v$; finally, let $\Phi[\bar{v}(n, b)] = v(n/3, a) = \mu(m, a)$, with $a = 20$, and $j = 0, 1, \ldots, 19$, representing the 20 amino acids of the polypeptide chain. Then

(i) DNA replication: $v = I(v)$ (55a)

(ii) DNA/RNA transcript: $\bar{v} = Tr(v)$ (55b)

(iii) Protein synthesis: $\mu = \Phi(\bar{v})$ (55c)

While the operations I and Tr require only trivial machines for the process of transcription, Φ is a recursive computation of the form

$$j(i) \equiv y(i) = y(i-1) + a^i f(x) \tag{56}$$

Using the suggested recursion [compare with Eq. (14)]:

$$y(i) = a^i f(x) + a^{i-1} f(x^*) + a^{i-2} f(x^{**}) \ldots$$

or

$$y(i) = \sum_{k=0}^{i} a^{i-k} f(x^{(k)*}) \tag{57}$$

and

$$y(m) \equiv \mu(m, a)$$

The function f is, of course, computed by the ribosome which reads the codon x, and synthesizes the amino acids which, in turn, are linked together by the recursion to a connected polypeptide chain.

Visualizing the whole process as the operations of a sequential finite state machine was probably more than just a clue in "breaking the genetic code" and identifying as the input state to this machine the triplet (u, v, w) of adjacent symbols in the $\bar{v}$-number representation of the messenger RNA.

A method for computing v-numbers of molecular sequences directly from properties of the generated structure was suggested by Pattee (1961). He used the concept of a sequential "shift register," i.e., in principle that of an autonomous tile. For computing periodic sequences in growing helical molecules, the computation for the next element to be attached to the helix is solely determined by the present and some earlier building block. No extraneous computing system is required.

If on a higher level of the hierarchical organization the neuron is taken as a functional unit, the examples are numerous in which it is seen as a recursive function computer. Depending on what is taken to be the "signal," a single pulse, an average frequency code, a latency code, a probability code (Bullock, 1968), etc., the neuron becomes an "all or nothing" device for computing logical functions (McCulloch and Pitts, 1943), a linear element (Sherrington, 1906), a logarithmic element, etc., by changing in essence only a single parameter characteristic for that neuron (Von Foerster, 1967b). The same is true for neural nets in which the recursion is achieved by loops or sometimes directly through recurring fibers. The "reverberating" neural net is a typical example of a finite state machine in its dynamic equilibrium.

In the face of perhaps a whole library filled with recorded instances in which the concept of the finite state machine proved useful, it may come as a surprise that on purely physical grounds these systems are absurd. In order to keep going they must be nothing less than perpetual motion machines. While this is easily accomplished by a mathematical object, it is impossible for an object of reality. Of course, from a heuristical point of view it is irrelevant whether or not a model is physically realizable, as long as it is self-consistent and an intellectual stimulus for further investigations.

However, when the flow of energy between various levels of organization is neglected, and the mechanisms of energy conversion and transfer are ignored, difficulties arise in matching descriptive parameters of functional units on one level to those of higher or lower levels. For instance, a relation between the code of a particular nuclear RNA molecule and, say, the pulse frequency code at the same neuron cannot be established, unless mechanisms of energy transfer are considered. As long as the question as to what keeps the organism going and how this is done is not asked, the gap between functional units on different levels of organization remains open. Can it be closed by thermodynamics?

Three different kinds of molecular mechanisms that offer themselves readily for this purpose will be briefly discussed. All of them make use of various forms of energy as radiation (vh), potential energy (V, structure), work ($p\Delta v$), and heat ($k\Delta T$), and its various conversions from one form to another.

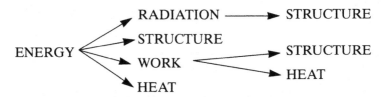

We remain in the terminology of finite state machines and classify the three kinds of mechanism according to their inputs and their outputs, dropping, however, for the moment all distinctions of forms of energy, except that of potential energy (structure) as distinct from all other forms (energy).

(i) Molecular store: Energy in,
          Energy out.
(ii) Molecular computer: Energy in,
          Structure out.
(iii) Molecular carrier: Energy and structure in,
          Energy out.

These three cases will now be briefly reviewed.

B. Molecular Store

Probably the most obvious, and hence perhaps the oldest, approach to link macroscopic behavior, as for example, the forgetting of nonsense syllables (Ebbinghaus, 1885), with the quantum mechanical decay of the available large number of excited metastable states in macromolecules, assumes no further analyzable "elementary impressions" that are associated with a molecule's meta-stable state (Von Foerster, 1948; Von Foerster, 1949). By a nondestructive read-out they can be transferred to another molecule, and a record of these elementary impressions may either decay or else grow, depending on whether the product of the quantum decay time constant with the scanning rate of the read-out is either smaller or else larger than unity. While this model gives good agreement between macroscopic variables such as forgetting rates, temperature dependence of conceived lapse of time (Hoagland, 1951; Hoagland, 1954), and such microscopic variables as binding energies, electron orbital frequencies, it suffers the malaise of all recording schemes, namely, it is unable to infer anything from the accumulated records. Only if an inductive inference machine which computes the appropriate behavior functions is attached to this record can an organism survive (Von Foerster *et al.*, 1968). Hence, one may abandon speculations about systems that just record specifics, and contemplate those that compute generalizations.

C. Molecular Computer

The good match between macroscopic and microscopic variables of the previous model suggests that this relation should be pursued further. Indeed, it can be shown (Von Foerster, 1969) that the energy intervals between excited meta-stable states are so organized that the decay times in the lattice vibration band correspond to neuronal pulse intervals, and their energy levels to a polarization potential of from 60 mV to 150 mV. Consequently, a pulse train of various pulse intervals will "pump" such a molecule up into higher states of excitation, depending on its initial condition. However, if the excitation level reaches about 1.2 eV, the molecule undergoes configurational changes with life spans of 1 day or longer. In this "structurally charged" state it may now participate in various ways in altering the transfer function of a neuron, either transmitting its energy to other molecules or facilitating their reaction. Since in this model undirected electrical potential energy is used to cause specific structural change, it is referred to as "energy in—structure out." This, however, gives rise to a concept of molecular computation, the result of which is deposition of energy on a specific site of utilization. This is the content of the next and last model.

D. Molecular Carrier

One of the most widely used principles of energy dissemination in a living organism is that of separation of sites of synthesis and utilization. The general method employed in this transfer is a cyclic operation that involves one or many molecular carriers which are "charged" at the site where environmental energy can be absorbed, and are "discharged" where this energy must be used. Charging and discharging is usually accomplished by chemical modifications of the basic carrier molecules. One obvious example of the directional flow of energy and the cyclic flow of matter is, of course, the complementarity of the processes of photosynthesis and respiration

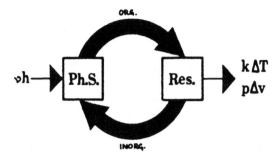

FIGURE 9. Directional flow of energy and cyclic flow of matter in photosynthesis coupled with respiration.

(Fig. 9). Light energy, vh, breaks the stable bonds of inorganic oxides and transforms them into energetically charged organic molecules. These, in turn, are burned up in the respiratory process, releasing the energy in the form of work, $p\Delta v$, or heat, $k\Delta T$, at the site of utilization and return again as inorganic oxides to the site of synthesis.

Another example is the extremely involved way in which in the mitochondria the uphill reaction is accomplished. This reaction not only synthesizes adenosine triphosphate (ATP) by coupling a phosphate group to adenosine diphosphate (ADP), but also charges the ATP molecule with considerable energy which is effectively released during muscular contraction; the contraction process converts ATP back again into ADP by losing the previously attached phosphate group.

Finally, the messenger RNA may be cited as an example of separate sites for synthesis and utilization, although in this case the energetics are as yet not so well established as in the other cases. Here, apparently it is structure which is to be transferred from one place to another, rather than energy.

Common in all these processes is the fact that during synthesis not only a releasable package of energy, ΔE, is put on this molecular carrier but also an address label saving where to deliver the package. This address requires an additional amount of organization, $-\Delta H$, (negentropy), in order to locate its destination. Hence we have the crucial condition

$$\frac{\Delta E}{\Delta H} < 0 \tag{58}$$

which says "for high energy have a low entropy, and for low energy have a high entropy." This is, of course, contrary to the usual course of events in which these two quantities are coupled with each other in a positive relationship.

It can be shown, however, that if a system is composed of constituents which in the ground state are separated, but when "excited" hang together by "reasonably stable" metastable states, it fulfills the crucial condition above (Von Foerster, 1964).

Let

$$V = \pm\left(Ae^{-x/k} + B\sin\frac{2\pi x}{p} \right) \tag{59}$$

with

$$A/B \gg 1 \quad \text{and} \quad \kappa/p \gg 1$$

be the potential distribution in two one-dimensional linear "periodic crystals," C^+ and C^-, where the $\pm$ refer to corresponding cases. The essential difference between these two linear structures which can be envisioned as linear distributions of electric charges changing their sign (almost) periodically is that energy is required to put "crystal" C^+ together, while for "crystal" C^- about the same energy is required to decompose it into its constituents. These linear lattices have metastable equilibria at

$$C^+ \rightarrow x_1, x_3, x_5 \ldots$$

$$C^- \rightarrow x_0, x_2, x_4 \ldots$$

which are solutions of

$$e^{x/k} \cos\frac{2\pi x}{p} = \frac{1}{2\pi}\frac{Ap}{B\kappa} \approx 1$$

These states are protected by an energy threshold which lets them stay in this state on the average of amount an time

$$\tau = \tau_0 e^{\Delta v/kT} \tag{60}$$

where τ_0^{-1} is an electron orbital frequency, and ΔV is the difference between the energies at the valley and the crest of the potential wall $[\pm\Delta V_n = V(x_n) - V(x_{n+1})]$.

In order to find the entropy of this configuration, we solve the Schrödinger equation (given in normalized form)

$$\psi'' + \psi[\lambda - V(x)] = 0 \tag{61}$$

for its eigenvalues λ_i and eigenfunctions ψ_i, ψ_i^*, which, in turn, give the probability distribution for the molecule being in the ith eigenstate:

$$\left(\frac{dp}{dx}\right)_i = \psi_i \cdot \psi_i^* \tag{62}$$

with, of course,

$$\int_{-\infty}^{+\infty} \psi_i \cdot \psi_i^* dx = 1 \tag{63}$$

whence we obtain the entropy

$$H_i = -\int_{-\infty}^{+\infty} \psi_i \cdot \psi_i{}^* \ln \psi_i \cdot \psi_i{}^* \tag{64}$$

for the ith eigenstate.

It is significant that indeed for the two crystals C^+ and C^- the change in the ratio of energy to entropy for charging ($\Delta E = e(V(x_n) - V(x_{n+2}))$) goes into opposite directions:

$$C^- \rightarrow \left(\frac{\Delta E}{\Delta H}\right)^- > 0$$

$$C^+ \rightarrow \left(\frac{\Delta E}{\Delta H}\right)^+ < 0$$

This shows that the two crystals are quite different animals: one is dead (C^-), the other is alive (C^+).

IV. Summary

In essence this paper is a proposal to restore the original meaning of concepts like memory, learning, behavior, etc. by seeing them as various manifestations of a more inclusive phenomenon, namely, cognition. An attempt is made to justify this proposition and to sketch a conceptual machinery of apparently sufficient richness to describe these phenomena in their proper extension. In its most concise form the proposal was presented as a search for mechanisms within living organisms that enable them to turn their environment into a trivial machine, rather than a search for mechanisms in the environment that turn the organisms into trivial machines.

This posture is justified by realizing that the latter approach—when it succeeds—fails to account for the mechanisms it wishes to discover, for a trivial machine does not exhibit the desired properties; and when it fails does not reveal the properties that made it fail.

Within the conceptual framework of finite state machines, the calculus of recursive functionals was suggested as a descriptive (phenomenological) formalism to account for memory as potential awareness of previous interpretations of experiences, hence for the origin of the *concept* of "change," and to account for transitions in domains that occur when going from "facts" to "description of facts" and—since these in turn are facts too—to "descriptions of descriptions of facts" and so on.

Elementary finite function machines can be strung together to form linear or two-dimensional tesselations of considerable computational flexibility and complexity. Such tesselations are useful models for aggregates of interacting functional units at various levels in the hierarchical organization of organisms. On the molecular level, for instance, a stringlike tesselation coiled to a helix may compute itself (self-replication) or, in conjunction with other elements, compute other molecular functional units (synthesis).

While in the discussion of descriptive formalisms the concept of recursive functionals provides the bridge for passing through various descriptive domains, it is the concept of energy transfer connected with entropic change that links operationally the functional units on various organizational levels. It is these links, conceptual or operational, which are the prerequisites for interpreting structures and function of a living organism seen as an autonomous self-referring organism. When these links are ignored, the concept of "organism" is void, and its unrelated pieces becomes trivialities or remain mysteries.

Acknowledgments. Some of the ideas and results presented in this article grew out of work jointly sponsored by the Air Force Office of Scientific Research under Grant AF-OSR 7–67, by the Office of Education under Grant OEC-1-7-071213-4557, and by the Air Force Office of Scientific Research under Grant AF 49 (638)-1680.

References

Ashby, W. R., 1956, "An Introduction to Cybernetics," Chapman and Hall, London.

Ashby, W. R., 1962, The Set Theory of Mechanisms and Homeo-stasis, Technical Report 7, NSF Grant 17414, Biological Computer Laboratory, Electrical Engineering Department, University of Illinois, Urbana, 44 pp.

Ashby, W. R., and Walker, C., 1966, On temporal characteristics of behavior in certain complex systems, *Kybernetik* 3:100.

Bullock, T. H., 1968, Biological sensors, *in* "Vistas in Science" (D. L. Arm, ed.) pp. 176–206, University of New Mexico, Albuquerque.

Ebbinghaus, H., 1885, "Über das Gedächtnis: Untersuchungen zur experimentellen Psychologie," Drucker & Humbold, Leipzig.

Eccles, J. C., Ito, M., and Szentagothai, J., 1967, "The Cerebellum as a Neuronal Machine," Springer-Verlag, New York.

Estes, W. K., 1959, The statistical approach to learning theory, *in* "Psychology: A Study of a Science, I/2" (S. Koch, ed.) pp. 380–491, McGraw-Hill, New York.

Fitzhugh, H. S. II, 1963, Some considerations of polystable systems, *IEEE Transactions* 7:1.

Gill, A., 1962, "Introduction to the Theory of Finite State Machines," McGraw-Hill, New York.

Gunther, G., 1967, Time, timeless logic and self-referential systems, *in* "Interdisciplinary Perspectives of Time" (R. Fischer, ed.) pp. 396–406, New York Academy of Sciences, New York.

Hoagland, H., 1951, Consciousness and the chemistry of time, *in* "Problems of Consciousness Tr. First Conf." (H. A. Abramson, ed.) pp. 164–198, Josiah Macy Jr. Foundation, New York.

Hoagland, H., 1954, (A remark), *in* "Problems of Consciousness Tr. Fourth Conf." (H. A. Abramson, ed.) pp. 106–109, Josiah Macy Jr. Foundation, New York.

John, E. R., Shimkochi, M., and Bartlett, F., 1969, Neural readout from memory during generalization, *Science* 164:1534.

Konorski, J., 1962, The role of central factors in differentiation, *in* "Information Processing in the Nervous System" (R. W. Gerard and J. W. Duyff, eds.) Vol. 3, pp. 318–329, Excerpta Medica Foundation, Amsterdam.

Lettvin, J. Y., Maturana, H. R., McCulloch, W. S., and Pitts, W., 1959, What the frog's eye tells the frog's brain, *Proc. I.R.E.* 47:1940.

Löfgren, L., 1962, Kinematic and tesselation models of self-repair, *in* "Biological Prototypes and Synthetic Systems" (E. E. Bernard and M. R. Kare, eds.) pp. 342–369, Plenum Press, New York.

Löfgren, L., 1968, An axiomatic explanation of complete self-reproduction, *Bull. of Math. Biophysics* 30(3):415.

Logan, F. A., 1959, The Hull-Spence approach, *in* "Psychology: A Study of a Science, I/2" (S. Koch, ed.) pp. 293–358, McGraw-Hill, New York.

McCulloch, W. S., and Pitts, W., 1943, A logical calculus of the ideas immanent in nervous activity, *Bull. of Math. Biophysics* 5:115.

Maturana, H. R., 1969, Neurophysiology of cognition, *in* "Cognition—A Multiple View" (P. L. Garvin, ed.) in press, Spartan Books, New York.

Maturana, H. R., Uribe, G., and Frenk, S., 1968, A Biological Theory of Relativistic Colour Coding in the Primate Retina, Supplemento No. 1, Arch. Biologia y Med. Exp., University of Chile, Santiago, 30 pp.

Pask, G., 1962, A proposed evolutionary model, *in* "Principles of Self-Organization" (H. Von Foerster and G. W. Zopf, Jr., eds.) pp. 229–254, Pergamon Press, New York.

Pask, G., 1968, A cybernetic model for some types of learning and mentation, *in* "Cybernetic Problems in Bionics" (H. L. Oestreicher and D. R. Moore, eds.) pp. 531–586, Gordon & Breach, New York.

Pattee, H. H., 1961, On the origin of macro-molecular sequences, *Biophys. J.* 1:683.

Pitts, W., and McCulloch, W. S., 1947, How we know universals; the perception of auditory and visual forms, *Bull. of Math. Biophysics* 9:127.

Selfridge, O. G., 1962, The organization of organization, *in* "Self-Organizing Systems" (M. C. Yovits, G. T. Jacoby and G. D. Goldstein, eds.) pp. 1–8, Spartan Books, New York.

Sherrington, C. S., 1906, "Integrative Action of the Nervous System," Yale University Press, New Haven.

Skinner, B. F., 1959, A case history in scientific method, *in* "Psychology: A Study of a Science, I/2" (S. Koch, ed.) pp. 359–379, McGraw-Hill, New York.

Ungar, G., 1969, Chemical transfer of learning, *in* "The Future of the Brain Sciences" (S. Bogoch, ed.) pp. 373–374, Plenum Press, New York.

Von Foerster, H., 1948, "Das Gedächtnis; Eine quantenmechanische Untersuchung," F. Deuticke, Vienna.

Von Foerster, H., 1949, Quantum mechanical theory of memory, *in* "Cybernetics, Transactions of the Sixth Conference" (H. Von Foerster, ed.) pp. 112–145, Josiah Macy Jr. Foundation, New York.

Von Foerster, H., 1964, Molecular bionics, *in* "Information Processing by Living Organisms and Machines" (H. L. Oestreicher, ed.) pp. 161–190, Aerospace Medical Division, Dayton.

Von Foerster, H., 1965, Memory without record, *in* "The Anatomy of Memory" (D. P. Kimble, ed.) pp. 388–433, Science and Behavior Books, Palo Alto.

Von Foerster, H., 1966, From stimulus to symbol, *in* "Sign, Image, Symbol" (G. Kepes, ed.) pp. 42–61, George Braziller, New York.

Von Foerster, H., 1967a, Biological principles of information storage and retrieval, *in* "Electronic Handling of Information: Testing and Evaluation" (A. Kent *et al.*, eds.) pp. 123–147, Academic Press, London.

Von Foerster, H., 1967b, Computation in neural nets, *Currents Mod. Biol.* 1:47.

Von Foerster, H., 1969, What is memory that it may have hindsight and foresight as well?, *in* "The Future of the Plain Sciences" (S. Bogoch, ed.) pp. 19–64, Plenum Press, New York.

von Foerster, H., Inselberg, A., and Weston, p., 1968, Memory and inductive inference, *in* "Cybernetic Problems in Bionics" (H. L. Oestreicher and D. R. Moore, eds.) pp. 31–68, Gordon & Breach, New York.

von Neumann, J., 1966, "The Theory of Self-Reproducing Automata," (A. Burks, ed.) University of Illinois Press, Urbana.

Walker, C., 1965, A Study of a Family of Complex Systems, An Approach to the Investigation of Organism's Behavior, Technical Report 5, AF-OSR Grant 7-65, Biological Computer Laboratory, Electrical Engineering Department, University of Illinois, Urbana, 251 pp.

Werner, G., 1969, The topology of the body representation in the somatic afferent pathways, *in* "The Neurosciences, II' in press, Rockefeller University Press, New York.

Weston, P., 1964, Noun chain trees, unpublished manuscript.

5
Thoughts and Notes on Cognition*

HEINZ VON FOERSTER
University of Illinois

Thoughts

Projecting the image of ourselves into things or functions of things in the outside world is quite a common practice. I shall call this projection "anthropomorphization." Since each of us has direct knowledge of himself, the most direct path of comprehending X is to find a mapping by which we can see ourselves represented by X. This is beautifully demonstrated by taking the names of parts of one's body and giving these names to things which have structural or functional similarities with these parts: the "head" of a screw, the "jaws" of a vise, the "teeth" of a gear, the "lips" of the cutting tool, the "sex" of electric connectors, the "legs" of a chair, a "chest" of drawers, etc.

Surrealists who were always keen to observe ambivalences in our cognitive processes bring them to our attention by pitching these ambivalences against semantic consistencies: the legs of a chair (Fig. 1[2]), a chest of drawers (Fig. 2[3]), etc.

At the turn of the century, animal psychologists had a difficult time in overcoming functional anthropomorphisms in a zoology populated with animals romanticized with human characteristics: the "faithful" dog, the "valiant" horse, the "proud" lion, the "sly" fox, etc. Konrad Lorenz, the great ornithologist, was chased from Vienna when he unwisely suggested controlling the population of the overbreeding, underfed, and tuberculosis-

I am deeply indebted to Humberto Maturana, Gotthard Gunther,[1] and Ross Ashby for their untiring efforts to enlighten me in matters of life, logic, and large systems, and to Lebbeus Woods for supplying me with drawings that illustrate my points better than I could do with words alone. However, should there remain any errors in exposition or presentation, it is I who am to blame and not these friends who have so generously contributed their time.

Some of the ideas expressed in this paper grew from work sponsored jointly by the Air Force Office of Scientific Research under Grants AFOSR 7-67 and AF 49(638)-1680, and by the Office of Education under Grant OEC-1-7-071213-4557.
* This article was originally published in *Cognition: A Multiple View*, P. Gavin (ed.), Spartan Books, New York, pp. 25–48 (1970).

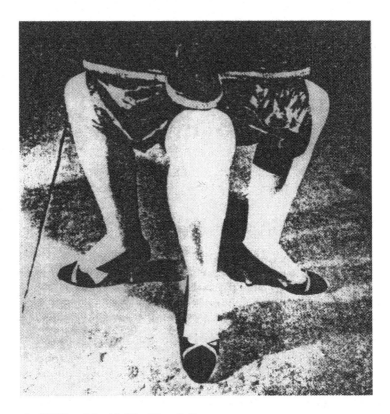

FIGURE 1. "L'Ultra Meuble" by Kurt Seligman.

carrying pigeons of the city by importing falcons which would raid the pigeons' nests for eggs. The golden heart of the Viennese could not stand the thought of "pigeon infanticide." Rather, they fed the pigeons twice as much. When Lorenz pointed out that the result of this would be twice as many underfed and tuberculosis-carrying pigeons, he had to go, and fast!

Of course, in principle there is nothing wrong with anthropomorphizations; in most cases they serve as useful algorithms for determining behavior. In trying to cope with a fox it is an advantage to know he is "sly," that is, he is a challenge to the brain rather than to the muscles.

Today, with most of us having moved to the big cities, we have lost direct contact with the animal world, and pieces of steel furniture with some functional properties, the computers, are becoming the objects of our endearments and, consequently, are bestowed now with romanticizing epithets. Since we live today, however, in an era of science and technology rather than in one of emotion and sentimentality, the endearing epithets for our machines are not those of character but of intellect. Although it is quite possible, and perhaps even appropriate to talk about a "proud IBM 360–50 system," the "valiant 1800," or the "sly PDP 8," I have never observed

FIGURE 2. "City of Drawers" by Salvador Dali.

anyone using this style of language. Instead, we romanticize what appears to be the intellectual functions of the machines. We talk about their "memories," we say that these machines store and retrieve "information," they "solve problems," "prove theorems," etc. Apparently, one is dealing here with quite intelligent chaps, and there are even some attempts made to design an A. I. Q., "an artificial intelligence quotient" to carry over into this new field of "artificial intelligence" with efficacy and authority the misconceptions that are still today quite popular among some prominent behaviorists.

While our intellectual relationship with these machines awaits clarification, in the emotional sphere we seem to do all right. I wish to make this comment as a footnote to Madeleine Mathiot's delightful observations in this volume* about various degrees of "awesomeness" associated with the referential genders "it," "he," and "she." She develops a three-valued logical place-value system in which the nonhuman "it" carries no reference to awesomeness either in the negative (absence) or else in the affirmative (presence), while the human "he" and "she" indeed carry reference to awesomeness, the masculine "he" referring to its absence, the feminine "she," of course, to its presence.

When in the early fifties at the University of Illinois ILLIAC II was built, "it" was the referential gender used by all of us. The computer group that

_____

* See Chapter 11.

now works on ILLIAC III promises that "he" will be operative soon. But ILLIAC IV reaches into quite different dimensions. The planners say that when "she" will be switched on, the world's computing power will be doubled.

Again, these anthropomorphisms are perfectly all right inasmuch as they help us establish good working relations with these tools. Since most of the people I know in our computer department are heterosexual males, it is clear that they prefer the days and nights of their work spent with a "she," rather than with an "it."

However, in the last decade or so something odd and distressing developed, namely, that not only the engineers who work with these systems gradually began to believe that those mental functions whose names were first metaphorically applied to some machine operations are indeed residing in these machines, but also some biologists—tempted by the absence of a comprehensive theory of mentation—began to believe that certain machine operations which unfortunately carried the *names* of some mental processes are indeed functional isomorphs of these operations. For example, in the search for a physiological basis of memory, they began to look for neural mechanisms which are analogues of electromagnetic or electrodynamic mechanisms that "freeze" temporal configurations (magnetic tapes, drums, or cores) or spatial configurations (holograms) of the electromagnetic field so that they may be inspected at a later time.

The delusion, which takes for granted a functional isomorphism between various and distinct processes that happen to be called by the same name, is so well established in these two professions that he who follows Lorenz's example and attempts now to "de-anthropomorphize" machines and to "de-mechanize" man is prone to encounter antagonisms similar to those Lorenz encountered when he began to "animalize" animals.

On the other hand, this reluctance to adopt a conceptual framework in which apparently separable higher mental faculties as, for example, "to learn," "to remember," "to perceive," "to recall," "to predict," etc., are seen as various manifestations of a single, more inclusive phenomenon, namely, "cognition," is quite understandable. It would mean abandoning the comfortable position in which these faculties can be treated in isolation and thus can be reduced to rather trivial mechanisms. Memory, for instance, contemplated in isolation is reduced to "recording," learning to "change," perception to "input," etc. In other words, by separating these functions from the totality of cognitive processes one has abandoned the original problem and now searches for mechanisms that implement entirely different functions that may or may not have any semblance with some processes that are, as Maturana* pointed out, subservient to the maintenance of the integrity of the organism as a functioning unit.

---

* See Chapter 1, pages 3–23.

Perhaps the following three examples will make this point more explicit.

I shall begin with "memory." When engineers talk about a computer's "memory" they really don't mean a computer's memory, they refer to devices, or systems of devices, for recording electric signals which when needed for further manipulations can be played back again. Hence, these devices are stores, or storage systems, with the characteristic of all stores, namely, the conservation of quality of that which is stored at one time, and then is retrieved at a later time. The content of these stores is a record, and in the pre-semantic-confusion times this was also the name properly given to those thin black disks which play back the music recorded on them. I can see the big eyes of the clerk in a music shop who is asked for the "memory" of Beethoven's *Fifth Symphony*. She may refer the customer to the bookstore next door. And rightly so, for memories of past experiences do not reproduce the causes for these experiences, but—by changing the domains of quality—transform these experiences by a set of complex processes into utterances or into other forms of symbolic or purposeful behavior. When asked about the contents of my breakfast, I shall not produce scrambled eggs, I just say, "scrambled eggs." It is clear that a computer's "memory" has nothing to do with such transformations, it was never intended to have. This does not mean, however, that I do not believe that these machines may eventually write their own memoirs. But in order to get them there we still have to solve some unsolved epistemological problems before we can turn to the problem of designing the appropriate software and hardware.

If "memory" is a misleading metaphor for recording devices, so is the epithet "problem solver" for our computing machines. Of course, they are no problem solvers, because they do not have any problems in the first place. It is *our* problems they help us solve like any other useful tool, say, a hammer which may be dubbed a "problem solver" for driving nails into a board. The danger in this subtle semantic twist by which the responsibility for action is shifted from man to a machine lies in making us lose sight of the problem of cognition. By making us believe that the issue is how to find solutions to some well defined problems, we may forget to ask first what constitutes a "problem," what is its "solution," and—when a problem is identified—what makes us want to solve it.

Another case of pathological semantics—and the last example in my polemics—is the widespread abuse of the term "information." This poor thing is nowadays "processed," "stored," "retrieved," "compressed," "chopped," etc., as if it were hamburger meat. Since the case history of this modern disease may easily fill an entire volume, I only shall pick on the so-called "information storage and retrieval systems" which in the form of some advanced library search and retrieval systems, computer based data processing systems, the nationwide Educational Resources Information Center (ERIC), etc., have been seriously suggested to serve as analogies for the workings of the brain.

Of course, these systems do not store information, they store books, tapes, microfiche or other sorts of documents, and it is again these books, tapes, microfiche or other documents that are retrieved which only if looked upon by a human mind may yield the desired information. Calling these collections of documents "information storage and retrieval systems" is tantamount to calling a garage a "transportation storage and retrieval system." By confusing *vehicles* for potential information with *information*, one puts again the problem of cognition nicely into one's blind spot of intellectual vision, and the problem conveniently disappears. If indeed the brain were seriously compared with one of these storage and retrieval systems, distinct from these only by its quantity of storage rather than by quality of process, such a theory would require a demon with cognitive powers to zoom through this huge system in order to extract from its contents the information that is vital to the owner of this brain.

Dificile est satiram non scribere. Obviously, I have failed to overcome this difficulty, and I am afraid that I will also fail in overcoming the other difficulty, namely, to say now what cognition *really* is. At this moment, I even have difficulties in relating my feelings on the profoundness of our problem, if one cares to approach it in its full extension. In a group like ours, there are probably as many ways to look at it as there are pairs of eyes. I am still baffled by the mystery that when Jim, a friend of Joe, hears the noises that are associated with reading aloud from the black marks that follow.

ANN IS THE SISTER OF JOE

—or just sees these marks—knows that indeed Ann is the sister of Joe, and, *de facto*, changes his whole attitude toward the world, commensurate with his new insight into a relational structure of elements in this world.

To my knowledge, we do not yet understand the "cognitive processes" which establish this insight from certain sensations. I shall not worry at this moment whether these sensations are caused by an interaction of the organism with objects in the world or with their symbolic representations. For, if I understood Dr. Maturana correctly, these two problems, when properly formulated, will boil down to the same problem, namely, that of cognition *per se*.

In order to clarify this issue for myself, I gathered the following notes which are presented as six propositions labeled $n = 1 \rightarrow 6$. Propositions numbered $n.1$, $n.2$, $n.3$, etc., are comments on proposition numbered n. Propositions numbered $n.m1$, $n.m2$, etc., are comments on proposition $n.m$, and so on.

Here they are.

Notes

1 A living organism, Ω, is a bounded, autonomous unit whose functional and structural organization is determined by the interaction of its contiguous elementary constituents.

1.1 The elementary constituents, the cells, are, in turn, bounded, functional, and structural units, however, they are not necessarily autonomous.

1.11 Autonomy of cells is progressively lost with increasing differentiation in organisms of ascending complexity which, on the other hand, provides the appropriate "organic environment" for these units to maintain their structural and functional integrity.

1.2 A living organism, Ω, is bounded by a closed orientable surface. Topologically this is equivalent to a sphere with an even number $2p$ of holes which are connected in pairs by tubes. The number p is called the genus of the surface.

1.21 Should the histological distinction between ectoderm and endoderm be maintained, then a surface of genus $p = (s + t)/2$ is equivalent to a sphere with s surface holes which are connected through a network or tubes with t T-branches. Ectoderm is then represented by the surface of the sphere, endoderm by the lining of the tubes (Fig. 3).

1.3 Any closed orientable surface is metrizable. Hence, each point on this surface can be labeled by the two coordinates α, β, of a geodesic coordinate system that may be chosen to cover the surface conveniently. One of the properties of a geodesic coordinate system is that it is locally Cartesian.

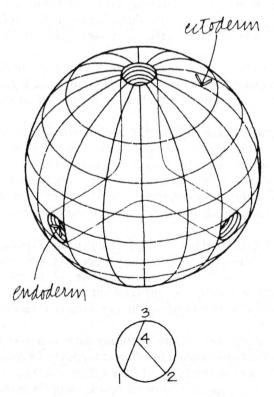

FIGURE 3. Closed orientable surface of genus $p = 2$ ($s = 3, t = 1$ [$s + t$]/2 = 4/2 = 2].

Surface coordinates α, β, will be referred to as the "proprietory coordinates," denoted by the single symbol ξ.

1.31 If to the vicinity of each surface point ξ the Gaussian curvature γ is given, then the totality of triples α, β, γ, determines the shape Γ of the surface ($\gamma = \gamma[\alpha,\beta]$).

1.32 Since a living organism is bounded by a closed orientable surface, an appropriate geodesic coordinate system can be drawn on the surface of this organism at an arbitrary "rest state," and each surface element (ectodermal or endodermal cell) can be labeled according to the proprietory coordinates ξ of its location.

1.33 A cell c_ξ so labeled shall carry its label under subsequent distortions of the surface (continuous distortions), and even after transplants to locations ξ' (discontinuous distortions).

1.331 The geodesic coordinates on the surface of the organism can be mapped onto a topologically equivalent unit sphere ($R = 1$) so that to each point ξ and its vicinity on the organism corresponds precisely one point λ and its vicinity on the unit sphere. Consequently, each cell c_ξ on the surface of the organism has an image c_λ on the surface of the unit sphere.

1.332 It is clear that surface distortions of the organism, even transplants of cells from one location to another, are not reflected by any changes on the surface of this sphere. The once established map remains invariant under such transformations, hence this sphere will be referred to as the "representative body sphere" (Fig. 3, or appropriate modifications with $p > 2$).

1.34 Since the volume enclosed by a closed orientable surface is metrizable, all that has been said ($1.3 \rightarrow 1.332$) for surface points ξ and cell c_ξ holds for volume points ζ and cells c_ζ with representative cells c_μ in the body sphere.

1.4 The organism, Ω, is supposed to be embedded in an "environment" with fixed Euclidean metric, with coordinates a, b, c, or x for short, in which its position is defined by identifying three environmental points x_1, x_2, and x_3 with three surface points ξ_1, ξ_2, and ξ_3 of the organism. Conversely, the representative body sphere is embedded in a "representative environment" with variable non-Euclidean metric, and with the other conditions *mutatis mutandis*.

1.41 The two pairs of figures (Figs. 4a and 4b and Figs. 5a and 5b) illustrate the configuration of the proprietory space, ξ, of an organism (fish-like creature) as seen from an Euclidean environment (4a and 5a), and the configuration of the non-Euclidean environment as seen from the unit sphere (4b, 5b) for the two cases in which the organism is at rest (4a, 4b) and in motion (5a, 5b).

2 Phylogenetically as well as ontogenetically the neural tube develops from the ectoderm. Receptor cells r_ξ are differentiated ectodermal cells c_ξ. So are the other cells deep in the body which participate in the transmission of signals (neurons) n_ζ, the generation of signals (proprioceptors) p_ζ, as well

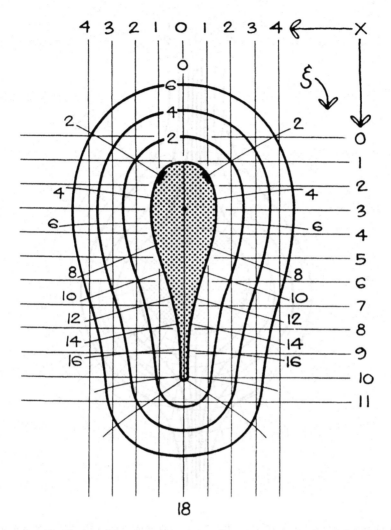

FIGURE 4a. Geodesics of the proprietory coordinate system of an organism Ω at rest $[\delta\Gamma = 0]$ embedded in an environment with Euclidean metric.

as those (effectors) e_ζ which cause by their signaling specialized fibers (muscles) m_ζ to contract, thus causing changes $\delta\Gamma$ in the shape of the organism (movement).

2.1 Let A be an agent of amount A distributed in the environment and characterized by a distribution function of its concentration (intensity) over a parameter p:

$$S(x, p) = \frac{d^2 A}{dx\, dp} = \left(\frac{da}{dp} \right)_x$$

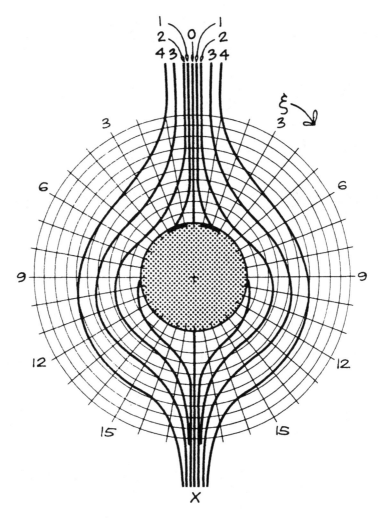

FIGURE 4b. Geodesics (circles, radii) of the proprietory coordinate system with respect to the representative body sphere embedded in an environment with non-Euclidean metric corresponding to the organism at rest (Fig. 4a).

with

$$\int_0^\infty S(x, p)dp = a_x \equiv \frac{dA}{dx}.$$

2.11 Let $s(\xi,p)$ be the (specific) sensitivity of receptor r_ξ with respect to parameter p, and ρ_ξ be its response activity:

$$s(\xi, p) = k\frac{dp}{da} = k\left(\frac{d\rho}{dp}\right)_\xi \bigg/ \left(\frac{da}{dp}\right)_\xi$$

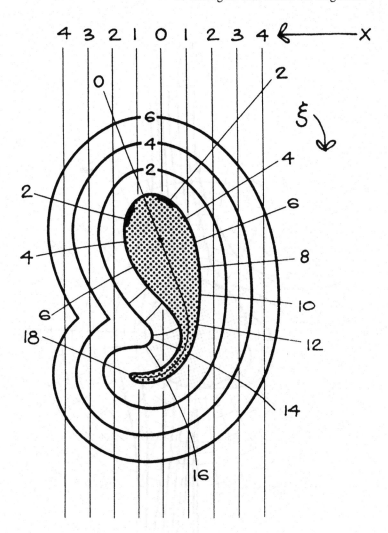

FIGURE 5a. Geodesics of the proprietory coordinate system of an organism Ω in motion ($\delta\Gamma \neq 0$) embedded in an environment with Euclidean metric.

with

$$\int_0^\infty s(\xi, p) \, dp = 1,$$

and k being a normalizing constant.

2.12 Let x and ξ coincide. The response ρ_ξ of receptor r_ξ to its stimulus S_ξ is now

$$\rho_\xi = \int_0^\infty S(\xi, p) \cdot s(\xi, p) \, dp = F(a_\xi).$$

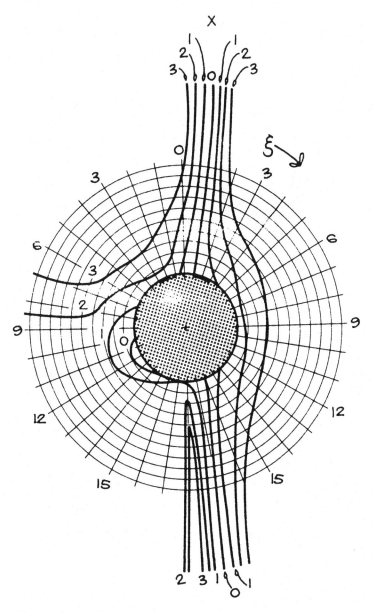

FIGURE 5b. Geodesics (circles, radii) of the proprietory coordinate system with respect to the representative body sphere embedded in an environment with non-Euclidean metric corresponding to the organism in motion (Fig. 5a).

2.2 This expression shows that neither the modality of the agent, nor its parametric characteristic, nor reference to environmental point x is encoded in the receptor's response, solely some clues as to the presence of a stimulant for receptor r_ξ are given by its activity ρ_ξ.

2.21 Since all receptors of an organism respond likewise, it is clear that organisms are incapable of deriving any notions as to the "variety of environmental features," unless they make reference to their own body by utilizing the geometrical significance of the label ξ of receptor r_ξ which reports: "so and so much ($\rho = \rho_1$) is at *this* place on my body ($\xi = \xi_1$)."

2.22 Moreover, it is clear that any notions of a "sensory modality" cannot arise from a "sensory specificity," say a distinction in sensitivity regarding different parameters p_1 and p_2, or in different sensitivities s_1 and s_2 for the same parameter p, for all these distinctions are "integrated out" as seen in expression 2.12. Consequently, these notions can only arise from a distinction of the body-oriented loci of sensation ξ_1 and ξ_2. (A pinch applied to the little toe of the left foot is felt not in the brain but at the little toe of the left foot. Dislocating one eyeball by gently pushing it aside displaces the environmental image of this eye with respect to the other eye.)

2.23 From this it becomes clear that all inferences regarding the environment of Ω must be computed by operating on the distribution function ρ_ξ. (It may also be seen that these operations ω_{ij} are in some sense coupled to various sensitivities $s_i[\xi,p_j]$.)

2.24 This becomes even more apparent if a physical agent in the environment produces "actions at a distance."

2.241 Let $g(x,p)$ be the environmental distribution of sources of the agent having parametric variety (p); let R be the distance between any point x in the environment and a fixed point x_0; and let $\Phi(R)$ be the distance function by which the agent loses its intensity. Moreover, let the point ξ_0 on the body of an organism coincide with x_0, then the stimulus intensity for receptor r_{ξ_0} is (Fig. 6):

$$S(x_0,p)_{\text{postition}} = \int_{\text{sensory field}} \Phi(x - x_0) g(x, p) dx$$

and its response is

$$\rho_{\xi_0,\text{position}} \equiv F(S_{\xi_0,\text{position}})$$

compare with 2.12).

2.242 Again this expression shows suppression of all spatial clues, save for self-reference expressed by the bodily location ξ_0 of the sensation *and* by the position of the organism as expressed by the limits of the integral which can, of course, only be taken over the sensory field "seen" by receptor cell r_{ξ_0} (Fig. 6).

2.25 Since ρ_ξ gives no clues as to the kind of stimulant (p), it must be either ξ, the place of origin of the sensation, or the operation $\omega(\rho_\xi)$, or both that establish a "sensory modality."

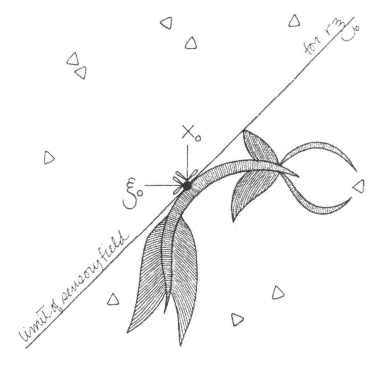

FIGURE 6. Geometry of the sensory field for a specific sensor $r_{\xi 0}$ susceptible to an agent Δ distributed over environmental space.

2.251 In some cases it is possible to compute the spatial distribution of an agent from the known distribution of its effects along a closed surface of given shape. For instance, the spatial distribution of an electrical potential V_x has a unique solution by solving Laplace's equation

$$\Delta V = 0$$

for given values V_ξ along a closed orientable surface (electric fish). Other examples may be cited.

2.252 In some other cases it is possible to compute the spatial distribution of an agent from its effects on just two small, but distinct regions on the body. For instance, the (Euclidean, 3-D) notion of "depth" is computed by resolving the discrepancy of having the "same scene" represented as different images on the retinas of the two eyes in binocular animals (Fig. 7). Let $L(x,y)$ be a postretinal network which computes the relation "x is left of y." While the right eye reports object "a" to be to the left of "b," ($L_r[a,b]$), the left eye gives the contradictory report of object "b" being to the left of "a," ($L_1[b,a]$). A network B which takes cognizance of the different origin of signals coming from cell groups $\{r_\xi\}_r$ and $\{r_\xi\}_1$ to the right

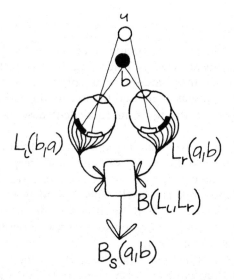

FIGURE 7. Computation of "depth" by resolving a sensory discrepancy in binocular vision; (L) networks computing the relation "x is left of y"; (B) networks computing the relation "x is behind y".

and left side of the animal's body computes with $B(L_r,L_1)$ a new "dimension," namely, the relation $B\ (a,b)$: "a is behind b with respect to me" (subscript $s \equiv$ "self").

2.253 The results of these computations always imply a relation (geometrical or otherwise) of the organism to its environment, as indicated by the relative notion of "behind"; or to itself, as indicated by the absolute notions of "my left eye" or "my right eye." This is the origin of "self-reference."

2.254 It is clear that the burden of these computations is placed upon the operations ω which compute on the distribution functions ρ_ξ.

2.26 For any of such operations to evolve, it is necessary that changes in sensation $\delta\rho_\xi$ are compared with causes of these changes that are controlled by the organism.

3 In a stationary environment, anisotropic with respect to parameters p_i, a movement $\delta\Gamma$ of the organism causes a change of its sensations $\delta\rho_\xi$. Hence we have

$$(\text{movement}) \rightarrow (\text{change in sensation})$$

but not necessarily

$$(\text{change in sensation}) \rightarrow (\text{movement}).$$

3.1 The terms "change in sensation" and "movement" refer to experiences by the organism. This is evidenced by the notation employed here which describes these affairs ρ_ξ, μ_ζ, purely in proprietory coordinates ξ and ζ. (Here μ_ζ has been used to indicate the activity of contractile elements m_ζ. Consequently μ_ζ is an equivalent description of $\delta\Gamma$: $\mu_\zeta \rightarrow \delta\Gamma$.)

3.11 These terms have been introduced to contrast their corresponding notions with those of the terms "stimulus" and "response" of an organism which refer to the experiences of one who *observes* the organism and not to those of the organism itself. This is evidenced by the notation employed here which describes these affairs S_x, $\delta\Gamma_x$ in terms of environmental coordinates x. This is correct insofar as for an observer 0 the organism Ω is a piece of environment.

3.111 From this it is clear that "stimulus" cannot be equated with "change in sensation" and likewise "response" not with "movement." Although it is conceivable that the complex relations that undoubtedly hold between these notions may eventually be established when more is known of the cognitive processes in both the observer and the organism.

3.112 From the *non sequitur* established under proposition 3, it follows *a fortiori*:

$$\text{not necessarily: (stimulus)} \rightarrow (\text{response}).$$

3.2 The presence of a perceptible agent of weak concentration may cause an organism to move toward it (approach). However, the presence of the same agent in strong concentration may cause this organism to move away from it (withdrawal).

3.21 This may be transcribed by the following schema:

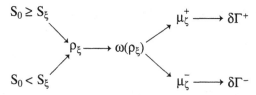

where the (+) and (−) denote approach and withdrawal respectively.

3.211 This schema is the minimal form for representing

a) "environment" [S]
b) "internal representation of environment" ($\omega[\rho_\xi]$)
c) "description of environment" ($\delta\Gamma^+,\delta\Gamma^-$).

4 The logical structure of descriptions arises from the logical structure of movements; "approach" and "withdrawal" are the precursors for "yes" and "no."

4.1 The two phases of elementary behavior, "approach" and "with-drawal," establish the operational origin of the two fundamental axioms of two-valued logic, namely, the "law of the excluded contradiction": $X \& \overline{X}$ (not: X *and* not-X); and the "law of the excluded middle": $X \vee \overline{X}$ (X *or* not-X); (Fig. 8).

4.2 We have from Wittgenstein's *Tractatus*,[4] proposition 4.0621:

. . . it is important that the signs "p" and "non-p" *can* say the same thing. For it shows that nothing in reality corresponds to the sign "non."

The occurrence of negation in a proposition is not enough to characterize its sense (non-non-$p = p$).

4.21 Since nothing in the environment corresponds to negation, nega-tion as well as all other "logical particles" (inclusion, alternation, implica-tion, etc.) must arise within the organism as a consequence of perceiving the relation of itself with respect to its environment.

4.3 Beyond being logical affirmative or negative, descriptions can be true or false.

4.31 We have from Susan Langer, *Philosophy in a New Key*[5]:

The use of signs is the very first manifestation of mind. It arises as early in biolog-ical history as the famous "conditioned reflex," by which a concomitant of a stim-ulus takes over the stimulus-function. The concomitant becomes a *sign* of the condition to which the reaction is really appropriate. This is the real beginning of mentality, for here is the birthplace of *error*, and herewith of *truth*.

4.32 Thus, not only the logical structure of descriptions but also their truth values are coupled to movement.

4.4 Movement, $\delta\Gamma$, is internally represented through operations on peripheral signals generated by:

a) proprioceptors, p_ζ:

$$\delta\Gamma \to \pi_\zeta \to \omega(\pi_\zeta),$$

b) sensors, r_ξ:

$$\delta\Gamma \to \delta\rho_\xi \to \omega(\rho_\zeta);$$

and movement is initiated by operations on the activity v_ζ of central elements n_ζ,

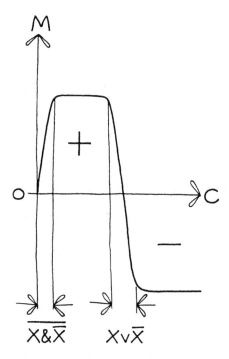

FIGURE 8. The laws of "excluded contradiction" (X & $\overline{X}$) and of "excluded middle" (X v $\overline{X}$) in the twilight zones between no motion ($M = 0$) and approach (+), and between approach (+) and withdrawal (−) as a function of the concentration (C) of a perceptible agent.

c)   intent:

$$\omega(v_\zeta) \to \mu_\zeta \to \delta\Gamma.$$

4.41 Since peripheral activity implies central activity

$$\rho_\xi \to v_\xi \leftarrow \pi_\zeta$$

we have

$$\omega(v_\zeta) \longrightarrow \delta\Gamma$$
$$\uparrow \qquad\qquad \downarrow$$
$$\delta\Gamma \longleftarrow \omega(v_\zeta) \,.$$

4.411 From this it is seen that a conceptualization of descriptions of (the internal representation of) the environment arises from the conceptualization of potential movements. This leads to the contemplation of expressions having the form

$$\omega^{(n)}(\delta\Gamma_1, \omega^{[n-1]}[\delta\Gamma_2, \omega^{(n-2)}(\dots[\rho_\xi])]),$$

that is "descriptions of descriptions of descriptions . . ." or, equivalently, "representations of representations of representations. . . ."

5 The information associated with an event E is the formation of operations ω which control this event's internal representation $\omega(\rho_\xi)$ or its description $\delta\Gamma$.

5.1 A measure of the number of choices of representations ($\omega_i[E]$) or of descriptions ($\delta\Gamma_i[E]$) of this event—or of the probabilities p_i of their occurrence—is the "amount of information" of this event with respect to the organism Ω. ($H[E,\Omega] = -\log_2 p_i$, that is, the negative mean value* of all the [$\log_2 p_i$]).

5.11 This shows that information is a relative concept. And so is H.

5.2 The class of different representations $\omega \equiv (\omega_i[E])$ of an event E determines an equivalence class for different events $(E_i[\omega]) \equiv E$. Hence, a measure of the number of events (E_i) which constitute a cognitive unit, a "category E"[6]—or of the probabilities p_i of their occurrence—is again the "amount of information", H, received by an observer upon perceiving the occurrence of one of these events.

5.21 This shows that the amount of information is a number depending on the choice of a category, that is, of a cognitive unit.

5.3 We have from a paper by Jerzy Konorski:[7]

It is not so, as we would be inclined to think according to our introspection, that the receipt of information and its utilization are two separate processes which can be

---

* The mean value of a set of quantities, x_i, whose probability of occurrence of p_i is given by $\overline{X}_i = \Sigma x_i \cdot p_i$.

combined one with the other in any way; on the contrary, information and its utilization are inseparable constituting, as a matter of fact, one single process.

5.31 These processes are the operations ω, and they are implemented
in the structural and functional organization of nervous activity.

5.4 Let v_i be the signals traveling along single fibers, i, and $v^{(1)}$ be the
outcome of an interaction of N fibers ($i = 1, 2 \ldots N$):

$$v^{(1)} = F^{(1)}(v_1, v_2, \ldots v_N) \equiv F^{(1)}([v_i]).$$

5.41 It is profitable to consider the activity of a subset of these fibers
as determiner for the functional interaction of the remaining ones ("inhibition" changes the functional interaction of "facilitatory" signals). This can
be expressed by a formalism that specifies the functions computed on the
remaining fibers:

$$v^{(1)} = f^{(1)}_{[v_j]}([v_i]), \quad j \neq i.$$

The correspondence between the values v of the row vector (v_j) and the
appropriate functions $f_v^{(1)}$ constitutes a functional for the class of functions
$f^{(1)}_{[v_j]}$.

5.411 This notation makes it clear that the signals themselves may
be seen as being, in part, responsible for determining the operations being
performed on them.

5.42 The mapping that establishes this correspondence is usually
interpreted as being the "structural organization" of these operations, while
the set of functions so generated as being their "functional organization."

5.421 This shows that the distinction between structural and functional organization of cognitive processes depends on the observer's point
of view.

5.43 With N fibers being considered, there are 2^N possible interpretations (the set of all subsets of N) of the functional and structural organization of such operations. With all interpretations having the same likelihood,
the "uncertainty" of this system regarding its interpretability is $H = \log_2 2^N$
$= N$ bits.

5.5 Let $v_i^{(1)}$ be the signals traveling along single fibers, i, and $v^{(2)}$ be the
outcome of an interaction of N_1 such fibers ($i = 1, 2, \ldots N_1$):

$$v^{(2)} = F^{(2)}([v_i^{(1)}])$$

or, recursively from 5.4:

$$v^{(k)} = F^{(k)}(F^{(k-1)}[F^{(k-2)}(\ldots F^{[1]}[v_i])]).$$

5.51 Since the $F^{(k)}$ can be interpreted as functionals $f^{(k)}_{[v_i]}$, this leads to
a calculus of recursive functionals for the representation of cognitive
processes ω.

5.511 This becomes particularly significant if $v_i^{(k-t)}$ denotes the activity of fiber, i, at a time interval t prior to its present activity $v_i^{(k)}$. That is, the
recursion in 5.5 can be interpreted as a recursion in time.

5.52 The formalism of recursive functionals as employed in 5.5 for representing cognitive processes, ω, is isomorphic to the lexical definition structure of nouns. Essentially, a noun signifies a class, $cl^{(1)}$, of things. When defined, it is shown to be a member of a more inclusive class, $cl^{(2)}$, denoted by a noun which, in turn, when defined is shown to be a member of a more inclusive class, $cl^{(3)}$, and so on [pheasant → bird → animal → organism → thing]:

$$cl^{(n)} = \left(cl_{i_{n-1}}^{[n-1]} \left[cl_{i_{n-2}}^{[n-2]} (\ldots [cl_{i_1}^{(1)}]) \right] \right)$$

where the notation (e_i) stands for a class composed of elements e_i, and subscripted subscripts are used to associate these subscripts with the corresponding superscripts.

5.521 The highest order n* in this hierarchy of classes is always represented by a single, undefined term "thing," "entity," "act," etc., which refers to basic notions of being able to perceive at all.

5.6 Cognitive processes create descriptions of, that is information, about the environment.

6 The environment contains no information. The environment is as it is.

References

1. G. Gunther, "Cybernetic Ontology and Transjunctional Operations," *Self-Organizing Systems*, ed. M. C. Yovits *et al.* (Washington, D.C.: Spartan Books, 1962).
2. A. H. Barr, Jr. (ed.), *Fantastic Art, Dada, Surrealism* (3rd. ed.; New York: The Museum of Modern Art, 1947), p. 156.
3. M. Jean, *Histoire de la painture Surrealiste* (Edition du Seuil; Paris: 1959), p. 284.
4. L. Wittgenstein, *Tractatus Logico Philosophicus* (New York: Humanities Press, 1961).
5. S. Langer, *Philosophy in a New Key* (New York: new American Library, 1951).
6. H. Quastler, "A Primer in Information Theory," *Information Theory in Biology*, ed. H. P. Yockey *et al.* (New York: Pergamon Press, 1958), pp. 3–49.
7. J. Konorski, "The Role of Central Factors in Differentiation," *Information Processing in the Nervous System*, 3, eds. R. W. Gerard and J. W. Duyff (Amsterdam: Excerpta Medica Foundation, 1962), pp. 318–29.

6
Responsibilities of Competence*

HEINZ VON FOERSTER
Biological Computer Laboratory, University of Illinois

At our last Annual Symposium I submitted to you a theorem to which Stafford Beer referred on another occasion as "Heinz Von Foerster's Theorem Number One". As some of you may remember, it went as follows:

"The more profound the problem that is ignored, the greater are the chances for fame and success."

Building on a tradition of a single instance, I shall again submit a theorem which, in all modesty, I shall call "Heinz Von Foerster's Theorem Number Two". It goes as follows:

"The hard sciences are successful because they deal with the soft problems; the soft sciences are struggling because they deal with the hard problems."

* Adapted from the keynote address at the Fall Conference of the American Society for Cybernetics, Dec. 9, 1971, in Washington, D.C. Published in the *Journal of Cybernetics*, *2* (2), pp. 1–6, (1972).

Should you care to look closer, you may discover that Theorem 2 could serve as a corollary to Theorem 1. This will become obvious when we contemplate for a moment the method of inquiry employed by the hard sciences. If a system is too complex to be understood it is broken up into smaller pieces. If they, in turn, are still too complex, they are broken up into even smaller pieces, and so on, until the pieces are so small that at least one piece can be understood. The delightful feature of this process, the method of reduction, "reductionism", is that it inevitably leads to success.

Unfortunately, the soft sciences are not blessed with such favorable conditions. Consider, for instance, the sociologist, psychologist, anthropologist, linguist, etc. If they would reduce the complexity of the system of their interest, i.e., society, psyche, culture, language, etc., by breaking it up into smaller parts for further inspection they would soon no longer be able to claim that they are dealing with the original system of their choice. This is so, because these scientists are dealing with essentially nonlinear systems whose salient features are represented by the *interactions* between whatever one may call their "parts" whose properties in isolation add little, if anything, to the understanding of the workings of these systems when each is taken as a whole. Consequently, if he wishes to remain in the field of his choice, the scientist who works in the soft sciences is faced with a formidable problem: he cannot afford to loose sight of the full complexity of his system, on the other hand it becomes more and more urgent that his problems be solved. This is not just to please him. By now it has become quite clear that his problems concern us all. "Corruption of our society", "psychological disturbances", "cultural erosion", the "breakdown of communication", and all the other of these "crises" of today are our problems as well as his. How can we contribute to their solution?

My suggestion is that we apply the *competences* gained in the hard sciences—and not the method of reduction—to the solution of the hard problems in the soft sciences. I hasten to add that this suggestion is not new at all. In fact, I submit that it is precisely *Cybernetics* that interfaces hard competence with the hard problems of the soft sciences. Those of us who witnessed the early development of cybernetics may well remember that before Norbert Wiener created that name for our science it was referred to as the study of "Circular-Causal and Feedback Mechanisms in Biological and Social Systems", a description it carried even years after he wrote his famous book. Of course, in his definition of Cybernetics as the science of "communication and control in the animal and the machine" Norbert Wiener went one step further in the generalization of these concepts, and today "Cybernetics" has ultimately come to stand for the science of *regulation* in the most general sense.

Since our science embraces indeed this general and all-pervasive notion, why then, unlike most of our sister sciences, do we not have a patron saint or a diety to bestow favors on us in our search for new insights, and who protects our society from evils from without as well as from within? Astronomers and physicists are looked after by Urania; Demeter patronizes agriculture; and various Muses help the various arts and sciences. But who helps Cybernetics?

One night when I was pondering this cosmic question I suddenly had an apparition. Alas, it was not one of the charming goddesses who bless the other arts and sciences. Clearly, that funny little creature sitting on my desk must be a demon. After a while he started to talk. I was right. "I am Maxwell's Demon", he said. And then he disappeared.

When I regained my composure it was immediately clear to me that nobody else but this respectable demon could be our patron, for Maxwell's Demon is *the paradigm for regulation.*

As you remember, Maxwell's Demon regulates the flow of molecules between two containers in a most *unnatural* way, namely, so that heat flows from the cold container to the hotter, as opposed to the natural course of events where without the demon's interference heat always flows from the hot container to the colder.

I am sure you also remember how he proceeds: He guards a small aperture between the two containers which he opens to let a molecule pass whenever a fast one comes from the cool side or a slow one comes from the hot side. Otherwise he keeps the aperture closed. Obviously, by this maneuver he gets the cool container becoming cooler, and the hot container getting hotter, thus apparently upsetting the Second Law of Thermodynamics. Of course, we know by now that while he succeeds in obtaining this perverse flow of heat, the Second Law remains untouched. This is because of his need for a flashlight to determine the velocity of the upcoming molecules. Were he at thermal equilibrium with one of the containers he couldn't see a thing: he is part of a black body. Since he can do his antics only as long as the battery of his flashlight lasts, we must include into the system with an active demon not only the energy of the two containers, but also that of the battery. The entropy gained by the battery's decay is not completely compensated by the negentropy gained from the increased disparity of the two containers.

The moral of this story is simply that while our demon cannot beat the Second Law, he can, by his regulatory activity, retard the degradation of the available energy, i.e., the growth of entropy, to an arbitrary slow rate.

This is indeed a very significant observation because it demonstrates the paramount importance of regulatory mechanisms in living organisms. In this context they can be seen as manifestations of Maxwell's Demon, retarding continuously the degradation of the flow of energy, that is, retarding the increase of entropy. In other words, as regulators living organisms are "entropy retarders".

Moreover, as I will show in a moment, Maxwell's Demon is not only an entropy retarder and a paradigm for regulation, but he is also a functional isomorph of a Universal Turing Machine. Thus, the three concepts of regulation, entropy retardation, and computation constitute an interlaced conceptual network which, for me, is indeed the essence of Cybernetics.

I shall now briefly justify my claim that Maxwell's Demon is not only the paradigm for regulation but also for computation.

When I use the term "computation" I am not restricting my self to specific operations as, for instance, addition, multiplication, etc. I wish to interpret "computation" in the most general sense as a mechanism, or "algorithm", for *ordering*. The ideal, or should I say the most general, representation of such mechanism is, of course, a Turing Machine, and I shall use this machine to illuminate some of the points I wish to make.

There are two levels on which we can think of "ordering". The one is when we wish to make a description of a given arrangement of things. The other one when we wish to re-arrange things according to certain descriptions. It will be obvious at once that these two operations constitute indeed the foundations for all that which we call "computation".

Let A be a particular arrangement. Then this arrangement can be computed by a universal Turing machine with a suitable initial tape expression which we whall call a "description" of A: $D(A)$. The length $L(A)$ of this description will depend on the alphabet (language) used. Hence, we may say that a language α_1 reveals more order in the arrangement A than another language α_2, if and only if the length $L_1(A)$ of the suitable initial tape description for computing A is shorter than $L_2(A)$, *or mutatis mutandis*.

This covers the first level of above, and leads us immediately to the second level.

Among all suitable initial tape descriptions for an arrangement A_1 there is a shortest one: $L^*(A_1)$. If A_1 is re-arranged to give A_2, call A_2 to be of a higher order than A_1 if and only if the shortest initial tape description $L^*(A_2)$ is shorter than $L^*(A_1)$, or *mutatis mutandis*.

This covers the second level of above, and leads us to a final statement of perfect ordering (computation).

Among all arrangements A_i there is one, A^*, for which the suitable initial tape description is the shortest $L^*(A^*)$.

I hope that with these examples it has become clear that living organisms (replacing now the Turing machine) interacting with their environment (arrangements) have several options at their disposal: (i) they may develop "languages" (sensors, neural codes, motor organs, etc.) which "fit" their given environment better (reveal more order); (ii) they may change their surroundings until it "fits" their constitution; and (iii), they may do both. However, it should be noted that whatever option they take, it will be done by computation. That these computations are indeed functional isomorphs of our demon's activity is now for me to show.

The essential function of a Turing machine can be specified by five operations:

(i) *Read* the input symbol x.
(ii) *Compare* x with z, the intenal state of the machine.
(iii) *Write* the appropriate output symbol y.
(iv) *Change* the internal state z to the new state z'.
(v) *Repeat* the above sequence with a new input state x'.

Similarly, the essential function of Maxwell's Demon can be specified by five operations equivalent to those above:

(i) *Read* the velocity v of the upcoming molecule M.
(ii) *Compare* $(mv^2/2)$ with the mean energy $\langle mv^2/2 \rangle$ (temperature T) of, say, the cooler container (internal state T).
(iii) *Open* the aperture if $(mv^2/2)$ is greater than $\langle mv^2/2 \rangle$; otherwise keep it closed.
(iv) *Change* the internal state T to the new (cooler) state T'.
(v) *Repeat* the above sequence with a new uncoming molecule M'.

Since the translation of the terms occurring in the correspondingly labeled points is obvious, with the presentation of these two lists I have completed my proof.

How can we make use of our insight that Cybernetics is the science of regulation, computation, ordering, and entropy retardation? We may, of course, apply our insight to the system that is generally understood to be the *cause célèbre* for regulation, computation, ordering, and entropy retardation, namely, the human brain.

Rather than following the physicists who order their problems according to the number of *objects* involved ("The one-body problem", "The two-body problem", "The three-body problem", etc.), I shall order our problems according to the number of *brains* involved by discussing now "The one-brain problem", "The two-brain problem", "The many-brain problem", and "The all-brain problem".

1. The Single-Brain Problem: The Brain Sciences

It is clear that if the brain sciences do not want to degenerate into a physics or chemistry of living—or having once lived—tissue they must develop a theory of the brain: T(B). But, of course, this theory must be written by a brain: B(T). This means that this theory must be constructed so as to write itself T(B(T)).

Such a theory will be distinct in a fundamental sense from, say, physics which addresses itself to a (not quite) successful description of a "subject-less world" in which even the observer is not supposed to have a place. This leads me now to pronounce my Theorem Number Three:

"The Laws of Nature are written by man. The laws of biology must write themselves."

In order to refute this theorem it is tempting to invoke Gödel's Proof of the limits of the Entscheidungsproblem in systems that attempt to speak of themselves. But Lars Löfgren and Gotthard Günther have shown that self-explanation and self-reference are concepts that are untouched by Gödel's arguments. In other words, a science of the brain in the above sense is, I claim, indeed a legitimate science with a legitimate problem.

2. The Two-Brain Problem: Education

It is clear that the majority of our established educational efforts is directed toward the trivialization of our children. I use the term "trivialization" exactly as used in automata theory, where a trivial machine is characterized by its fixed input-output relation, while in a non-trivial machine (Turing machine) the output is determined by the input *and* its internal state. Since our educational system is geared to generate predictable citizens, its aim is to amputate the bothersome internal states which generate unpredictability and novelty. This is most clearly demonstrated by our method of examination in which only questions are asked for which the answers are known (or defined), and are to be memorized by the student. I shall call these questions "illegitimate questions".

Would it not be fascinating to think of an educational system that de-trivializes its students by teaching them to ask "legitimate questions", that is, questions for which the answers are unknown?

3. The Many-Brain Problem: Society

It is clear that our entire society suffers from a severe dysfunction. On the level of the individual this is painfully felt by apathy, distrust, violence, disconnectedness, powerlessness, alienation, and so on. I call this the "participatory crisis", for it excludes the individual from participating in the social process. The society becomes the "system", the "establishment" or what have you, a depersonalized Kafkanesque ogre of its own ill will.

It is not difficult to see that the essential cause for this dysfunction is the absence of an adequate input for the individual to interact with society. The so-called "communication channels", the "mass media" are only one-way: they talk, but nobody can talk back. The feedback loop is missing and, hence, the system is out of control. What cybernetics could supply is, of course, a universally accessible social input device.

4. The All-Brain Problem: Humanity

It is clear that the single most distressing characteristic of the global system "mankind" is its demonstrated instability, and a fast approaching singularity. As long as humanity treats itself as an open system by ignoring the signals of its sensors that report about its own state of affairs, we shall approach this singularity with no breaks whatsoever. (Lately I began to wonder whether the information of its own state can reach all elements in time to act should they decide to listen rather than fight.)

The goal is clear: we have to close the system to reach a stable population, a stable economy, and stable resources. While the problem of constructing a "population servo" and an "economic servo" can be solved with the mental resources on this planet, for the stability of our material resources we are forced by the Second Law of Thermodynamics to turn to extra-planetary sources. About $2 \cdot 10^{14}$ kilowatts solar radiation are at our disposal. Wisely used, this could leave our earthy, highly structured, invaluable organic resources, fossilized or living, intact for the use and enjoyment of uncounted generations to come.

If we are after fame and success we may ignore the profundity of these problems in computation, ordering, regulation, and entropy retardation. However, since we as cyberneticians supposedly have the competence to attack them, we may set our goal above fame and success by quietly going about their solution. If we wish to maintain our scientific credibility, the first step to take is to apply our competence to ourselves by forming a global society which is not so much *for* Cybernetics as it *functions* cybernetically. This is how I understand Dennis Gabor's exhortation in an earlier issue: "Cyberneticians of the world, unite!" Without communication there is no regulation; without regulation there is no goal; and without a goal the concept of "society" or "system" becomes void.

Competence implies responsibilities. A doctor must act at the scene of the accident. We can no longer afford to be the knowing spectators at a global disaster. We must share what competence we have through communication and cooperation in working together through the problems of our time. This is the only way in which we can fulfill our social and individual responsibilities as cyberneticians who should practice what they preach.

7
Perception of the Future and the Future of Perception*

HEINZ VON FOERSTER
University of Illinois, Urbana, Illinois

Abstract

"The definition of a problem and the action taken to solve it largely depend on the view which the individuals or groups that discovered the problem have of the system to which it refers. A problem may thus find itself defined as a badly interpreted output, or as a faulty output of a faulty output device, or as a faulty output due to a malfunction in an otherwise faultless system, or as a correct but undesired output from a faultless and thus undesirable system. All definitions but the last suggest corrective action; only the last definition suggests change, and so presents an unsolvable problem to anyone opposed to change" (Herbert Brün, 1971).

Truisms have the disadvantage that by dulling the senses they obscure the truth. Almost nobody will become alarmed when told that in times of continuity the future equals the past. Only a few will become aware that from this follows that in times of socio-cultural change the future will *not* be like the past. Moreover, with a future not clearly perceived, we do not know how to act with only one certainty left: if we don't act ourselves, we shall be acted upon. Thus, if we wish to be subjects, rather than objects, what we see now, that is, our perception, must be foresight rather than hindsight.

Epidemic

My colleagues and I are, at present, researching the mysteries of cognition and perception. When, from time to time, we look through the windows of our laboratory into the affairs of this world, we become more and more distressed by what we now observe. The world appears to be in the grip of a fast-spreading disease which, by now, has assumed almost global dimen-

---

* This article is an adaptation of an address given on March 29, 1971, at the opening of the Twenty-fourth Annual Conference on World Affairs at the University of Colorado, Boulder, Colorado, U.S.A. Reprinted from *Instructional Science, 1* (1), pp. 31–43 (1972). With kind permission from Kluwer Academic Publishers.

sions. In the individual the symptoms of the disorder manifest themselves by a progressive corruption of his faculty to perceive, with corrupted language being the pathogene, that is, the agent that makes the disease so highly contagious. Worse, in progressive stages of this disorder, the afflicted become numb, they become less and less aware of their affliction.

This state of affairs makes it clear why I am concerned about perception when contemplating the future, for:

if we can't perceive,
we can't perceive of the future
and thus, we don't know how to act now.

I venture to say that one may agree with the conclusion. If one looks around, the world appears like an anthill where its inhabitants have lost all sense of direction. They run aimlessly about, chop each other to pieces, foul their nest, attack their young, spend tremendous energies in building artifices that are either abandoned when completed, or when maintained, cause more disruption than was visible before, and so on. Thus, the conclusions seem to match the facts. Are the premises acceptable? Where does perception come in?

Before we proceed, let me first remove some semantic traps, for—as I said before—corrupt language is the pathogene of the disease. Some simple perversions may come at once to mind, as when "incursion" is used for "invasion," "protective reaction" for "aggression," "food denial" for "poisoning men, beasts, and plants," and others. Fortunately, we have developed some immunity against such insults, having been nourished with syntactic monstrosities as "X is better" without ever saying "than what." There are, however, many more profound semantic confusions, and it is these to which I want to draw your attention now.

There are three pairs of concepts in which one member of these pairs is generally substituted for the other so as to reduce the richness of our conceptions. It has become a matter of fact to confuse process with substance, relations with predicates, and quality with quantity. Let me illustrate this with a few examples out of a potentially very large catalogue, and let me at the same time show you the paralytic behavior that is caused by this conceptual dysfunction.

Process/Substance

The primordial and most proprietary processes in any man and, in fact, in any organism, namely "information" and "knowledge," are now persistently taken as commodities, that is as substance. Information is, of course, the process by which knowledge is acquired, and knowledge is the processes that integrate past and present experiences to form new activities, either as nervous activity internally perceived as thought and will, or externally perceivable as speech and movement (Maturana, 1970, 1971; Von Foerster, 1969, 1971).

Neither of these processes can be "passed on" as we are told in phrases like ". . . Universities are depositories of Knowledge which is passed on from generation to generation . . . ," etc., for *your* nervous activity is just *your* nervous activity and, alas, not *mine*.

No wonder that an educational system that confuses the process of creating new processes with the dispensing of goods called "knowledge" may cause some disappointment in the hypothetical receivers, for the goods are just not coming: there are no goods.

Historically, I believe, the confusion by which knowledge is taken as substance comes from a witty broadsheet printed in Nuremberg in the Sixteenth Century. It shows a seated student with a hole on top of his head into which a funnel is inserted. Next to him stands the teacher who pours into this funnel a bucket full of "knowledge," that is, letters of the alphabet, numbers and simple equations. It seems to me that what the wheel did for mankind, the Nuremberg Funnel did for education: we can now roll faster down the hill.

Is there a remedy? Of course, there is one! We only have to perceive lectures, books, slides and films, etc., not as *information* but as *vehicles* for potential information. Then we shall see that in giving lectures, writing books, showing slides and films, etc., we have not solved a problem, we just created one, namely, to find out in which context can these things be seen so that they create in their perceivers new insights, thoughts, and actions.

Relation/Predicate

Confusing relations with predicates has become a political pastime. In the proposition "spinach is green," "green" is a predicate; in "spinach is good," "good" is a relation between the chemistry of spinach and the observer who tastes it. He may refer to his relation with spinach as "good." Our mothers, who are the first politicians we encounter, make use of the semantic ambiguity of the syntactic operator "is" by telling us "spinach *is* good" as if they were to say "spinach is *green*."

When we grow older we are flooded with this kind of semantic distortion that could be hilarious if it were not so far reaching. Aristophanes could have written a comedy in which the wisest men of a land set out to accomplish a job that, in principle, cannot be done. They wish to establish, once and for all, all the properties that define an obscene object or act. Of course, "obscenity" is not a property residing within things, but a subject-object relationship, for if we show Mr. X a painting and he calls it obscene, we know a lot about Mr. X but very little about the painting. Thus, when our lawmakers will finally come up with their imaginary list, we shall know a lot about them, but their laws will be dangerous nonsense.

"Order" is another concept that we are commanded to see in things rather than in our perception of things. Of the two sequences A and B,

A: 1, 2, 3, 4, 5, 6, 7, 8, 9
B: 8, 5, 4, 9, 1, 7, 6, 3, 2

sequence A is seen to be ordered while B appears to be in a mess, until we are told that B has the same beautiful order as A, for B is in alphabetical order (eight, five, four, . . .). "Everything has order once it is understood" says one of my friends, a neurophysiologist, who can see order in what appears to me at first the most impossible scramble of cells. My insistence here to recognize "order" as a subject-object relation and not to confuse it with a property of things may seem too pedantic. However, when it comes to the issue "law and order" this confusion may have lethal consequences. "Law and order" is no issue, it is a desire common to all; the issue is "which laws and what order," or, in other words, the issue is "justice and freedom."

Castration

One may dismiss these confusions as something that can easily be corrected. One may argue that what I just did was doing that. However, I fear this is not so; the roots are deeper than we think. We seem to be brought up in a world seen through descriptions by others rather than through our own perceptions. This has the consequence that instead of using language as a tool with which to express thoughts and experience, we accept language as a tool that determines our thoughts and experience.

It is, of course, very difficult to prove this point, for nothing less is required than to go inside the head and to exhibit the semantic structure that reflects our mode of perception and thinking. However, there are now new and fascinating experiments from which these semantic structures can be inferred. Let me describe one that demonstrates my point most dramatically.

The method proposed by George Miller (1967) consists of asking independently several subjects to classify on the basis of similarity of meaning a number of words printed on cards (Fig. 1). The subject can form as many classes as he wants, and any number of items can be placed in each class. The data so collected can be represented by a "tree" such that the branch-points further away from the "root" indicate stronger agreement among the subjects, and hence suggest a measure of similarity in the meaning of the words for this particular group of subjects.

Figure 2 shows the result of such a "cluster analysis" of the 36 words of Fig. 1 by 20 adult subjects ("root" on the left). Clearly, adults classify according to syntactic categories, putting nouns in one class (bottom tree), adjectives in another (next to bottom tree), then verbs, and finally those little words one does not know how to deal with.

FIGURE 1. Example of 36 words printed on cards to be classified according to similarity in meaning.

The difference is impressive when the adults' results are compared with the richness of perception and imagery of children in the third and fourth grade when given the same task (Fig. 3). Miller reflects upon these delightful results:

"Children tend to put together words that might be used in talking about the same thing—which cuts right across the tidy syntactic boundaries so important to adults. Thus all twenty of the children agree in putting the verb 'eat' with the noun 'apple'; for many of them 'air' is 'cold'; the foot' is used to 'jump; you 'live' in a 'house'; 'sugar' is 'sweet'; and the cluster of doctor,' 'needle,' 'suffer,' 'weep,' and 'sadly' is a small vignette in itself."

What is wrong with our education that castrates our power over language? Of the many factors that may be responsible I shall name only one that has a profound influence on our way of thinking, namely, the misapplication of the "scientific method."

Scientific Method

The scientific method rests on two fundamental pillars:

(i) Rules observed in the past shall apply to the future. This is usually referred to as the principle of conservation of rules, and I have no doubt that you are all familiar with it. The other pillar, however, stands in the shadow of the first and thus is not so clearly visible:
(ii) Almost everything in the universe shall be irrelevant. This is usually referred to as the principle of the necessary and sufficient cause, and what

ADULTS

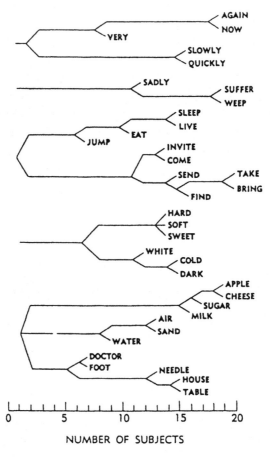

NUMBER OF SUBJECTS

FIGURE 2. Cluster analysis of the 36 words of Fig. 1 classified by 20 adult subjects. Note that syntactic categories are faithfully respected, while semantic relations are almost completely ignored.

it demands is at once apparent when one realizes that "relevance" is a triadic relation that relates a set of propositions $(P_1, P_2, \ldots)$ to another set of propositions $(Q_1, Q_2, \ldots)$ in the mind (M) of one who wishes to establish this relation. If P are the causes that are to explain the perceived effects Q, then the principle of necessary and sufficient cause forces us to reduce our perception of effects further and further until we have hit upon the necessary and sufficient cause that produces the desired effect: everything else in the universe shall be irrelevant.

It is easy to show that resting one's cognitive functions upon these two pillars is counter-productive in contemplating any evolutionary process, be

GRADES 3 AND 4

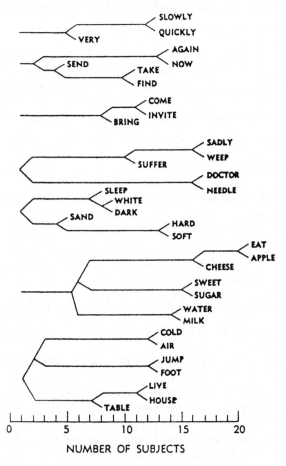

FIGURE 3. The same 36 words of Figs. 1 and 2 classified by children in the third and fourth grade. Note the emergence of meaningful cognitive units, while syntactic categories are almost completely ignored.

it the growing up of an individual, or a society in transition. In fact, this was already known by Aristotle who distinguished two kinds of cause, one the "efficient cause," the other the "final cause," which provide us with two distinct explanatory frameworks for either inanimate matter, or else living organisms, the distinction being that the efficient cause *precedes* its effect while the final cause *succeeds* its effect. When striking with a match the treated surface of a matchbook, the striking is the (efficient) cause for the match to ignite. However, the cause for my striking the match is my wish to have it ignited (final cause).

Perhaps, with this distinction, my introductory remarks may appear much clearer. Of course, I had in mind the final cause when I said that if we can perceive of the future (the match being ignited), we know how to act now (strike!). This leads me immediately to draw a conclusion, namely:

At any moment we are free to act toward the future we desire.

In other words, the future will be as we wish and perceive it to be. This may come as a shock only to those who let their thinking be governed by the principle that demands that only the rules observed in the past shall apply to the future. For those the concept of "change" is inconceivable, for change is the process that obliterates the rules of the past.

Quality/Quantity

In order to protect society from the dangerous consequences of change, not only a whole branch of business has emerged, but also the Government has established several offices that busy themselves in predicting the future by applying the rules of the past. These are the Futurists. Their job is to confuse quality with quantity, and their products are "future scenarios" in which the qualities remain the same, only the quantities change: more cars, wider highways, faster planes, bigger bombs, etc. While these "future scenarios" are meaningless in a changing world, they have become a lucrative business for entrepreneurs who sell them to corporations that profit from designing for obsolescence.

With the diagnosis of the deficiency to perceive qualitative change, that is, a change of our subject-object and subject-subject relationships, we are very close to the root of the epidemic that I mentioned in my opening remarks. An example in neurophysiology may help to comprehend the deficiency that now occurs on the cognitive level.

The visual receptors in the retina, the cones and the rods, operate optimally only under certain conditions of illumination. Beyond or below this condition we suffer a loss in acuity or in color discrimination. However, in the vertebrate eye the retina almost always operates under these optimal conditions, because of the iris that contracts or dilates so as to admit under changing conditions of brightness the same amount of light to the receptors. Hence, the scenario "seen" by the optic nerve has always the same illumination independent of whether we are in bright sunshine or in a shaded room. How, then, do we know whether it is bright or shady?

The information about this datum resides in the regulator that compares the activity in the optic nerve with the desired standard and causes the iris to contract when the activity is too high, and to dilate when it is too small. Thus, the information of brightness does not come from inspecting the scenario—it appears always to be of similar brightness—it comes from an inspection of the regulator that suppresses the perception of change.

There are subjects who have difficulties in assessing the state of their regulator, and thus they are weak in discriminating different levels of brightness. They are called "dysphotic." They are the opposite of photographers who may be called "photic," for they have a keen sense of brightness discrimination. There are subjects who have difficulties in assessing the regulators that maintain their identity in a changing world. I shall call individuals suffering from this disorder "dysgnostic," for they have no way of knowing themselves. Since this disorder has assumed extraordinary dimensions, it has indeed been recognized at the highest national level.

As you all know, it has been observed that the majority of the American people cannot speak. This is interpreted by saying that they are "silent"; I say they are *mute*. However, as you all know very well, there is nothing wrong with the vocal tract of those who are mute: the cause of their muteness is deafness. Hence, the so-called "silent majority" is *de facto* a "deaf majority."

However, the most distressing thing in this observation is that there is again nothing wrong with their auditory system; they could hear if they wanted to: but they don't want to. Their deafness is voluntary, and in others it is their blindness.

At this point proof will be required for these outrageous propositions. *TIME Magazine* (1970) provides it for me in its study of Middle America.

There is the wife of a Glencoe, Illinois lawyer, who worries about the America in which her four children are growing up: "I want my children to live and grow up in an America as I *knew* it," [note the principle of conservation of rule where the future equals the past] "where we were proud to be citizens of this country. I'm damned sick and tired of *listening* to all this nonsense about how awful America is." [Note voluntary deafness.]

Another example is a newspaper librarian in Pittsfield, Massachusetts, who is angered by student unrest: "Every time I see protestors, I say, 'Look at those creeps.'" [Note reduction of visual acuity.] "But then my 12-year-old son says, 'They're not creeps. They have a perfect right to do what they want.'" [Note the un-adult-erated perceptual faculty in the young.]

The tragedy in these examples is that the victims of "dysgnosis" not only do not know that they don't see, hear, or feel, they also do not want to.

How can we rectify this situation?

Trivialization

I have listed so far several instances of perceptual disorders that block our vision of the future. These symptoms collectively consitute the syndrome of our epidemic disease. It would be the sign of a poor physician if he were to go about relieving the patient of these symptoms one by one, for the elimination of one may aggrevate another. Is there a single common denominator that would identify the root of the entire syndrome?

To this end, let me introduce two concepts, they are the concepts of the "trivial" and the "non-trivial" machine. The term "machine" in this context refers to well-defined functional properties of an abstract entity rather than to an assembly of cogwheels, buttons and levers, although such assemblies may represent embodiments of these abstract functional entities.

A trivial machine is characterized by a one-to-one relationship between its "input" (stimulus, cause) and its "output" (response, effect). This invariable relationship is "the machine." Since this relationship is determined once and for all, this is a deterministic system; and since an output once observed for a given input will be the same for the same input given later, this is also a predictable system.

Non-trivial machines, however, are quite different creatures. Their input-output relationship is not invariant, but is determined by the machine's previous output. In other words, its previous steps determine its present reactions. While these machines are again deterministic systems, for all practical reasons they are unpredictable: an output once observed for a given input will most likely be not the same for the same input given later.

In order to grasp the profound difference between these two kinds of machines it may be helpful to envision "internal states" in these machines. While in the trivial machine only one internal state participates always in its internal operation, in the non-trivial machine it is the shift from one internal state to another that makes it so elusive.

One may interpret this distinction as the Twentieth Century version of Aristotle's distinction of explanatory frameworks for inanimate matter and living organisms.

All machines we construct and buy are, hopefully, trivial machines. A toaster should toast, a washing machine wash, a motorcar should predictably respond to its driver's operations. In fact, all our efforts go into one direction, to create trivial machines or, if we encounter non-trivial machines, to convert them into trivial machines. The discovery of agriculture is the discovery that some aspects of Nature can be trivialized: If I till today, I shall have bread tomorrow.

Granted, that in some instances we may be not completely successful in producing ideally trivial machines. For example, one morning turning the starter key to our car, the beast does not start. Apparently it changed its internal state, obscure to us, as a consequence of previous outputs (it may have exhausted its gasoline supply) and revealed for a moment its true nature of being a non-trivial machine. But this is, of course, outrageous and this state of affairs should be remedied at once.

While our pre-occupation with the trivialization of our environment may be in one domain useful and constructive, in another domain it is useless and destructive. Trivialization is a dangerous panacea when man applies it to himself.

Consider, for instance, the way our system of education is set up. The student enters school as an unpredictable "non-trivial machine." We don't know what answer he will give to a question. However, should he succeed in this system the answers he gives to our questions must be known. They are the "right" answers:

Q: "When was Napoleon born?"
A: "1769"
Right!
Student → Student
but
Q: "When was Napoleon born?"
A: Seven years before the Declaration of Independence."
Wrong!
Student → Non-student

Tests are devices to establish a measure of trivialization. A perfect score in a test is indicative of perfect trivialization: the student is completely predictable and thus can be admitted into society. He will cause neither any surprises nor any trouble.

I shall call a question to which the answer is known an "illegitimate question." Wouldn't it be fascinating to contemplate an educational system that would ask of its students to answer "legitimate questions" that is questions to which the answers are unknown (H. Brün in a personal communication). Would it not be even more fascinating to conceive of a society that would establish such an educational system? The necessary condition for such an utopia is that its members perceive one another as autonomous, non-trivial beings. Such a society shall make, I predict, some of the most astounding discoveries. Just for the record, I shall list the following three:

1. "Education is neither a right nor a privilege: it is a necessity."
2. "Education is learning to ask legitimate questions."

A society who has made these two discoveries will ultimately be able to discover the third and most utopian one:

3. "A is better off when B is better off."

From where we stand now, anyone who seriously makes just one of those three propositions is bound to get into trouble. Maybe you remember the story Ivan Karamazov makes up in order to intellectually needle his younger brother Alyosha. The story is that of the Great Inquisitor. As you recall, the Great Inquisitor walks on a very pleasant afternoon through his town, I believe it is Salamanca; he is in good spirits. In the morning he has burned at the stakes about a hundred and twenty heretics, he has done a good job, everything is fine. Suddenly there is a crowd of people in front of him, he moves closer to see what's going on, and he sees a stranger who is putting his hand onto a lame person, and that lame one can walk. Then a blind girl is brought before him, the stranger is putting his hand on her eyes, and she can see. The Great Inquisitor knows immediately who He is, and he says to his henchmen: "Arrest this man." They jump and arrest this man and put Him into jail. In the night the Great Inquisitor visits the stranger in his cell and he says: "Look, I know who You are, troublemaker. It took us one thousand and five hundred years to straighten out the troubles you have sown. You know very well that people can't make decisions by themselves. You know very well people can't be free. *We* have to make their decisions. *We* tell them who they are to be. You know that very well. Therefore, I shall burn You at the stakes tomorrow." The stranger stands up, embraces the Great Inquisitor and kisses him. The Great Inquisitor walks out, but, as he leaves the cell, he does not close the door, and the stranger disappears in the darkness of the night.

Let us remember this story when we meet those troublemakers, and let "Let there be vision: and there was light."

References

Brün, H. (1971). "Technology and the Composer," in Von Foerster, H., ed., *Interpersonal Relational Networks*. pp. 1/10. Cuernavaca: Centro Intercultural de Documentacion.

Maturana, H. R. (1970). "Biology of Cognition" BCL Report No. 9.0, Biological Laboratory, Department of Electrical Engineering, University of Illinois, Urbana, 93 pp.

Maturana, H. R. (1971). "Neurophysiology of Cognition," in Garvin, P., ed., *Cognition, A Multiple View*, pp. 3–23. New York: Spartan Books.

Miller, G. A. (1967). "Psycholinguistic Approaches to the Study of Communication," in Arm, D. L., ed., *Journeys in Science*, pp. 22–73. Albuquerque: Univ. New Mexico.

TIME Magazine. (1970). "The Middle Americans," (January 5).

Von Foerster, H. (1969). "What is Memory that It May Have Hindsight and Foresight as well?," in Bogoch, S., ed., *The Future of the Brain Sciences*, pp. 19–64. New York: Plenum Press.

Von Foerster, H. (1971). "Thoughts and Notes on Cognition," in Garvin, P., ed., *Cognition, A Multiple View*, pp. 25–48. New York: Spartan Books.

8
On Constructing a Reality*

HEINZ VON FOERSTER

Draw a distinction!

G. Spencer Brown[1]

The Postulate

I AM SURE YOU remember the plain citizen Jourdain in Molière's *Le Bourgeois Gentilhomme* who, nouveau riche, travels in the sophisticated circles of the French aristocracy and who is eager to learn. On one occasion his new friends speak about poetry and prose, and Jourdain discovers to his amazement and great delight that whenever he speaks, he speaks prose. He is overwhelmed by this discovery: "I am speaking Prose! I have always spoken Prose! I have spoken Prose throughout my whole life!"

A similar discovery has been made not so long ago, but it was neither of poetry nor of prose—it was the environment that was discovered. I remember when, perhaps ten or fifteen years ago, some of my American friends came running to me with the delight and amazement of having just made a great discovery: "I am living in an Environment! I have always lived in an Environment! I have lived in an Environment throughout my whole life!"

However, neither M. Jourdain nor my friends have as yet made another discovery, and that is when M. Jourdain speaks, may it be prose or poetry, it is he who invents it, and, likewise, when we perceive our environment, it is we who invent it.

Every discovery has a painful and a joyful side: painful, while struggling with a new insight; joyful, when this insight is gained. I see the sole purpose of my presentation to minimize the pain and maximize the joy for those who have not yet made this discovery; and for those who have made it, to

---

* This article is an adaptation of an address given on April 17, 1973, to the Fourth International Environmental Design Research Association Conference at the College of Architecture, Virginia Polytechnic Institute, Blacksburg, Virginia. Originally published in *Environmental Design Research*, Vol. 2, F.E. Preiser (ed.), Dowden, Hutchinson & Ross, Stroudberg, pp. 35–46 (1973).

let them know they are not alone. Again, the discovery we all have to make for ourselves is the following postulate.

The Environment as We Perceive It Is Our Invention

The burden is now upon me to support this outrageous claim. I shall proceed by first inviting you to participate in an experiment; then I shall report a clinical case and the results of two other experiments. After this I will give an interpretation, and thereafter a highly compressed version of the neurophysiological basis of these experiments and my postulate of before. Finally, I shall attempt to suggest the significance of all that to aesthetical and ethical considerations.

Experiments

The Blind Spot

Hold book with right hand, close left eye, and fixate star of Figure 1 with right eye. Move book slowly back and forth along line of vision until at an appropriate distance (from about 12 to 14 inches) round black spot disappears. With star well focused, spot should remain invisible even if book is slowly moved parallel to itself in any direction.

This localized blindness is a direct consequence of the absence of photo receptors (rods or cones) at that point of the retina, the "disk," where all fibers leading from the eye's light-sensitive surface converge to form the optic nerve. Clearly, when the black spot is projected onto the disk, it cannot be seen. Note that this localized blindness is not perceived as a dark blotch in our visual field (seeing a dark blotch would imply "seeing"), but this blindness is not perceived at all, that is, neither as something present, nor as something absent: Whatever is perceived is perceived "blotchless."

Scotoma

Well-localized occipital lesions in the brain (e.g., injuries from high-velocity projectiles) heal relatively fast without the patient's awareness of any perceptible loss in his vision. However, after several weeks motor dysfunction in the patient becomes apparent, for example, loss of control of arm or leg movements of one side or the other. Clinical tests, however, show that there is nothing wrong with the motor system, but that in some cases

FIGURE 1.

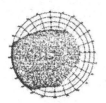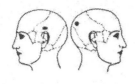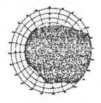

FIGURE 2.

there is substantial loss (Fig. 2) of a large portion of the visual field (*scotoma*).[9] A successful therapy consists of blind-folding the patient over a period of one to two months until he regains control over his motor system by shifting his "attention" from (nonexistent) visual clues regarding his posture to (fully operative) channels that give direct postural clues from (proprioceptive) sensors embedded in muscles and joints. Note again absence of perception of "absence of perception," and also the emergence of perception through sensorimotor interaction. This prompts two metaphors: Perceiving is doing, and If I don't see I am blind, I am blind; but if I see I am blind, I see.

Alternates

A single word is spoken once into a tape recorder and the tape smoothly spliced (without click) into a loop. The word is repetitively played back with high rather than low volume. After one or two minutes of listening (from 50 to 150 repetitions), the word clearly perceived so far abruptly changes into another meaningful and clearly perceived word: an "alternate." After ten to thirty repetitions of this first alternate, a sudden switch to a second alternate is perceived, and so on.[6] The following is a small selection of the 758 alternates reported from a population of about 200 subjects who were exposed to a repetitive playback of the single word *cogitate: agitate, annotate, arbitrate, artistry, back and forth, brevity, ça d'était, candidate, can't you see, can't you stay, Cape Cod you say, card estate, cardiotape, car district, catch a tape, cavitate, cha cha che, cogitate, computate; conjugate, conscious state, counter tape, count to ten, count to three, count yer tape, cut the steak, entity, fantasy, God to take, God you say, got a data, got your pay, got your tape, gratitude, gravity, guard the tit, gurgitate, had to take, kinds of tape, majesty, marmalade.*

Comprehension*

Into the various stations of the auditory pathways in a cat's brain microelectrodes are implanted that allow a recording (electroencephalogram)

* Literally, *con* = together; *prehendere* = to seize, grasp.

Figures

3: Session 3, Trial 1   4: Session 3, Trial 13

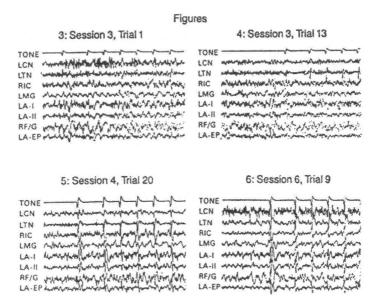

5: Session 4, Trial 20   6: Session 6, Trial 9

FIGURES 3–6

from the nerve cells first to receive auditory stimuli (cochlea nucleus, CN) up to the auditory cortex.[10] The cat so prepared is admitted into a cage that contains a food box whose lid can be opened by pressing a lever. However, the lever–lid connection is operative only when a short single tone (here C_6, which is about 1000 hertz) is repetitively presented. The cat has to learn that C_6 "means" food. Figures 3–6 show the pattern of nervous activity at eight ascending auditory stations and at four consecutive stages of this learning process.[10] The cat's behavior associated with the recorded neural activity is for "random search" in Figure 3, "inspection of lever" in Figure 4, "lever pressed at once" in Figure 5, and "walking straight toward lever (full comprehension)" in Figure 6. Note that no tone is perceived as long as this tone is uninterpretable (Figs. 3,4; pure noise), but the whole system swings into action with the appearance of the first "beep" (Figs. 5,6; noise becomes signal), when sensation becomes comprehensible, when *our* perception of "beep, beep; beep" is in the *cat's* perception "food, food, food."

Interpretation

In these experiments I have cited instances in which we see or hear what is not "there," or in which we do not see or hear what is "there" unless coordination of sensation and movement allows us to "grasp" what appears to be there. Let me strengthen this observation by citing now the "principle of undifferentiated encoding":

The response of a nerve cell does *not* encode the physical nature of the agents that caused its response. Encoded is only "how much" at this point on my body, but not "what."

Take, for instance, a light-sensitive receptor cell in the retina, a "rod" that absorbs the electromagnetic radiation originating from a distant source. This absorption causes a change in the electrochemical potential in the rod, which will ultimately give rise to a periodic electric discharge of some cells higher up in the postretinal networks (see below, Fig. 15), with a period that is commensurate with the intensity of the radiation absorbed, but without a clue that it was electromagnetic radiation that caused the rod to discharge. The same is true for any other sensory receptor, may it be the taste buds, the touch receptors, and all the other receptors that are associated with the sensations of smell, heat and cold, sound, and so on: They are all "blind" as to the quality of their stimulation, responsive only as to their quantity.

Although surprising, this should not come as a surprise, for indeed "out there" there is no light and no color, there are only electromagnetic waves; "out there" there is no sound and no music, there are only periodic variations of the air pressure; "out there" there is no heat and no cold, there are only moving molecules with more or less mean kinetic energy, and so on. Finally, for sure, "out there" there is no pain.

Since the physical nature of the stimulus—its *quality*—is not encoded into nervous activity, the fundamental question arises as to how does our brain conjure up the tremendous variety of this colorful world as we experience it any moment while awake, and sometimes in dreams while asleep. This is the "problem of cognition," the search for an understanding of the cognitive processes.

The way in which a question is asked determines the way in which an answer may be found. Thus it is upon me to paraphrase the "problem of cognition" in such a way that the conceptual tools that are today at our disposal may become fully effective. To this end let me paraphrase ($\rightarrow$) "cognition" in the following way:

$$\text{cognition} \rightarrow \text{computing a reality}$$

With this I anticipate a storm of objections. First, I appear to replace one unknown term *cognition*, with three other terms, two of which, *computing* and *reality*, are even more opaque than the definiendum, and with the only definite word used here being the indefinite article *a*. Moreover, the use of the indefinite article implies the ridiculous notion of other realities besides "the" only and one reality, our cherished Environment; and finally I seem to suggest by "computing" that everything, from my wristwatch to the galaxies; is merely computed, and is not "there." Outrageous!

Let me take up these objections one by one. First, let me remove the semantic sting that the term *computing* may cause in a group of women and men who are more inclined toward the humanities than to the sciences.

Harmlessly enough, computing (from *com-putare*) literally means to reflect, to contemplate (*putare*) things in concert (*com*), without any explicit reference to numerical quantities. Indeed, I shall use this term in this most general sense to indicate any operation (not necessarily numerical) that transforms, modifies, rearranges, orders, and so on, observed physical entities ("objects") or their representations ("symbols"). For instance, the simple permutation of the three letters A,B,C, in which the last letter now goes first—C,A,B—I shall call a computation; similarly the operation that obliterates the commas between the letters—CAB—and likewise the semantic transformation that changes CAB into *taxi*, and so on.

I shall now turn to the defense of my use of the indefinite article in the noun phrase *a reality*. I could, of course, shield myself behind the logical argument that solving for the general case, implied by the *a*, I would also have solved any specific case denoted by the use of *the*. However, my motivation lies much deeper. In fact, there is a deep hiatus that separates the *the* school of thought from the *a* school of thought in which, respectively, the distinct concepts of "confirmation" and "correlation" are taken as explanatory paradigms for perceptions. The *the* school: My sensation of touch is *confirmation* for my visual sensation that here is a table. The *a* school: My sensation of touch in *correlation* with my visual sensation generate an experience that I may describe by "here is a table."

I am rejecting the *the* position on epistemological grounds, for in this way the whole problem of cognition is safely put away in one's own cognitive blind spot: Even its absence can no longer be seen.

Finally one may rightly argue that cognitive processes do not compute wristwatches or galaxies, but compute at best *descriptions* of such entities. Thus I am yielding to this objection and replace my former paraphrase by

cognition → computing descriptions of a reality

Neurophysiologists, however, will tell us[4] that a description computed on one level of neural activity, say, a projected image on the retina, will be operated on again on higher levels, and so on, whereby some motor activity may be taken by an observer as a "terminal description," for instance, the utterance, "Here is a table." Consequently, I have to modify this paraphrase again to read

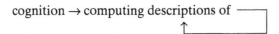

where the arrow turning back suggests this infinite recursion of descriptions of descriptions, etc. This formulation has the advantage that one unknown, namely, "reality," is successfully eliminated. Reality appears only implicit as the operation of recursive descriptions. Moreover, we may take advantage of the notion that computing descriptions is nothing else but computations.

Hence

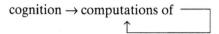

$$\text{cognition} \rightarrow \text{computations of} \underset{\uparrow\rule{0pt}{0pt}}{\boxed{}}$$

In summary, I propose to interpret cognitive processes as never-ending recursive processes of computation, and I hope that in the following *tour de force* of neurophysiology I can make this interpretation transparent.

Neurophysiology

Evolution

In order that the principle of recursive computation be fully appreciated as being the underlying principle of all cognitive processes—even of life itself, as one of the most advanced thinkers in biology assures me[5]—it may be instructive to go back for a moment to the most elementary—or as evolutionists would say, to very "early"—manifestations of this principle. These are the "independent effectors," or independent sensorimotor units, found in protozoa and metazoa distributed over the surface of these animals (Fig. 7). The triangular portion of this unit, protruding with its tip from the surface, is the sensory part; the onion-shaped portion, the contractile motor part. A change in the chemical concentration of an agent in the immediate vicinity of the sensing tip, and "perceptible" by it, causes an instantaneous contraction of this unit. The resulting displacement of this or any other unit by change of shape of the animal or its location may, in turn, produce perceptible changes in the agent's concentration in the vicinity of these units, which, in turn, will cause their instantaneous contraction, and so on. Thus we have the recursion

$$\boxed{\rightarrow \text{change of sensation} \rightarrow \text{change of shape} }$$

Separation of the sites of sensation and action appears to have been the next evolutionary step (Fig. 8). The sensory and motor organs are now connected by thin filaments, the "axons" (in essence degenerated muscle fibers having lost their contractility), which transmit the sensor's perturbations to its effector, thus giving rise to the concept of a "signal": See something here, act accordingly there.

The crucial step, however, in the evolution of the complex organization of the mammalian central nervous system (CNS) appears to be the appear-

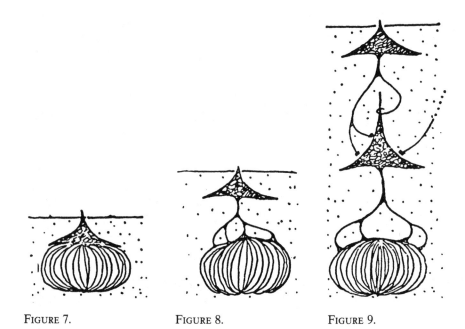

FIGURE 7.   FIGURE 8.   FIGURE 9.

ance of an "internuncial neuron," a cell sandwiched between the sensory and the motor unit (Fig. 9). It is, in essence, a sensory cell, but specialized so as to respond only to a universal "agent," namely, the electrical activity of the afferent axons terminating in its vicinity. Since its present activity may affect its subsequent responsivity, it introduces the element of computation in the animal kingdom and gives these organisms the astounding latitude of nontrivial behaviors. Having once developed the genetic code for assembling an internuncial neuron, to add the genetic command *repeat* is a small burden indeed. Hence, I believe, it is now easy to comprehend the rapid proliferation of these neurons along additional vertical layers with growing horizontal connections to form those complex interconnected structures we call "brains."

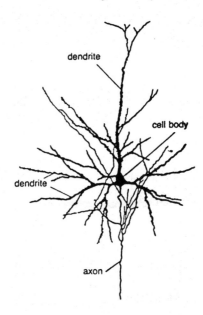

FIGURE 10.

The Neuron

The neuron, of which we have more than 10 billion in our brain, is a highly specialized single cell with three anatomically distinct features (Fig. 10): (1) the branch-like ramifications stretching up and to the side, the "dendrites"; (2) the bulb in the center housing the cell's nucleus, the "cell body"; and (3), the "axon," the smooth fiber stretching downward. Its various bifurcations terminate on dendrites of another (but sometimes—recursively—on the same) neuron. The same membrane that envelops the cell body forms also the tubular sheath for dendrites and axon, and causes the inside of the cell to be electrically charged against the outside with about $^1/_{10}$ of a volt. If in the dendritic region this charge is sufficiently perturbed, the neuron "fires" and sends this perturbation along its axon to its termination, the synapses.

Transmission

Since these perturbations are electrical, they can be picked up by "microprobes," amplified and recorded. Figure 11 shows three examples of periodic discharges from a touch receptor under continuous stimulation, the low frequency corresponding to a weak stimulus, the high frequency to a strong stimulus. The magnitude of the discharge is clearly everywhere the same, the pulse frequency representing the stimulus intensity, but the intensity only.

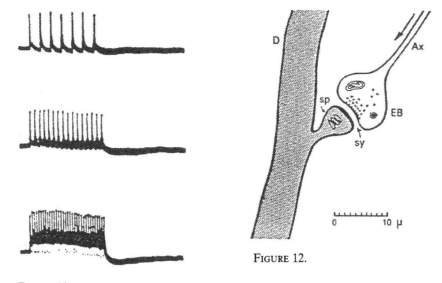

FIGURE 11.

FIGURE 12.

Synapse

Figure 12 sketches a synaptic junction. The afferent axon (Ax), along which the pulses travel, terminates in an end bulb (EB), which is separated from the spine (sp) of a dendrite (D) of the target neuron by a minute gap (sy), the "synaptic gap." (Note the many spines that cause the rugged appearance of the dendrites in Fig. 10). The chemical composition of the "transmitter substances" filling the synaptic gap is crucial in determining the effect an arriving pulse may have on the ultimate response of the neuron: Under certain circumstances it may produce an "inhibitory effect" (cancellation of

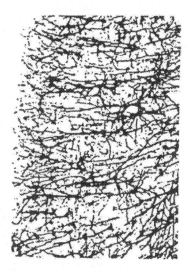

FIGURE 13.

another simultaneously arriving pulse), in others a "facilitory effect" (augmenting another pulse to fire the neuron). Consequently, the synaptic gap can be seen as the "microenvironment" of a sensitive tip, the spine, and with this interpretation in mind we may compare the sensitivity of the CNS to changes of the *internal* environment (the sum total of all microenvironments) to those of the *external* environment (all sensory receptors). Since there are only 100 million sensory receptors, and about 10,000 billion synapses in our nervous system, we are 100 thousand times more receptive to changes in our internal than in our external environment.

The Cortex

In order that one may get at least some perspective on the organization of the entire machinery that computes all perceptual, intellectual, and emotional experiences, I have attached Figure 13,[7] which shows a magnified section of about 2 square millimeters of a cat's cortex by a staining method that stains only cell body and dendrites, and of those only 1% of all neurons present. Although you have to imagine the many connections among these neurons provided by the (invisible) axons, and a density of packing that is 100 times that shown, the computational power of even this very small part of a brain may be sensed.

Descartes

This perspective is a far cry from that held, say, 300 years ago:[2]

If the fire A is near the foot B [Fig. 14], the particles of this fire, which as you know move with great rapidity, have the power to move the area of the skin of this foot

FIGURE 14.

that they touch; and in this way drawing the little thread, c, that you see to be attached at base of toes and on the nerve, at the same instant they open the entrance of the pore, d,e, at which this little thread terminates, just as by pulling one end of a cord, at the same time one causes the bell to sound that hangs at the other end. Now the entrance of the pore or little conduit, d,e, being thus opened, the animal spirits of the cavity F, enter within and are carried by it, partly into the muscles that serve to withdraw this foot from the fire, partly into those that serve to turn the eyes and the head to look at it, and partly into those that serve to advance the hands and to bend the whole body to protect it.

Note, however, that some behaviorists of today still cling to the same view,[8] with one difference only, namely, that in the meantime Descartes' "animal spirit" has gone into oblivion.

Computation

The retina of vertebrates, with its associated nervous tissue, is a typical case of neural computation. Figure 15 is a schematic representation of a mammalian retina and its postretinal network. The layer labeled 1 represents the array of rods and cones, and layer 2 the bodies and nuclei of these cells. Layer 3 identifies the general region where the axons of the receptors synapse with the dendritic ramifications of the "bipolar cells" (4) which, in turn, synapse in layer 5 with the dendrites of the ganglion cells" (6), whose

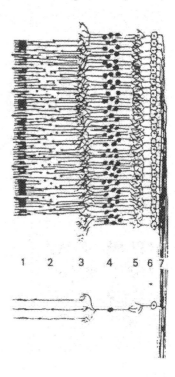

FIGURE 15.

activity is transmitted to deeper regions of the brain via their axons, which are bundled together to form the optic nerve (7). Computation takes place within the two layers labeled 3 and 5, that is, where the synapses are located. As Maturana has shown[3] it is there where the sensation of color and some clues as to form are computed.

Form computation: Take the two-layered periodic network of Figure 16, the upper layer representing receptor cells sensitive to, say, "light." Each of these receptors is connected to three neurons in the lower (computing) layer, with two excitatory synapses on the neuron directly below (symbolized by buttons attached to the body) and with one inhibitory synapse (symbolized by a loop around the tip) attached to each of the two neurons, one to the left and one to the right. It is clear that the computing layer will not respond to uniform light projected on the receptive layer, for the two excitatory stimuli on a computer neuron will be exactly compensated by the inhibitory signals coming from the two lateral receptors. This zero response will prevail under strongest and weakest stimulations as well as for slow or rapid changes of the illumination. The legitimate question may now arise: "Why this complex apparatus that doesn't do a thing?"

Consider now Figure 17, in which an obstruction is placed in the light path illuminating the layer of receptors. Again all neurons of the lower layer will remain silent, except the one at the edge of the obstruction, for it

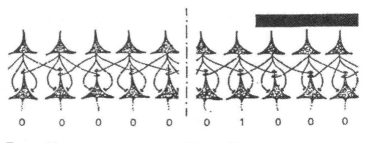

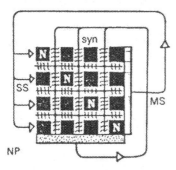

FIGURE 16.   FIGURE 17.

FIGURE 18.

receives two excitatory signals from the receptor above, but only one inhibitory signal from the sensor to the left. We now understand the important function of this net, for it computes any spatial *variation* in the visual field of this "eye," independent of the intensity of the ambient light and its temporal variations, and independent of place and extension of the obstruction.

Although all operations involved in this computation are elementary, the organization of these operations allows us to appreciate a principle of considerable depth, namely, that of the computation of abstracts, here the notion of "edge."

I hope that this simple example is sufficient to suggest to you the possibility of generalizing this principle in the sense that "computation" can be seen on at least two levels, namely, (1) the operations actually performed and (2) the organization of these operations represented here by the structure of the nerve net. In computer language (1) would again be associated with "operations," but (2) with the "program." As we shall see later, in "biological computers" the programs themselves may be computed on. This leads to the concepts of "metaprograms," "meta-metaprograms," and so on. This, of course, is the consequence of the inherent recursive organization of those systems.

FIGURE 19.

Closure

By attending to all the neurophysiological pieces, we may have lost the per-spective that sees an organism as a functioning whole. In Figure 18 I have put these pieces together in their functional context. The black squares labeled *N* represent bundles of neurons that synapse with neurons of other bundles over the (synaptic) gaps indicated by the spaces between squares. The sensory surface (SS) of the organism is to the left, its motor surface (MS) to the right, and the neuropituitary (NP), the strongly innervated master gland that regulates the entire endocrinal system, is the stippled lower boundary of the array of squares. Nerve impulses traveling horizon-tally (from left to right) ultimately act on the motor surface (MS) whose changes (movements) are immediately sensed by the sensory surface (SS), as suggested by the "external" pathway following the arrows. Impulses traveling vertically (from top to bottom) stimulate the neuropituitary (NP), whose activity release steroids into the synaptic gasp, as suggested by the wiggly terminations of the lines following the arrow, and thus modify the *modus operandi* of all synaptic junctures, hence the *modus operandi* of the system as a whole. Note the double closure of the system that now recursively operates not only on what it "sees," but on its operators as well. In order to make this twofold closure even more apparent I propose to wrap the diagram of Figure 18 around its two axes of circular symmetry until the artificial boundaries disappear and the torus (doughnut) in Figure 19 is obtained. Here the "synaptic gap" between the motor and sensory surfaces is the striated meridian in the front center, the neuropituitary the stippled equator. This, I submit, is the functional organization of a living organism in a (dough) nut shell.

The computations within this torus are subject to a nontrivial constraint, and this is expressed in the postulate of cognitive homeostais:

The nervous system is organized (or organizes itself) so that it computes a stable reality.

This postulate stipulates "autonomy," that is, "self-regulation," for every living organism. Since the semantic structure of nouns with the prefix *self-* becomes more transparent when this prefix is replaced by the noun, *autonomy* becomes synonymous with *regulation of regulation*. This is precisely what the doubly closed, recursively computing torus does: It regulates its own regulation.

Significance

It may be strange in times like these to stipulate autonomy, for autonomy implies responsibility: If I am the only one who decides how I act, then I am responsible for my action. Since the rule of the most popular game played today is to make someone else responsible for *my* acts—the name of the game is "heteronomy"—my arguments make, I understand, a most unpopular claim. One way of sweeping it under the rug is to dismiss it as just another attempt to rescue "solipsism," the view that this world is only in my imagination and the only reality is the imagining "I." Indeed, that was precisely what I was saying before, but I was talking only about a single organism. The situation is quite different when there are two, as I shall demonstrate with the aid of the gentleman with the bowler hat (Fig. 20).

He insists that he is the sole reality, while everything else appears only in his imagination. However, he cannot deny that his imaginary universe is populated with apparitions that are not unlike himself. Hence he has to

FIGURE 20.

concede that they themselves may insists that they are the sole reality and everything else is only a concoction of their imagination. In that case their imaginary universe will be populated with apparitions, one of which may be *he*, the gentleman with the bowler hat.

According to the principle of relativity, which rejects a hypothesis when it does not hold for two instances together, although it holds for each instance separately (Earthlings and Venusians may be consistent in claiming to be in the center of the universe, but their claims fall to pieces if they should ever get together), the solipsistic claim falls to pieces when besides me I invent another autonomous organism. However, it should be noted that since the principle of relativity is not a logical necessity—nor is it a proposition that can be proven to be either true or false—the crucial point to be recognized here is that I am free to choose either to adopt this principle or to reject. If I reject it, I am the center of the universe, my reality is my dreams and my nightmares, my language is monologue, and my logic monologic. If I adopt it, neither I nor the other can be the center of the universe. As in the heliocentric system, there must be a third that is the central reference. It is the relation between Thou and I, and this relation is *identity*:

$$\text{reality} = \text{community}$$

What are the consequences of all this in ethics and aesthetics?

The ethical imperative: Act always so as to increase the number of choices.

The aesthetical imperative: If you desire to see, learn how to act.

References

1. Brown, G. S. *Laws of Form*. Julian Press, New York, 1972, p. 3.
2. Descartes, R. *L'Homme*. Angot, Paris, 1664. Reprinted in *Oeuvres de Descartes*, Vol. 11. Adam and Tannery, Paris, 1957, pp. 119–209.
3. Maturana, H. R. A biological theory of relativistic colour coding in the primate retina. *Archivos de Biologia y Medicina Experimentales, Suplemento* 1, 1968.
4. Maturana, H. R. Neurophysiology of cognition. In *Cognition: A Multiple View* (P. Garvin, ed.). Spartan Press, New York, 1970, pp. 3–23.
5. Maturana, H. R. *Biology of Cognition*. University of Illinois, Urbana, Illinois, 1970.
6. Naeser, M. A., and Lilly, J. C. The repeating word effect: Phonetic analysis of reported alternatives. *Journal of Speech and Hearing Research*, 1971.
7. Sholl, D. A. *The Organization of the Cerebral Cortex*. Methuen, London, 1956.
8. Skinner, B. F. *Beyond Freedom and Dignity*. A. Knopf, New York, 1971.
9. Teuber, H. L. Neuere Betrachtungen über Sehstrahlung und Sehrinde. In *Das Visuelle System* (R. Jung and H. Kornhuber, eds.). Springer, Berlin, 1961, pp. 256–274.
10. Worden, F. G. EEG studies and conditional reflexes in man. In *The Central Nervous System and Behavior* (Mary A. B. Brazier, ed.). Josiah Macy, Jr., Foundation, New York, 1959, pp. 270–291.

9
Cybernetics of Epistemology*

HEINZ VON FOERSTER
Translated by Peter Werres

Summary

If "epistemology" is taken to be the theory of knowledge acquisition, rather than of knowledge per se, then—it is argued—the appropriate conceptual framework for such an epistemology is that of cybernetics, the only discipline that has given us a rigorous treatment of circular causality. The processes by which knowledge is acquired, i.e., the cognitive processes, are interpreted as computational algorithms which, in turn, are being computed. This leads to the contemplation of computations that compute computations, and so one, that is, of recursive computations with a regress of arbitrary depth.

From this point of view the activity of the nervous system, some experiments, the foundations of a future theory of behavior and its ethical consequences, are discussed.

Introduction

When I agreed to deliver my talk here in German, I had no idea what difficulties this would cause me. Over the last twenty years I have only thought and talked in English when it comes to scholarly matters. Many terms and research results were baptized in English after their conception and resist any translation efforts.

After vain attempts to turn my lecture into German, I have finally decided to step out of myself to look at the whole quilt of my thoughts as if they were a piece of tapestry and to describe to you, to the best of my ability, the figures, ornaments and symbols woven into its fabric. If in the process I appear to get lost in labyrinthine syntactical constructions, then this will not constitute

* Lecture given at Cybernetics and Bionics, the Fifth Congress of the German Society of Cybernetics (DSK) on March 28, 1973 in Nuremberg, Germany. Published in *Kybernetic und Bionik*, W.D. Keidel, W. Handler & M. Spring (eds.), Oldenburg; Munich, pp. 27–46 (1974). Reprinted with permission.

some form of affectedness, but the moaning of a rusty piece of machinery. Perhaps you already stumbled when reading the title of my lecture, "Cybernetics of Epistemology." Inwardly you were convinced that what I really had meant was "An Epistemology of Cybernetics." originally, this had actually been the case. But over the course of some thinking, it became clear to me that not only an epistemology of cybernetics, but any epistemology claiming completeness will be some form of cybernetic theory.

The essential contribution of cybernetics to epistemology is the ability to change an open system into a closed system, especially as regards the closing of a linear, open, infinite causal nexus into closed, finite, circular causality.

Here, perhaps, a historic footnote may be in order. Several years before Norbert Wiener called our area of study "Cybernetics," there were yearly symposia in New York where a group of scholars from different fields (who were close to Wiener) congregated to talk about common problems. The subject was "Circular Causal and Feedback Mechanisms in Biological and Social Systems".[1-5]

First of all, the idea of closed circular causality has the pleasant characteristic that the cause for an effect in the present can be found in the past if one cuts the circle at one spot, and that the cause lies in the future if one does the cutting at the diametrically opposed spot. Closed circular causality, thus, bridges the gap between effective and final cause, between motive and purpose.

Secondly, by closing the causal chain one also appears to have gained the advantage of having gotten rid of a degree of uncertainty: no longer does one have to concern oneself with the starting conditions—as they are automatically supplied by the end conditions. To be sure, this is the case, but the matter is anything but simple: only certain values of those conditions provide a solution for the processes within the circle; the problem has become an "Eigen-value" problem.

What also causes complication is that now the suspicion will be raised that the whole matter of circular causality might be mere logical mischief. We already know this from the theory of logical inference—the infamous vicious cycle: cause becomes effect and effect becomes cause.

It is my intent not only to liberate the "*circulus vitiosus*" from its bad reputation,[6] but to raise it to the honorable position of a "*circulus creativus*", a creative cycle.

I want to start with two preliminary propositions. For the first one I will use the two expressions "sensorium" and "motorium." With sensorium I mean the system of conscious sensations, and with motorium I mean controlled sequences of motion.

My first proposition:

The meaning of the signals of the sensorium are determined by the motorium; and the meaning of the signals of the motorium are determined by the sensorium.

That means that information—not in the sense of information theory, but in its everyday meaning—has its origin in this creative circle. Only that has meaning which I can "grasp."

My second preliminary proposition deals with the problem of a complete and closed theory of the brain. If any one of us mortals ever deals with this problem he will, without any doubt, use his brain. This observation is at the basis of my second preliminary proposition.

My second preliminary proposition:

The laws of physics, the so-called "laws of nature," can be described by us. The laws of brain functions—or ever more generally—the laws of biology, must be written in such a way that the writing of these laws can be deducted from them, i.e., they have to write themselves.

Let me now return to my theme, namely, that an epistemology is, for all practical purposes, a cybernetics. This will become clear at once if "episte-mology" is understood as a "theory of knowledge acquisition," rather than a theory of knowledge.

While in German the creative process of becoming knowledgeable is indicated by the "*generativ*" or "*creativ*" prefix "*er...*", to augment static knowledge (Kenntnis) by new knowledge Kenntnis becomes Er-Kenntnis. The English version of this process is borrowing from Greek, the root "*g n*" which indicates emergence (genesis) as well as perception (cognition) to be a creative processes.

I would like now to replace the phrase "knowledge acquisition" with the term "cognition," and suggest for this process an operational definition which preserves its semantic essence and, moreover, enables us to make use of contemporary conceptual tools:

"Computing" in this context is not at all restricted to the numerical domain, but is taken in its general sense as "contemplating (*putare*) things together (*com...*)". Some skeptics most likely will begin raising their eyebrows. "Why 'a' reality—why not 'the' reality?" one may ask. After all, we are here—the Cybernetics Symposium, the Meistersingerhalle in Nuremberg, the physical universe—how could there be any other reality?

Indeed, a deep epistemological abyss separates the two views, which are distinguished by using, in one case, the definite article "the", and in the other, the indefinite article "a". The distinction here arises from two fundamentally different positions regarding "reality." When we hold that different independent observations are *confirmations*, we talk about *the* reality. However, we can take the position that only through *correlating* different independent observations, realities are emerging.

The first case: my visual sense tells me, here stands a lectern; my sense of touch confirms that. Also, our chairman, Dr. Kupfmuller, would say so, if I were to ask him.

The second case: my visual sense tells me there stands something; my sense of touch tells me there stands something, and I hear Dr. Kupfmuller say, "There stands a lectern." The correlation between these three independent observations permits me to say: "There stands a lectern."

In the first case, we assume that with each new independent observation we confirm the correct perception of the previous observation of the reality. In the second case we use Ockham's Razor to shave off unnecessary assumptions and allow a reality to emerge through the correlation of our sensations.

In any case, I intend to use the indefinite article, as it represents the general case: clearly *the* reality is a special case of *a* reality.

We have barely laid the indefinite article to rest when we are already confronted with a new issue. What is "computing" supposed to mean? I could by no means claim in all seriousness that the lectern, my wrist watch, or the Andromeda Nebula is being computed by me. At the most, one could say that a "description of reality" is computed, because with my verbal references ("lectern", "wristwatch", "Andromeda"), I have just demonstrated that certain sequences of motion of my body combined with certain hissing and grunting sounds, permitted listeners to interpret these as a description.

Neurophysiologists may in this context object that incoming signals have to undergo many steps of modification before we experience a verbal message. First the retina provides a two-dimensional projection of the exterior world which one may call a "description of the first order." The next post-retinal networks then offer to the ganglia cells a modified description of this description; thus a "description of the second order." And so it goes on via the various stations of computing all the way to descriptions of higher and highest orders. We can thus modify the second version of my proposition in the following way:

$$Cognition \rightarrow computing\ of\ a\ description$$

This paraphrasing has two advantages: the contention "the reality" or "a reality" has disappeared, as now there is no more talk about reality. Secondly, we can use the insight that computing a description is nothing but computing. This way we reach a final paraphrasing of the forever renewed process of knowledge acquisition, i.e.:

$$Cognition \rightarrow computations\ of$$

I, therefore, interpret knowledge, or the process of knowledge acquisition, as recursive computations.

At this point I could follow two different paths: I could now talk about the characteristics of recursive functions and of the characteristics of machinery to compute these functions. In machine language, I would then talk about cascades of compiler languages and about the theory of metaprograms. In this case, naturally, the concept of the Touring Machine would be the ideal conceptual tool. It becomes, for instance, very clear that the structure of the quadruples of a Touring Machine computing the code number of another machine or its own cannot be confused with the structure of the computed tape descriptions.[7] Or neurophysiologically expressed: in order say "lectern" or to know that a lectern is located here, I do not have to have the letters l-e-c-t-e-r-n inscribed into my brain, nor does a tiny representation of a lectern have to be located somewhere internally of me. Rather, I need a structure which computes for me the different manifestations of a description. But all of this belongs in the session on "Artificial Intelligence" and here we are in a session on "Biocybernetics of the Central Nervous System." I thus consider it more fitting to talk about the effects of my earlier propositions as regards activities in the central nervous system.

Let us first note the immensity of the problem with which we are confronted. For this purpose let us recall the law of undifferentiated encoding.

The principle of undifferentiated encoding:

The states of a nerve cell do not encode the nature of the cause of its activity. (Encoded is only "this much at this part of my body." but not "what.")

One example: one cell in the retina in a certain moment absorbs a stream of photons of x amount per second. In the process it creates an electro-chemical potential which is a function of the magnitude of the stream of photons, which means that the "how much" is being encoded: but the signals which are the cause of this potential neither provide any indication that photons were the cause of activity, nor do they tell us of which frequencies the stream of photons consisted. Exactly the same is true for all other cells of sensory perception. Take, for instance, the hair cells in the cochlea of Meisner's touch bodies, or the papillae of taste, or whatever cells you may choose. In none of them is the quality of the cause of activity encoded, only the quantity. An indeed, "out there" there is no light and there are no colors—there are electromagnetic waves; "out there" there are no sounds and no music—there are longitudinal periodic pressure waves; "out there" there is no heat or cold—there is a higher or lower median molecular kinetic energy, and so, and so forth, and quite certainly there is no pain "out there."

Then the fundamental question is: how do we experience the world in its overwhelming multiplicity when as incoming data we only have: first, the intensity of stimuli, and second, the bodily coordinates of the source of stimuli, i.e. stimulation at a certain point of my body?

Given that the qualities of sensory impression are not encoded in the receptive apparatus, it is clear that the central nervous system is organized in such a way that it computes these qualities from this meager input.

We (at least I) know very little about these operations. I am, therefore, already looking forward to several lectures that will deal with this problem, especially those of Donald McKay, Horst Mittelstaedt, and Hans Lukas Teuber, who will discuss their famous reaference-principle. I myself will have to limit myself to a few hints regarding the nature of these operations.

Let me first use a pedagogical artifice in order to formulate the problem of the computing of perceptual richness in such a way that in this new perspective the issue is seen from the viewpoint of the perceiving organism and not—as usual—from the viewpoint of an observer who, seduced by his own perception, always thinks he knows how "out there" looks, and who then tries to figure out with micropipets in the nervous system of an organism how "that of the outside" will look "in the inside." Epistemologically viewed, this constitutes a case of cheating, as the observer, so to speak, peeks sideways for "answers" (his own world view) which he then will compare will some cellular states of activity, from which alone our organism will have to piece together its view of the world. However, how it does this, that is the problem.[8]

It is clear that every organism has a finite volume which is bounded by a closed surface, a volume which is thread by an intricate system of "tubes." the latter surfacing at several places, let's say at "s" number of places. Ontogenetically speaking, the surface is defined by the ectoderm and contains all sensitive endorgans, while the interior is defined by the endoderm. Topologically viewed, such a surface constitutes an orientable two-dimensional manifold of genus $p = (s + t)/2$, with "t" indicating the number of "T-connections" within the system of tubes. According to a well-known law of topology, every closed and orientable surface of finite order is metrizable, i.e. we can superimpose on the surface of each organism a geodetic coordinate system, which in the immediate vicinity of each point is Euclidian. I will call this system of coordinates "proprietary," and I shall represent the two coordinates ξ_1, ξ_2, which define a given point at the surface of this Representative Unit Sphere with a single symbol ξ. This way I assure that each sensory cell is given unambiguously a pair of coordinates that uniquely apply to this organism.

According to an equally well-known law of topology, every orientable surface of "p" is identical with the surface of a sphere of the same, which means that we can show the sensitive surface of any organism on a sphere—the "Representative Unit Sphere" [Figure 1]—so that each sensory cell of the organism corresponds to a point on the Unit Sphere and vice versa. It can easily be shown that the same considerations can also be used for the interior.

It is clear that the Representative Unit Share remains invariant to all deformations and movements of the organism. In other words, the proprietary coordinates are principal invariants.

However, seen from an observer's point of view, the organism is embedded in an Euclidian coordinate system (the "participatory" coordinate system $\kappa_1, \kappa_2, \kappa_3, \ldots$ or κ, for short, as above, and he may well interpret the relation $\xi = F(\kappa)$ as the "Formfunction", F, or by also considering temporal (t) variations $\xi = B(\kappa,t)$ as the Behavior Function B.

[Note: For non-deformable endo- or exo-skeletal) organisms is, of course, $F(x) = B(x,t) = 0$. Moreover, because of (approximate) incompressibility we have:

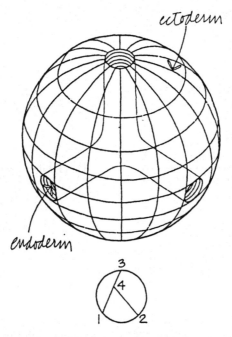

FIGURE 1. Closed orientable surface of genus $p = 2$, $s = 3$, $t = 1$, $[s + t]/2 = 4/2 = 2$.

This situation for a "fishlike" creature is sketched in Fig. 2a. The proprietary coordinates, ξ, of the fish are here superimposed over the participatory coordinates, κ, of the observer.

The pedagogical artifice of which I spoke before consists of transforming the system of coordinates in such a way that the proprietary coordinates become Euclidian everywhere, which means that the world of the fish has to be mapped onto the Representative Unit Sphere (Fig. 2b; ξ are polar coordinates.) With this, however, the world "out there" is no longer Euclidian, as can be seen from the divergence and convergence of the vertical x coordinates.

A motion of the fish as seen by the observer (Figure 3a) means for the fish that he has changed the metric of his environment by tensing his muscles (see Fig. 3b); he himself, naturally, has remained the same (invariance of self-reference) is expressed by the Representative Unit Sphere.

Please note the line of vision from eye to tail, which is by the observer interpreted as a straight line (Fig. 3a) which, however, has to be computed as such by the fish, who sees his tail through the curved geodetic of Fig. 3b. For this computation he only has available to himself the activation intensity along his surface. Although I do not want to show here in all rigor that these data are insufficient to compute the nature of his "environment," I nevertheless hope that it becomes clear that only by correlating the motor activity of the organism with the resulting changes in its sense organs makes it possible for it to interpret these neural activity uniquely. In somewhat

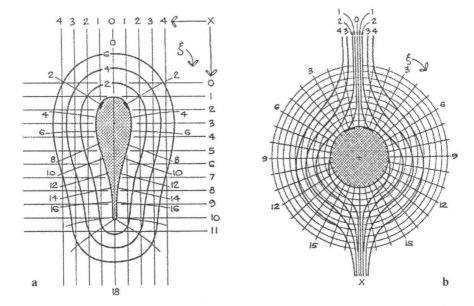

FIGURE 2. (Fig. 2a): Geodetic coordinates of the proprietary system of coordinates, ξ, of an organism at rest, imbedded in its environment with Euclidian metric. (Fig. 2b): Geodesics (circles, radii) of the proprietary coordinate system with respect to the representative body sphere embedded in an environment with non-Euclidean metric corresponding to the organism at rest (Fig. 2a).

different form, the Eigen-value problem resurfaces, which I had mentioned earlier and about which I will talk again later.

At present we are attempting to establish in experiments, once and for all, the proposition that the motorium provides the interpretation for the sensorium and that the sensorium provides the interpretation for the motorium.

There are extremely interesting experiments with infants in which one can show that gaining perceptual multitude is directly correlated to the manipulation of certain suited objects,[9,10,11] but unfortunately one cannot talk to infants, or, more correctly, one can talk to them, but one does not understand their answers. With adults there is the problem that the sensory motor system is already so well integrated that it is very hard to separate what was learned earlier and is now carried over into the experimental situation from what is assimilated as "new" during the experiment, unless one goes into a completely "new dimension," the access to which remains principally closed to our earlier experiences.

This dimension would, for instance, be the fourth spacial dimension. All of us, sooner or later in our lives, have had the bitter experience that it is extremely difficult, perhaps impossible, to squeeze oneself into the fourth

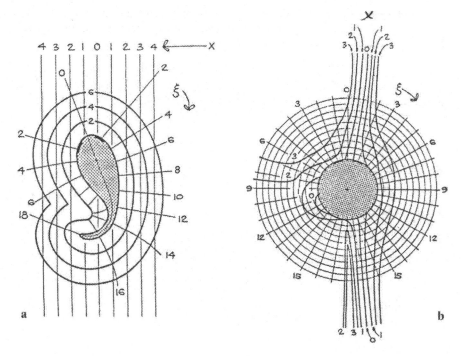

FIGURE 3. (Fig. 3a): Geodetic coordinates of the proprietary system of coordinates, ξ, of an organism in motion, imbedded in its environment with Euclidian metric. (Fig. 3b): Geodesics (circles, radii) of the proprietary coordinate system with respect to the representative body sphere embedded in an environment with non-Euclidean metric corresponding to the organism in motion (Fig. 3a).

dimension. However, every point of our three-dimensional space is an open door to the entry into the fourth dimension, but no matter how much we stretch and twist, we remain stuck in the all-too-well-known three dimensions.

At the Biological Computer Lab, we, therefore, asked ourselves: would the principle of the origins of knowledge acquisition, as mutuality of "Gnosis" and "Episteme," of Sensorium and Motorium, not gain immensely in plausibility once we are able to show that the fourth dimension becomes much more perceivable if it can also be *grasped*, which means that we can touch and handle four-dimensional objects?[12-13]

Thanks to the friendly computers (which do not know of any dimensionality) this can be carried out by giving to a test person manipulators which are on line with a fast computer. According to the position of the manipulators, the computer will compute two related two-dimensional projections of a four-dimensional body on the screen of a picture tube which—viewed by a person through a stereoscope—will be seen as three-dimensional projections of this body floating in space.

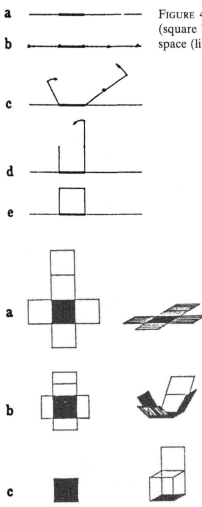

a ——————————   FIGURE 4. Step-by-step construction of a 2-D cube (square by folding three 1-D cubes from the source space (line) around their boundaries (points).

b

c

d

e

a

b

c

FIGURE 5. Step by step construction of a 3-D cube (cube) by folding five 2-D cubes (squares) out of the source space (plane) around their boundaries (edges).

The construction of (n + 1) = dimensional objects out of corresponding objects of lower dimensionality can, without difficulty, be developed with a recursive formula which tells how the (n + 1)-dimensional entity is constructed from an analog n-dimensional entity. In Figures 4–6 it is show how, for example, a four-dimensional "cube" (Tesseract) is step-by-step developed from a one-dimensional "cube'. The recursive injunction is:

(i) Add one n-dimensional cube each to the 2n boundaries of the n-dimensional "base cube", and for completion, add to one of these a "lid cube."

(ii) Fold the thus added cubes around the 2n boundaries of the base cube into the next higher dimension until they stand perpendicular, i.e., until their normal projections disappear.

FIGURE 6. Step by step construction of a 4-D cube (Tesseract) by folding seven 3-D cubes (cubes) out of the source space (space) around their boundaries (edges). Compare Fig. 6a with Salvador Dali's painting "Crucifixion (Corpus Hypercubus)" in the New York metropolitan Museum as, for instance, reproduced in.[17]

a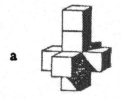

b

c

(iii) For completion, fold the lid cube around its connected boundary so that all unconnected boundaries fall into place.

In Figures 4a and 4b, injunction (i) is carried out for the construction of a two-dimensional "cube" (square) from four one-dimensional "cubes" (line segments); 4c and 4d realize injunction (ii); and the last step of constructing this structure by closing the lid is depicted in 4d and 4e.

For the construction of three- and four-dimensional cubes, the three necessary steps are sketched in Figures 5 and 6. It is clear that perpendicular projections of such structures into the spaces from where they come do not offer anything new (perpendicular projections of a 3D cube onto a plane will show only a square.) Not until these structures are tilted and obliquely projected do their higher manifolds become apparent.

This can be seen in Figure 7, where a Tesseract, with its base cube parallel to our space is obliquely projected into 3-space. (To be appreciated by a reader of this report, this projection, in turn, is projected onto the two-dimensionality of the plane of the paper.)

Figures 8a, b, and c respectively show the stereoscopic pairs of Tesseract (hypercube), the handed block from a Soma cube (a three-dimensional handed figure which can be made to change handedness by a four-space rotation), and a representation of a Klein bottle as two Mobius strips with corresponding points connected through four-space.

These objects can be appreciated in their three-dimensionality without a stereopticon when they are fixated with crossed eyes: fixate your index finger held between eye and plane of the paper at about 12 inches away. Observe the convergence of the two stereo pictures into one, then transfer your gaze from finger to paper, maintaining the crossed eyes. A 3D structure will float before your very eyes!

The way in which our experiment was conducted, either the person in charge of the experiment or the test-person can control, via two manipula-

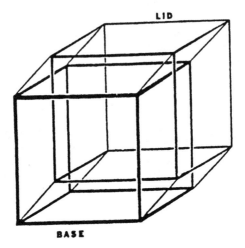

FIGURE 7. Oblique projection of a Tesseract (4-D cube) standing parallel to the space of the observer with its base cubes (base). (End cube = LID).

tors with three degrees of freedom, the projections appearing on the screen, whereby the right hand controls the three rotations in the xy, sz, and yz planes of our space, and the left hand the three in the fourth dimension, namely wx, wy, wz. Although the experiments have not yet been concluded, I can already report that the realization that these strangely changing entities are nothing but the projections of one and the same object (a "generating invariant") is usually gained by most of the naive test-persons by a loud and enthusiastic, "Oh, I see," within 20–40 minutes, if they themselves are permitted to use the manipulators. A similar exclamation will not be heard by the same category of test-persons until four to eight hours of sessions if it is the instructor who works the manipulators (who, if asked, will patiently again and again explain the geometrical situation.)

Through further developing the apparatus, we hope to be able to show the importance of motor-sensory correlation for the processes of cognition even more clearly: entry into the fourth dimension will be carried out via isometric contractions of the muscles of the neck, the arm, or the upper part of the leg, i.e. through motor-participation not showing 3D, but only 4D consequences.

As a further hint as to how the term "recursive computation" may be understood, I will quote another neurological principle:

The state of activity of a nerve cell is exclusively determined by the (electro-chemical) conditions in its immediate vicinity (micro-environment) and by its (immediately) proceeding own state of activity: there is no neurological "action at a distance."

It is obvious that we are not reacting to the table over there but to states of activity of our cells in the retina and of our proprioceptors which enable

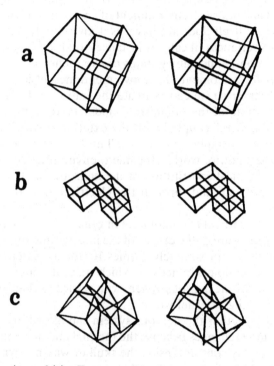

FIGURE 8. Projections of (a) a Tesseract, (b) a Soma cube, and (c) a Klein bottle, all tilted against the space of the observer.

us after certain operations in the Central Nervous System, to relate to the "tale over there." To be sure, this sounds trivial, but it is meant to make easier a transformation of the observation which, similar to physics, turns the statement "the moon is attracted by the earth" (theory of action at a distance) into the statement, "the moon moves in the gravitational field of the earth" (theory of action at the point). An immediate result of this shift in the point of view is that the usual distinction between "sensitive" and "switching" neurons—or, should you prefer, between the neurons of the peripheral and the central nervous system—will disappear, as every nerve cell is now to be considered a "sensitive" cell specifically reacting to its micro-environment. Yet, as the total sum of the micro-environments of all neurons of an organism constitutes its "entire environment," it is clear that only an outside observer has the privilege of distinguishing between an "exterior" and an "interior" environment of an organism. This is a privilege that the organism itself does not have, as it knows only one environment: that, which it experiences (it can, for instance, not differentiate between hallucinatory and non-hallucinatory states of experience.)

Let us keep, for a moment, the above-mentioned inside/outside distinction by the observer so we can estimate the impact of both environments on the nervous state of the organism. I understand every sensitive cell to be a "point of activity" coupled with the exterior environment ("exterior world"), and every synaptic spine of a cell in the CNS as a preferred point of activity of the "interior world", the micro-environment of which is determined by the chemical constitution of the neurotransmitters in the corresponding synaptic gap and through the electrical state of activity of the afferent axon.

If one takes the ratio of the amount of internal or external points of activity as a relative measure of the effects of the internal and the external world, one comes up with approximately 2 times 10^8 for the external and 2 times 10^{13} for the internal points of activity, which translates into a 100 000 times higher sensitivity of the nervous system with regard to changes of the internal than those of the external world.

The question whether the nervous system can afford this extraordinary sensitivity for inner changes because the thermal and hormonal parameters remain so incredibly constant inside the skull or whether synaptic proliferation is enhanced by this constancy, belongs, to the category of questions asking what came first, the chicken or the egg. In any case, this observation may be a hint for neuropharmcologists and psychiatrists that they are dealing with a highly sensitive system which may exhibit noticeable changes in its entire mode of operation with even minute changes in metabolism; to the information scientists it may tell that one is dealing with a computer, the program structure of which is modifiable by its activities.[14]

In Figure 9, the above-outlined overview of the organization of the nervous system is schematically reproduced. The black squares symbolize neuron bundles, which can have an effect on the next bundle via spaces— a collection of synaptic gaps. The flow of signals along the bundle runs from left to right, starting with the sensitive surface and terminating in the motor surface, the changes of which are fed back via the exterior world—the "motor-sensory synaptic gap"—to the sensory surface, thus closing the flow of signals through a circuit. A second circular flow of signals begins at the

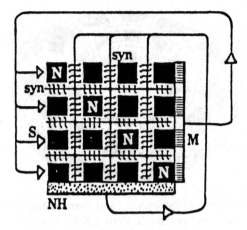

FIGURE 9. Flow of signals in the nervous system from the sensory surface (left boundary S) via bundles of nerves (black squares N) and synaptic gaps (syn) to the motor surface (right boundary M), which, in turn changes the stimulus distribution along the sensory surface; and, on the other hand, the signal flow from the neurohypophysis (lower boundary NH), whose activity modulates the composition of the steroids in the synapses and, hence, the operational modalities everywhere within the various bundles of neurons.

lower edge, schematizing the connection of the central nervous system with the neurohypophysis, which, in analogy to the motor surface, controls, via the vascular system—the "endocrine-operational synaptic gap"—the micro-environment of all synapses (shown by the minute threads in the spaces.)

If one, in this representation, interprets the length of the edge of a square with the number of points of activity in the neuron bundle belonging to it, we would have had to sketch the whole system with $10^5 \times 10^5$ squares in order to do justice to the much larger proportion of interior compared with exterior surface. A square would then have to be represented by a point with a diameter of about 1 micron if the whole diagram were to be exactly the same size as the one given here. In order to express this functional scheme geometrically, we can close circles of signals flowing in a right angle to one another by wrapping them around a vertical and a horizonal axis. A plane figure wrapped according to two right-angular axes is called a torus. Figure 10 shows a representation of this thought of the double closure of the stream of signals. The seam up front corresponds to the motor sensory synaptic gap, the horizontal seam, to the neurohypophysis.

This minimal diagram of the primal organization of an innervated being may also help see the problem which occurs if we attempt to deduce the procedures of computing a reality without the help of an observer who pretends to know both sides. In other words: If we wish to develop a consis-

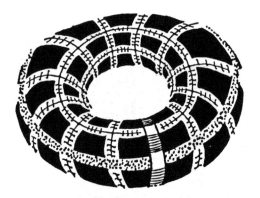

Figure 10. Double closure of the nervous and hormonal causal chain. Horizontal dotted line (equator) neurohypophysis. Vertically broken seam (meridian) motor-sensory "synaptic gap".

tent and complete theory of cognition—or of "observation"—based exclusively on recursive computations within the organism itself, without calling upon the help of a "second order" observer who tells us what he sees regarding the first order observer, and so on and so forth, up the never ending hierarchical ladder.[15,16]

As a general suggestion for researching this problem, I would postulate the following proposition:

The postulate of the epistemic homeostasis—

The nervous system as a whole is organized in such a way (organizes itself in such a way) that it computes a stable reality.

This makes it clear that here again, with "stable realities", we are dealing with an Eigen-value problem, and I could imagine that this observation may be of value in psychiatry.

Some may have seen in these remarks their existentialist basis. By means of the double closure of the circle of signals—or the complete closure of the causal nexus—I have done nothing more than stipulate the autonomy of each individual living being anew: the causes of my actions are not somewhere else or with somebody else—that would be heteronomy: the other is responsible. Rather, the causes of my actions are within myself: I am my own regulator! Frankl, Jaspers, or Buber would perhaps express it the following way: in each and every moment I can decide who I am.

And with this, the responsibility for who I am and how I act falls back to me; autonomy means responsibility; heteronomy means irresponsibility.

Here we see that the epistemological problems of ethics coincide to a larger degree with those of cybernetics, and thus we, in the field of cybernetics, have the responsibility to partake in the solution of the social and ethical problems of our times.

Postscript

I would like to thank the German Society for Cybernetics for enabling me to take part in this convention, the Department of Electrical Engineering of the University of Illinois for their general provision of time and assistance; and further my thanks go to Mr. Lebbeus Woods (Figures 1–3) and Rodney Clough (Figures 9 & 10) for their artistic contributions, and to Kathy Roberts who provided her secretarial expertise. For the English edition, special thanks go to Linda Goetz of Coast Lines Secretarial in Half Moon Bay, California, who transformed a chaotic manuscript into this easily readable and visually pleasing form.

Bibliography

1. Von Foerster, H. (ed.): Cybernetics: Circular causal and Feedback Mechanisms in Biological and Social Systems, Transactions of the Sixth Conference, Josia Macy Jr. Foundation, New York, 202 pp., (1949).
2. Von Foerster, H., Mead, Margaret and Teuber, H. L. (eds.): Cybernetics: Circular Causal and Feedback Mechanisms in Biological and Social Systems, Transactions of the Seventh Conference, Josia Macy Jr. Foundation, New York, 240 pp., (1950).
3. Von Foerster, H., Mead, Margaret and Teuber, H. L. (eds.): Cybernetics: Circular Causal and Feedback Mechanisms in Biological and Social Systems, Transactions of the Eighth Conference, Josia Macy Jr. Foundation, New York, 240 pp., (1951).
4. Von Foerster, H., Mead, Margaret and Teuber, H. L. (eds.): Cybernetics: Circular Causal and Feedback Mechanisms in Biological and Social Systems, Transactions of the Ninth Conference, Josia Macy Jr. Foundation, New York, 184 pp., (1953).
5. Von Foerster, H., Mead, Margaret and Teuber, H. L. (eds.): Cybernetics: Circular Causal and Feedback Mechanisms in Biological and Social Systems, Transactions of the Tenth Conference, Josia Macy Jr. Foundation, New York, 100 pp., (1955).
6. Katz, J.J.: The Problem of Induction and Its Solution. University of Chicago Press; Chicago, 126 pp., (1962).
7. Lofgren, L.: "Recognition of Order and Evolutionary System" in Computer and Information Sciences II, J. Tou (ed.), Academic Press; New York, 165–175, (1967).
8. Von Foerster, H.: "Thoughts and Notes on Cognition" in Cognition: A Multiple View, P. Garvin (ed.), Spartan Books; New York, 25–48, (1970).
9. Piaget, J. and Inhelder, B.: The Child's Conception of Space. Norton; New York, (1956).
10. Bower, T. G. R.: "The Object in the World of the Infant' in Scientific American, 225 (4), p. 30–38 (October 1971).
11. Witz, K.: Models of Sensory-Motor Schemes in Infants, Research Report, Department of Mathematics, University of Illinois, Urbana, 33 pp, (1972).

12. Arnold, P.: "Experiencing the Fourth Spatial Dimension" in Accomplishment Summary 70/71, BCL Report No. 71.2, The Biological Computer Laboratory, University of Illinois, Urban, pp. 201–215, (1971).

13. Arnold, P.: "A Proposal for a Study of the Mechanisms of Perception of, and Formation of Internal Representations of, the Spatial Fourth Dimension" in Accomplishment Summary 71/72, BCL Report No. 72.2, The Biological Computer Laboratory, University of Illinois, Urbana, pp. 223–235, (1972).

14. Von Foerster, H.: "Molecular Ethology: An Immodest Proposal for Semantic Clarification" in Molecular Mechanisms in Memory and Learning, G. Ungar (ed.), Plenum Press; New York, 213–248, (1970).

15. Maturana, H.: Biology of Cognition, BCL Report No. 9.0, The Biological Computer Laboratory, University of Illinois, Urbana, 95 pp., (1970).

16. Maturana, H.: "Neurophysiology of Cognition" in Cognition: A Multiple View, P. Garvin (ed.), Spartan Books; New York, 3–23, (1970).

17. Descharnes, R.: Die Welt Salvador Dalis, Edita Lausanne, Abb. p. 185 (1962).

10
Notes on an Epistemology for Living Things*

I. Problem

While in the first quarter of this century physicists and cosmologists were forced to revise the basic notions that govern the natural sciences, in the last quarter of this century biologists will force a revision of the basic notions that govern science itself. After that "first revolution" it was clear that the classical concept of an "ultimate science," that is an objective description of the world in which there are no subjects (a "subjectless universe"), contains contradictions.

To remove these one had to account for an "observer" (that is at least for one subject): (i) Observations are not absolute but relative to an observer's point of view (i.e., his coordinate system: Einstein); (ii) Observations affect the observed so as to obliterate the observer's hope for prediction (i.e., his uncertainty is absolute: Heisenberg).

After this, we are now in the possession of the truism that a description (of the universe) implies one who describes (observes it). What we need now is the description of the "describer" or, in other words, we need a theory of the observer. Since it is only living organisms which would qualify as being observers, it appears that this task falls to the biologist. But he himself is a living being, which means that in his theory he has not only to account for himself, but also for his writing this theory. This is a new state of affairs in scientific discourse for, in line with the traditional viewpoint which separates the observer from his observations, reference to this discourse was to be carefully avoided. This separation was done by no means because of

---

* This article is an adaptation of an address given on September 7, 1972, at the *Centre Royaumont pour un Science de L'homme*, Royaumont, France, on the occasion of the international colloquim "l'Unite de l'homme: invariants biologiques et universaux culturel." The French version of this address has been published under the title "Notes pour une epistemologie des objets vivants" in *L'Unite de L'Homme: Invariants Biologiques et Universaux Culturel*, Edgar Morin and Massimo Piattelli-Palmerini (eds), Editions de Seul, Paris, pp. 401–417 (1974).

excentricity or folly, for under certain circumstances inclusion of the observer in his descriptions may lead to paradoxes, to wit the utterance "I am a liar."

In the meantime however, it has become clear that this narrow restriction not only creates the ethical problems associated with scientific activity, but also cripples the study of life in full context from molecular to social organizations. Life cannot be studied *in vitro*, one has to explore it *in vivo*.

In contradistinction to the classical problem of scientific inquiry that postulates first a description-invariant "objective world" (as if there were such a thing) and then attempts to write its description, now we are challenged to develop a description-invariant "subjective world," that is a world which includes the observer: *This is the problem.*

However, in accord with the classic tradition of scientific inquiry which perpetually asks "How?" rather than "What?," this task calls for an epistemology of "How do we know?" rather than "What do we know?"

The following notes on an epistemology of living things address themselves to the "How?" They may serve as a magnifying glass through which this problem becomes better visible.

II. Introduction

The twelve propositions labeled 1, 2, 3, . . . 12, of the following 80 Notes are intended to give a minimal framework for the context within which the various concepts that will be discussed are to acquire their meaning. Since Proposition Number 12 refers directly back to Number 1, Notes can be read in a circle. However, comments, justifications, and explanations, which apply to these propositions follow them with decimal labels (e.g., "5.423") the last digit ("3") referring to a proposition labeled with digits before the last digit ("5.42"), etc. (e.g., "5.42" refers to "5.4," etc.).

Although Notes may be entered at any place, and completed by going through the circle, it appeared advisable to cut the circle between propositions "11" and "1," and present the notes in linear sequence beginning with Proposition 1.

Since the formalism that will be used may for some appear to obscure more than it reveals, a preview of the twelve propositions* with comments in prose may facilitate reading the notes.

*1. The environment is experienced as the residence of objects, stationary, in motion, or changing.***
Obvious as this proposition may look at first glance, on second thought one may wonder about the meaning of a "changing object." Do we mean the

---

* In somewhat modified form.
** Propositions appear in italics.

change of appearance of the same object as when a cube is rotated, or a person turns around, and we take it to be the same object (cube, person, etc.); or when we see a tree growing, or meet an old schoolmate after a decade or two, are they different, are they the same, or are they different in one way and the same in another? Or when Circe changes men into beasts, or when a friend suffers a severe stroke, in these metamorphoses, what is invariant, what does change? Who says that these were the same persons or objects?

From studies by Piaget[1] and others[2] we know that "object constancy" is one of many cognitive skills that are acquired in early childhood and hence are subject to linguistic and thus cultural bias.

Consequently, in order to make sense of terms like "biological invariants," "cultural universals," etc., the logical properties of "invariance" and "change" have first to be established.

As the notes procede it will become apparent that these properties are those of descriptions (representations) rather than those of objects. In fact, as will be seen, "objects" do owe their existence to the properties of representations.

To this end the next four propositions are developed.

2. The logical properties of "invariance" and "change" are those of representations. If this is ignored, paradoxes arise.
Two paradoxes that arise when the concepts "invariance" and "change" are defined in a contextual vacuum are cited, indicating the need for a formalization of representations.

3. Formalize representations R, S, regarding two sets of variables {x} and {t}, tentatively called "entities" and "instants" respectively.
Here the difficulty of beginning to talk about something which only later makes sense so that one can begin talking about it, is pre-empted by "tentatively," giving two sets of as yet undefined variables highly meaningful names, viz, "entities" and "instants," which only later will be justified.

This apparent deviation from rigor has been made as a concession to lucidity. Striking the meaningful labels from these variables does not change the argument.

Developed under this proposition are expressions for representations that can be compared. This circumvents the apparent difficulty to compare an apple with itself before and after it is peeled. However, little difficulties are encountered by comparing the peeled apple as it is *seen now* with the unpeeled apple as it is *remembered* to have been before.

With the concept "comparison," however an operation ("computation") on representations is introduced, which requires a more detailed analysis. This is done in the next proposition. From here on the term "computation" will be consistently applied to all operations (not necessarily numerical) that transform, modify, re-arrange, order, etc., either symbols (in the "abstract" sense) or their physical manifestations (in the "concrete" sense). This is done to enforce a feeling for the realizability of these operations in the structural and functional organization of either grown nervous tissue or else constructed machines.

4. Contemplate relations, "Rel," between representations, R, and S.
However, immediately a highly specific relation is considered, viz, an "Equivalence Relation" between two representations. Due to the structural properties of representations, the computations necessary to confirm or deny equivalence of representations are not trivial. In fact, by keeping track of the computational pathways for establishing equivalence, "objects" and "events" emerge as *consequences* of branches of computation which are identified as the processes of abstraction and memorization.

5. Objects and events are not primitive experiences. Objects and events are representations of relations.
Since "objects" and "events" are not primary experiences and thus cannot claim to have absolute (objective) status, their interrelations, the "environment," is a purely personal affair, whose constraints are anatomical or cultural factors. Moreover, the postulate of an "external (objective) reality" disappears to give way to a reality that is determined by modes of internal computations[3].

6. Operationally, the computation of a specific relation is a representation of this relation.
Two steps of crucial importance to the whole argument forwarded in these notes are made here at the same time. One is to take a computation for a representation; the second is to introduce here for the first time "recursions." By recursion is meant that on one occasion or another a function is substituted for its own argument. In the above Proposition 6 this is provided for by taking the computation of a relation between *representations* again as a representation.

While taking a computation for a representation of a relation may not cause conceptual difficulties (the punched card of a computer program which controls the calculations of a desired relation may serve as a adequate metaphor), the adoption of recursive expressions appears to open the door for all kinds of logical mischief.

However, there are means to avoid such pitfalls. One, e.g., is to devise a notation that keeps track of the order of representations, e.g., "the representation of a representation of a representation" may be considered as a third order representation, $R^{(3)}$. The same applies to relations of higher order, n: $Rel^{(n)}$.

After the concepts of higher order representations and relations have been introduced, their physical manifestations are defined. Since representation and relations are computations, their manifestations are "special purpose computers" called "representors" and "relators" respectively. The distinction of levels of computation is maintained by referring to such structures as n-th order representors (relators). With these concepts the possibility of introducing "organism" is now open.

7. A living organism is a third order relator which computes the relations that maintain the organism's integrity.
The full force of recursive expressions is now applied to a recursive definition of living organisms first proposed by H. R. Maturana[4,5] and further developed by him and F. Varela in their concept of "autopoiesis"[6].

As a direct consequence of the formalism and the concepts which were developed in earlier propositions it is now possible to account for an interaction between the internal representation of an organism of himself with one of another organism. This gives rise to a theory of communication based on a purely connotative "language." The surprising property of such a theory is now described in the eighth proposition.

8. A Formalism necessary and sufficient for a theory of communication must not contain primary symbols representing communicabilia (e.g., symbols, words, messages, etc.).
Outrageous as this proposition may look at first glance, on second thought however it may appear obvious that a theory of communication is guilty of circular definitions if it assumes communicabilia in order to prove communication.

The calculus of recursive expressions circumvents this difficulty, and the power of such expressions is exemplified by the (indefinitely recursive) reflexive personal pronoun "I." Of course the semantic magic of such infinite recursions has been known for some time, to wit the utterance "I am who I am"[7].

9. Terminal representations (descriptions) made by an organism are manifest in its movements; consequently the logical structure of descriptions arises from the logical structure of movements.
The two fundamental aspects of the logical structure of descriptions, namely their sense (affirmation or negation), and their truth value (true or false), are shown to reside in the logical structure of movement: approach and withdrawal regarding the former aspect, and functioning or dysfunctioning of the conditioned reflex regarding the latter.

It is now possible to develop an exact definition for the concept of "information" associated with an utterance. "Information" is a relative concept that assumes meaning only when related to the cognitive structure of the observer of this utterance (the "recipient").

10. The information associated with a description depends on an observer's ability to draw inferences from this description.
Classical logic distinguishes two forms of inference: deductive and inductive[8]. While it is in principle possible to make infallible deductive inferences

("necessity"), it is in principle impossible to make infallible inductive inferences ("chance"). Consequently, chance and necessity are concepts that do not apply to the world, but to our attempts to create (a description of) it.

11. The environment contains no information; the environment is as it is.

12. Go back to Proposition Number 1.

References

1. Piaget, J.: *The Construction of Reality in the Child*. Basic Books, New York, (1954).
2. Witz, K. and J. Easley: *Cognitive Deep Structure and Science Education* in *Final Report*, Analysis of Cignitive Behavior in Children; Curriculum Laboratory, University of Illinois, Urbana, (1972).
3. Castaneda, C.: *A Separate Reality*. Simon and Schuster, New York, (1971).
4. Maturana, H.: *Neurophysiology of Cognition* in *Cognition: A Multiple View*, P. Garvin (ed.), Spartan Books, New York, pp. 3–23, (1970).
5. Maturana, H.: *Biology of Cognition*, BCL Report No. 9.0, Biological Computer Laboratory, Department of Electrical Engineering, University of Illinois, Urbana, 95 pp., (1970).
6. Maturana, H. and F. Varela: *Autopoiesis*. Faculdad de Ciencias, Universidad de Chile, Santiago, (1972).
7. Exodus, 3, 14.
8. Aristotle: *Metaphysica*. Volume VIII of *The Works of Aristotle*, W. D. Ross (ed., tr.). The Clarendon Press, Oxford, (1908).

III. Notes

1.   The environment is experienced as the residence of objects, stationary, in motion, or changing.
1.1   "Change" presupposes invariance, and "invariance" change.

2.   The logical properties of "invariance" and "change" are those of representations. If this is ignored paradoxes arise.
2.1   The paradox of "invariance:"

THE DISTINCT BEING THE SAME

But it makes not sense to write $x_1 = x_2$ (why the indices?).
And $x = x$ says something about "=" but nothing about x.
2.2   The paradox of "change:"

THE SAME BEING DISTINCT

But it makes no sense to write $x \neq x$.

3.   Formalize the representations R, S, ... regarding two sets of variables x_i and t_j (i, j = 1, 2, 3 ...), tentatively called "entities" and "instants" respectively.
3.1   The representation R of an entity x regarding the instant t_1 is distinct from the representation of this entity regarding the instant t_2:

$$R(x(t_1)) \neq R(x(t_2))$$

3.2   The representation S of an instant t regarding the entity x_1 is distinct from the representation of this instant regarding the entity x_2:

$$S(t(x_1)) \neq S(t(x_2))$$

3.3   However, the comparative judgment ("distinct from") cannot be made without a mechanism that computes these distinctions.

3.4   Abbreviate the notation by

$$R(x_i(t_j)) \rightarrow R_{ij}$$
$$S(t_k(x_l)) \rightarrow S_{kl}$$

$(i, j, k, l = 1, 2, 3, \ldots)$

4.   *Contemplate relations Rel_μ between the representations R and S:*

$$Rel_\mu(R_{ij}, S_{kl})$$

$(\mu = 1, 2, 3, \ldots)$

4.1   Call the relation which obliterates the distinction $x_i \neq x_l$ and $t_j \neq t_k$ (i.e., $i = l$; $j = k$) the "Equivalence Relation" and let it be represented by:

$$Equ(R_{ij}, S_{ji})$$

4.11   This is a representation of a relation between two representations and reads:

"The representation R of an entity x_i retarding the instant t_j is equivalent to the representation S of an instant t_j regarding the entity x_i."

4.12   A possible linguistic metaphor for the above representation of the equivalence relation between two representations is the equivalence of "thing acting" (most Indo-European languages) with "act thinging" (some African languages) (cognitive duality). For instance:

"The horse gallops" $\leftrightarrows$ "The gallop horses"

4.2   The computation of the equivalence relation 4.1 has two branches:

4.21   One computes equivalences for x only

$$Equ(R_{ij}, S_{ki}) = Obj(x_j)$$

4.211   The computations along this branch of equivalence relation are called "abstractions:" *Abs.*

4.212   The results of this branch of computation are usually called "objects" (entities), and their invariance under various transformations $(t_j, t_k, \ldots)$ is indicated by giving each object a distinct but invariant label N_i ("Name"):

$$Obj(x_j) \rightarrow N_j$$

4.22   The other branch computes equivalences for t only:

$$Equ(R_{ij}, S_{jl}) \equiv Eve(t_j)$$

4.221   The computations along this branch of equivalence relation are called "memory:" *Mem.*

4.222   The results of this branch of computation are usually called "events" (instants), and their invariance under various transformations $(x_i, x_l, \ldots)$ is indicated by associating with each event a distinct but invariant label T_j ("Time"):

$$Eve(t_j) \rightarrow T_j$$

4.3   This shows that the concepts "objects," "event," "name," "time," "abstraction," "memory," "invariance," "change," generate each other.

From this follows the next proposition:

5.   *Objects and events are not primitive experiences. "Objects" and "Events" are representations of relations.*
5.1   A possible graphic metaphor for the complementarity of "object" and "event" is an orthogonal grid that is mutually supported by both (Fig. 1).
5.2   "Environment" is the representation of relations between "objects" and "events"

$$Env(Obj, Eve)$$

5.3   Since the computation of equivalence relations is not unique, the results of these computations, namely, "objects" and "events" are likewise not unique.
5.31   This explains the possibility of an arbitrary number of different, but internally consistent (language determined) taxonomies.
5.32   This explains the possibility of an arbitrary number of different, but internally consistent (culturally determined) realities.
5.4   Since the computation of equivalence relations is performed on primitive experiences, an external environment is not a necessary prerequisite of the computation of a reality.

6.   *Operationally, the computation Cmp(Rel) of a specific relation is a representation of this relation.*

$$R = Cmp(Rel)$$

6.1   A possible mathematical metaphor for the equivalence of a computation with a representation is, for instance, Wallis' computational algorithm for the infinite product:

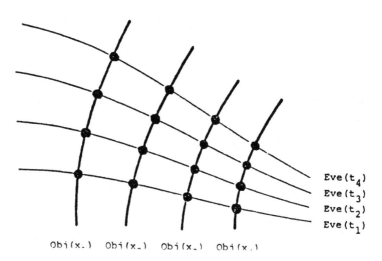

$$Eve(t_4)$$
$$Eve(t_3)$$
$$Eve(t_2)$$
$$Eve(t_1)$$

Obj(x.)   Obj(x.)   Obj(x.)   Obj(x.)

FIGURE 1. "Objects" creating "Events" and *vice versa.*

$$2 \cdot \frac{2}{1} \cdot \frac{2}{3} \cdot \frac{4}{3} \cdot \frac{4}{5} \cdot \frac{6}{5} \cdot \frac{6}{7} \ldots$$

Since this is one of many possible definitions of π (3.14159...), and π is a number, we may take π as a (numerical) representation of this computation.

6.2   Call representations of computations of relations "second order representations." This is clear when such a representation is written out fully:

$$R = Cmp(\mathrm{Rel}(\mathbf{R}_{ij}, \mathbf{S}_{kl})),$$

where $\mathbf{R}_{ij}$ and $\mathbf{S}_{kl}$ are, of course, "first order representations" as before (3.3).

6.21   From this notation it is clear that first order representations can be interpreted as zero-order relations (note the double indices on S and R).

6.22   From this notation it is also clear that higher order (n-th order) representations and relations can be formulated.

6.3   Call a physical mechanism that computes an n-th order representation (or an n-th order relation) an "n-th order representor" $RP^{(n)}$ (or "n-th order relator" $RL^{(n)}$) respectively.

6.4   Call the externalized physical manifestation of the result of a computation a "terminal representation" or a "description."

6.5   One possible mechanical metaphor for relator, relation, objects, and descriptions, is a mechanical desk calculator (the relator) whose internal structure (the arrangement of wheels and pegs) is a representation of a relation commonly called "addition:" Add (a, b; c). Given two objects, a = 5, b = 7, it computes a terminal representation (a description), c, of the relation between these two objects in digital, decadic, form:

$$\mathrm{Add}\,(5, 7; 12)$$

6.51   Of course, a machine with a different internal representation (structure) of the same relation Add (a, b; c), may have produced a different terminal representation (description), say, in the form of prime products, of this relation between the same objects:

$$\mathrm{Add}\,(5, 7; 2^2 \cdot 3^1)$$

6.6   Another possible mechanical metaphor for taking a computation of a relation as a representation of this relation is an electronic computer and its program. The program stands for the particular relation, and it assembles the parts of the machine such that the terminal representation (print-out) of the problem under consideration complies with the desired form.

6.61   A program that computes programs is called a "meta-program." In this terminology a machine accepting meta-programs is a second-order relator.

6.7   These metaphors stress a point made earlier (5.3), namely, that the computations of representations of objects and events is not unique.

6.8   These metaphors also suggest that my nervous tissue which, for instance, computes a terminal representation in the form of the following utterance: "These are my grandmother's spectacles" neither resembles my grandmother nor her spectacles; nor is there a "trace" to be found of either (as little as there are traces of "12" in the wheels and pegs of a desk calculator, or of numbers in a program). Moreover, my utterance "These are my grandmother's spectacles" should neither be confused with my grandmother's spectacles, nor with the program that computes this utterance, nor with the representation (physical manifestation) of this program.

6.81   However, a relation between the utterance, the objects, and the algorithms computing both, is computable (see 9.4).

7.   *A living organism Ω is a third-order relator ($\Omega = RL^{(3)}$) which computes the relations that maintain the organism's integrity [1] [2]:*

$$\Omega \; Equ \, [R(\Omega(\text{Obj})), S(\text{Eve}(\Omega))]$$

This expression is recursive in Ω.

7.1   An organism is its own ultimate object.

7.2   An organism that can compute a representation of this relation is self-conscious.

7.3   Amongst the internal representations of the computation of objects $\text{Obj}(x_i)$ within one organism Ω may be a representation $\text{Obj}(\Omega^*)$ of another organism Ω^*. Conversely, we may have in Ω^* a representation $\text{Obj}^*(\Omega)$ which computes Ω.

7.31   Both representations are recursive in Ω, Ω^* respectively. For instance, for Ω:

$$\text{Obj}^{(n)}(\Omega^{*(n-1)}\,(\text{Obj}^{*(n-1)}\,(\Omega^{(n-2)}(\text{Obj}^{(n-2)}(\dots\Omega^*))))).$$

7.32   This expression is the nucleus of a theory of communication.

8.   *A formalism necessary and sufficient for a theory of communication must not contain primary symbols representing "communicabilia" (e.g., symbols, words, messages, etc.).*

8.1   This is so, for if a "theory" of communication were to contain primary communicabilia, it would not be a theory but a technology of communication, taking communication for granted.

8.2   The nervous activity of one organism cannot be shared by another organism.

8.21   This suggests that indeed nothing is (can be) "communicated."

8.3   Since the expression in 7.31 may become cyclic (when $\text{Obj}^{(k)} = \text{Obj}^{(k-2i)}$), it is suggestive to develop a teleological theory of communication in which the stipulated goal is to keep $\text{Obj}(\Omega^*)$ invariant under perturbations by Ω^*.

8.31   It is clear that in such a theory such questions as: "Do you see the color of this object as I see it?" become irrelevant.

8.4   Communication is an observer's interpretation of the interaction between two organisms Ω_1, Ω_2.

8.41   Let $\text{Evs}_1 \equiv \text{Evs}\,(\Omega_1)$, and $\text{Evs}_2 \equiv \text{Evs}(\Omega_2)$, be *sequences* of events $\text{Eve}(t_j)$, ($j = 1, 2, 3, \dots$) with regard to two organisms Ω_1 and Ω_2 respectively; and let *Com* be an observer's (internal) representation of a relation between these sequences of events:

$$\text{OB}(Com(\text{Evs}_1, \text{Evs}_2))$$

8.42   Since either Ω_1 or Ω_2 or both can be observers ($\Omega_1 = \text{OB}_1; \Omega_2 = \text{OB}_2$) the above expression can become recursive in either Ω_1 or in Ω_2 or in both.

8.43   This shows that "communication" is an (internal) representation of a relation between (an internal representation of) oneself with somebody else.

$$R(\Omega^{(n+1)}, Com(\Omega^{(n)}, \Omega^*))$$

8.44   Abbreviate this by

$$C(\Omega^{(n)}, \Omega^*).$$

8.45   In this formalism the reflexive personal pronoun "I" appears as the indefinitely applied) recursive operator

$$Equ[\Omega^{(n+1)}C(\Omega^{(n)}, \Omega^{(n)})]$$

or in words:

"I am the observed relation between myself and observing myself."

8.46   "I" is a relator (*and* representor) of infinite order.

9.   *Terminal representations (descriptions) made by an organism are manifest in its movements; consequently, the logical structure of descriptions arises from the logical structure of movements.*
9.1   It is known that the presence of a perceptible agent of weak concentration may cause an organism to move toward it (approach). However, the presence of the same agent in strong concentration may cause this organism to move away from it (withdrawal).
9.11   That is "approach" and "withdrawal" are the precursors for "yes" or "no."
9.12   The two phases of elementary behavior, "approach" and "withdrawal," establish the operational origin of the two fundamental axioms of two-valued logic, namely, the "law of the excluded contradiction:"

$$\overline{x \& \overline{x}},$$

in words: "not: x *and* not-x;"
and the law of the excluded middle:

$$x \vee \overline{x},$$

in words: "x *or* not-x;" (see Fig. 2).
9.2   We have from Wittgenstein's Tractatus [3], proposition 6.0621: ". . . it is important that the signs "p" and non-p" *can* say the same thing. For it shows that nothing in reality corresponds to the sign "non."
The occurence of negation is a proposition is not enough to characterize its sense (non-non-p = p)."
9.21   Since nothing in the environment corresponds to negation, negation as well as all other "logical particles" (inclusion, alternation, implication, etc.) must arise within the organism itself.
9.3   Beyond being logical affirmative or negative, descriptions can be true or false.
9.31   We have from Susan Langer, *Philosophy in a New Key* [4]:

"The use of signs in the very first-manifestation of mind. It arises as early in biological history as the famous 'conditioned reflex,' by which a concomitant of a stimulus takes over the stimulus-function. The concomitant becomes a *sign* of the condition to which the reaction is really appropriate. This is the real beginning of mentality, for here is the birthplace of *error*, and herewith of *truth*."

9.32   Thus, not only the sense (yes or no) of descriptions but also their truth values (true or false) are coupled to movement (behavior).
9.4   Let D* be the terminal representation made by an organism Ω^*, and let it be observed by an organism Ω; let Ω's internal representation of this description be $D(\Omega, D^*)$; and, finally, let Ω's internal representation of his environment be $E(\Omega, E)$. Then we have:

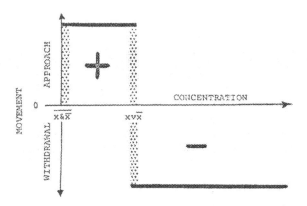

FIGURE 2. The laws of "excluded contradiction" ($\overline{x \,\&\, \overline{x}}$) and of "excluded middle" ($x \vee \overline{x}$) in the twilight zones between no motion (M = 0) and approach (+), and between approach (+) and withdrawal (−) as a function of the concentration (C) of a perceptible agent.

The domain of relations between D and E which are computable by Ω represents the "information" gained by Ω from watching Ω^*:

$$Inf(\Omega, D^*) \equiv \text{Domain } \text{Re}\, l_\mu(D, E)$$

($\mu = 1, 2, 3, \ldots$ m)

9.41   The logarithm (of base 2) of the number m of relations Rel_μ computable by Ω (or the negative mean value of the logarithmic probabilities of their occurance $\langle\log_2 p_i\rangle = \Sigma p_i \log_2 p_i;\, i = 1 \rightarrow m$) is the "amount of information, H" of the description D^* with respect to Ω:

$$H(D^*, \Omega) = \log_2 m$$

$$\left(\text{or } H(D^*, \Omega) = -\sum_1^m p_i \log_2 p_i \right)$$

9.42   This shows that information is a relative concept. And so is H.

9.5   We have from a paper by Jerzy Konorski[5]:

"... It is not so, as we would be inclined to think according to our introspection, that the receipt of information and its utilization are two separate processes which can be combined one with the other in any way; on the contrary, information and its utilization are inseparable constituting, as a matter of fact, one single process."

10.   The information associated with a description depends on an observer's ability to draw inferences from this description.
10.1 "Necessity" arises from the ability to make infallible deductions.
10.2 "Chance" arises from the inability to make infallible inductions.

11.   The environment contains no information. The environment is as it is.

12.   The environment is experienced as the residence of objects, stationary, in motion, or changing (Proposition 1).

References

1. Maturana, H.: *Neurophysiology of Cognition* in *Cognition: A Multiple View*, P. Garvin (ed.), Spartan Books, New York, pp. 3–23, (1970).
2. Maturana, H.: *Biology of Cognition*, BCL Report No. 9.0, Biological Computer Laboratory, Department of Electrical Engineering, University of Illinois, Urbana, 95 pp., (1970).
3. Wittgenstein, L.: *Tractatus Logico Philosophicus*, Humanities Press, New York, (1961).
4. Langer, S.: *Philosophy in a New Key*, New American Library, New York, (1951).
5. Konorski, J.: *The Role of Central Factors in Differentiation* in *Information Processing in the Nervous System*, R. W. Gerard and J. W. Duyff (eds.), Excerpta Medica Foundation, Amsterdam, 3, pp. 318–329, (1962).

11
Objects: Tokens for (Eigen-)Behaviors*

A seed, alas, not yet a flower, for Jean Piaget to his 80th birthday from Heinz von Foerster with admiration and affection.

I shall talk about notions that emerge when the organization of sensori-motor interactions (and also that of central processes (cortical-cerebellar-spinal, cortico-thalamic-spinal, etc.)) is seen as being essentially of circular (or more precisely of recursive) nature. Recursion enters these considerations whenever the changes in a creature's sensations are accounted for by its movements ($s_i = S(m_k)$), and its movements by its sensations ($m_k = M(s_j)$). When these two accounts are taken together, then they form "recursive expressions," that is, expressions that determine the states (movements, sensations) of the system (the creature) in terms of these very states ($s_i = S(M(s_j)) = SM(s_j); m_k = M(S(m_i)) = MS(m_i)$).

One point that with more time, effort and space could be made rigorously and not only suggestively as it has been made here, is that what is referred to as "objects" (GEGEN-STAENDE = "against-standers") in an observer-excluded (linear, open) epistemology, appears in an observer-included (circular, closed) epistemology as "tokens for stable behaviors" (or, if the terminology of Recursive Function Theory is used, as "tokens for Eigen-functions").

Of the many possible entries into this topic the most appropriate one for this occasion appears to me the (recursive) expression that forms the last line on page 63 of J. Piaget's *L'Equilibration des Structures Cognitives* (1975):

$$\text{Obs.O} \rightarrow \text{Obs.S} \rightarrow \text{Coord.S} \rightarrow \text{Coord.O} \rightarrow \text{Obs.O} \rightarrow \text{etc.}$$

This is an observer's account of an interaction between a subject S and an object (or a set of objects) O. The symbols used in this expression (defined on page 59 *op. cit.*) stand for (see also Fig. 1):

---

*This contribution was originally prepared for and presented at the University of Geneva on June 29, 1976, on occasion of Jean Piaget's 80th birthday. The French version of this paper appeared in *Hommage a Jean Piaget: Epistémologie génétique et équilibration*. B. Inhelder, R. Garcia, J. Voneche (eds.), Delachaux et Niestle Neuchatel (1977).

261

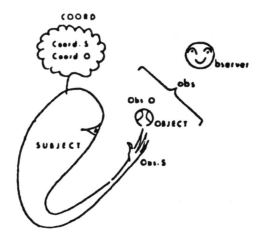

FIGURE 1.

Obs.S   "observables relatifs a l'action du sujet"
Obs.O   "observables relatifs aux objets"
Coord.S "coordinations inferentielles des actions (ou operations) du sujet"
Coord.O "coordinations inferentielles entre objets"
"etc."   "the (syntactic) injunction to iterate (with no limits specified) the sequence of these operations (HVF)"

For the sake of brevity (lucidity?) I propose to compress the symbolism of before even further, compounding all that is observed (i.e. Obs.O and Obs.S) into a single variable

$$obs,$$

and compounding coordinating operations that are performed by the subject (i.e. Coord.S and Coord.O) into a single operator

COORD.

COORD transforms, rearranges, modifies etc., the forms, arrangements, behaviors, etc., observed at one occasion (say, initially obs_0, and call it the "primary argument") into those observed at the next occasion, obs_1. Express the outcome of this operation through the equality:[1]

---

[1] By replacing the arrow "→", whose operational meaning is essentially to indicate a one-way (semantic) connectedness (e.g., "goes to," "implies," "invokes," leads to," etc.) between adjacent expressions, with the equality sign provides the basis for a calculus. However, in order that legitimate use of this sign can be made, the variables "obs_1" must belong to the same domain. The choice of domain is, of course left to the observer who may wish to express his observations in form of, for instance, numerical values, of vectors representing arrangements or geometrical configurations, or his observations of behaviors in form of mathematical functions (e.g., "equations of motion," etc.), or by logical propositions (e.g., McCulloch-Pitts" "TPE's" 1943 (i.e., Temporal Propositional Expressions), etc.).

$$obs_1 = COORD(obs_0).$$

While some relational fine structure is (clearly) lost in this compression, gained, however, may be an easiness by which the progression of events, suggested on the last lined page of 62 *op. cit.* and copied here can now be watched.

Obs. S(n) $\longrightarrow$ Coord. S(n) $\longleftrightarrow$ Obs. O(n) $\longleftarrow$ Coord. O(n)

Obs. S(n+1) $\longrightarrow$ Coord. S(n+1) $\longrightarrow$ Obs. O(n+1) $\longrightarrow$ Coord. O(n+1)

Obs. S(n+2) $\longrightarrow$ Coord. S(n+2) $\longleftrightarrow$ Obs. O(n+2) $\longleftarrow$ Coord. O(n+2)

etc.   etc.

Allow the operator COORD to operate on the previous outcome to give

$$obs_2 = COORD(obs_1) = COORD(COORD(obs_0)) \qquad (2)$$

and (recursively) after n steps $\qquad\qquad (obs_0)))..),$

$$obs_n = COORD(COORD(COORD(\cdots\cdots\cdots \mid n\ times\mid \qquad (3)$$
$$\underline{\qquad\qquad n\ times \qquad\qquad}$$

or by notational abbreviation

$$obs_n = COORD^{(n)}(obs_0). \qquad (4)$$

By this notational abbreviation it is suggested that also functionally

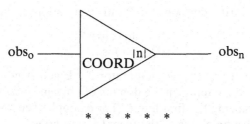

can be replaced by

$obs_0 \longrightarrow$ COORD$^{|n|}$ $\longrightarrow obs_n$

* * * * *

Let n grow without limit (n $\rightarrow \infty$):

$$obs_\infty = \lim_{n\to\infty} COORD^{(n)}(obs_0) \qquad (5)$$

or:

$$obs_\infty = COORD(COORD(COORD(COORD. \ldots \qquad (6)$$

Contemplate the above expression (6) and note:

(i) that the independent variable obs_0, the "primary argument" has disappeared (which may be taken as a signal that the simple connection between independent and dependent variables is lost in indefinite recursions, and that such expressions take on a different meaning).

(ii) that, because obs_∞ expresses an indefinite recursion of operators COORD onto operators COORD, any indefinite recursion within that expression can be replaced by obs_∞:

$$obs_\infty =$$

COORD(COORD(COORD(COORD(. . . .

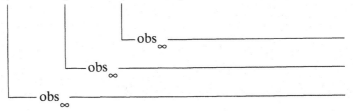

(iii) Hence:

$$obs_\infty = obs_\infty \qquad (7.0)$$
$$obs_\infty = COORD(obs_\infty) \qquad (7.1)$$
$$obs_\infty = COORD(COORD(obs_\infty)) \qquad (7.2)$$
$$obs_\infty = COORD(COORD(COORD(obs_\infty))) \qquad (7.3)$$
etc.

Note that while in this form the *horror infinitatis* of expression (6) has disappeared (all expressions in COORD are finite), a new feature has emerged, namely, that the dependent variable obs_∞ is, so to say, "self-depending" (or "self-defining," or "self-reflecting," etc., through the operator COORD).

Should there exist values $obs_{\infty i}$ that satisfy equations (7), call these values

"Eigen-Values"
$$obs_{\infty i} \equiv Obs_i \qquad (8)$$

(or "Eigen-Functions," "Eigen-Operators," "Eigen-Algorithms," "Eigen-Behaviors," etc., depending on the domain of obs) and denote these "Eigen-Values by capitalizing the first letter. (For examples see Appendix A).

Contemplate expressions of the form (7) and note:

(i) that Eigenvalues are discrete (even if the domain of the primary argument obs_0 is continuous).

This is so because any infinitesimal perturbation $\pm\epsilon$ from an Eigenvalue Obs_i (i.e., $Obs_i \pm \epsilon$) will disappear, as did all other values of obs, except those for which obs = Obs_i, and obs will be brought either back to Obs_i (*stable* Eigenvalue), or to another Eigenvalue Obs_j (*instable* Eigenvalue Obs_i).

In other words, Eigenvalues represent *equilibria*, and depending upon the chosen domain of the primary argument, these equilibria may be equilibrial values ("Fixed Points"), functional equilibria, operational equilibria, structural equilibria, etc.

(ii) that Eigenvalues Obs_i and their corresponding operators COORD stand to each other in a complementary relationship, the one implying the other, and vice versa; there the Obs_i represent the externally observable manifestations of the (introspectively accessible) cognitive computations (operations) COORD.

(iii) that Eigenvalues, because of their self-defining (or self-generating) nature imply topological "closure" ("circularity") (see Figures 2 and 3):

This state of affairs allows a symbolic re-formulation of expression (5);

$$\lim_{n\to\infty} COORD^{(n)} \equiv COORD\text{---}\!\!\!\boxed{}$$

that is, the snake eating its own tail: cognition computing its own cognitions.

$$* \quad * \quad * \quad * \quad *$$

Let there be, for a given operator COORD, at least three Eigenvalues

$$Obs_1, Obs_2, Obs_3,$$

FIGURE 2.

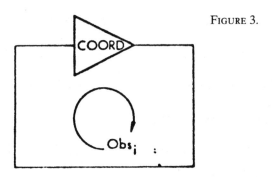

FIGURE 3.

and let there be an (albegraic) composition "*" such that

$$Obs_1 * Obs_2 = Obs_3, \tag{10}$$

then the coordinating operations COORD appear to coordinate the whole (i.e., the composition of the parts) as a composition of the apparent coordinations of the parts (see proof in Appendix B):

$$COORD(Obs_1 * Obs_2) = COORD(Obs_1) * COORD(Obs_2). \tag{11}$$

In other words, the coordination of compositions (i.e., the whole) corresponds to the composition of coordinations.

This is the condition for what may be called the "principle of cognitive continuity" (e.g., breaking pieces of chalk produces pieces of chalk).

This may be contrasted with the "principle of cognitive diversity" which arises when the Obs_i and the composition "*" are *not* the Eigenvalues and compositions complementing the coordination COORD′:

$$COORD'(Obs_1 * Obs_2) \neq COORD'(Obs_1) * COORD'(Obs_2), \tag{12}$$

and which says that the whole is neither more nor is it less than the sum of its parts: it is *different*. Moreover, the formalism in which this sentiment appears (expression (12)) leaves little doubt that it speaks neither of "wholes," nor of "parts" but of a subject's distinction drawn between two states of affairs which by an (other) observer may be seen as being not qualitatively, but only quantitatively distinct.

$$*\quad*\quad*\quad*\quad*$$

Eigenvalues have been found ontologically to be discrete, stable, separable and composable, while ontogenetically to arise as equilibria that determine themselves through circular processes. Ontologically, Eigenvalues and objects, and likewise, ontogenetically, stable behavior and the manifestation of a subject's "grasp" of an object cannot be distinguished. In both cases "objects" appear to reside exclusively in the subject's own experience of his sensori-motor coordinations; that is, "objects" appear to be exclusively subjective? Under which conditions, then, do objects assume "objectivity?"

Apparently, only when a subject, S_1, stipulates the existence of another subject, S_2, not unlike himself, who, in turn, stipulates the existence of still another subject, not unlike himself, who may well be S_1.

In this atomical social context each subject's (observer's) experience of his own sensori-motor coordination can now be referred to by a token of this experience, the "object," which, at the same time, may be taken as a token for the externality of communal space.

With this I have returned to the topology of closure

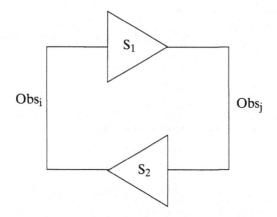

where equilibrium is obtained when the Eigenbehaviors of one participant generate (recursively) those for the other (see, for instance, Appendix Example A 2); where one snake eats the tail of the other as if it were its own, and where cognition computes its own cognitions through those of the other: here is the origin of ethics.

Appendix A

Examples:

A1. Consider the operator (linear transform) Op_1:

$$Op_1 = \text{"divide by two and add one"}$$

and apply it (recursively) to x_0, x_1, etc., (whose domains are the real numbers).

Choose an initial x_0, say $x_0 = 4$.

$$x_1 = Op_1(4) = \frac{4}{2} + 1 = 2 + 1 = 3;$$
$$x_2 = Op_1(3) = 2.500;$$
$$x_3 = Op_1(2.500) = 2.250;$$
$$x_4 = Op_1(2.250) = 2.125;$$
$$x_5 = Op_1(2.125) = 2.063;$$
$$x_6 = Op_1(2.063) = 2.031;$$
$$x_{11} = Op_1(x_{10}) = 2.001;$$
$$x_\infty = Op_1(x_\infty) = 2.000$$

Choose another initial value; say $x_0 = 1$

$$x_1 = Op_1(1) = 1.500;$$
$$x_2 = Op_1(1.500) = 1.750;$$
$$x_3 = Op_1(1.750) = 1.875;$$
$$x_8 = Op_1(x_7) = 1.996;$$
$$x_{10} = Op_1(x_9) = 1.999;$$
$$x_\infty = Op_1(x_\infty) = 2.000$$

And indeed:

$$\frac{1}{2} \cdot 2 + 1 = 2$$

$$Op_1(2) = 2$$

i.e., "2" is the (only eigenvalue of Op_1.

A2. Consider the operator Op_2:

$$Op_2 = \exp(\cos \quad).$$

There are three eigenvalues, two of which imply each other ("bi-stability"), and the third one being instable:

$$Op_2(2.4452\ldots) = 0.4643\ldots$$
$$\qquad\qquad\qquad\qquad\qquad \text{stable}$$
$$Op_2(0.4643\ldots) = 2.4452\ldots$$
$$Op_2(1.3029\ldots) = 1.3092\ldots \quad \text{instable}$$

This means that:

$$Op_2^{(2)}(2.4452\ldots) = 2.4452 \text{ stable}$$

$$Op_2^{(2)}(0.4643\ldots) = 0.4643 \text{ stable}$$

A3. Consider the differential operator Op_3:

$$Op_3 = \frac{d}{dx}.$$

The eigenfunction for this operator is the exponential function "exp:"

$$Op_3(exp) = exp$$

i.e.,

$$\frac{de^x}{dx} = e^x$$

The generalizations of this operator are, of course, all differential equation, integral equations, integro-differential equations, etc., which can be seen at once when these equations are re-written in operator form, say:

$$F(Op_3^{(n)}, Op_3^{(n-1)}. \ldots, f) = 0$$

Of course, these operators, in turn, may be eigenvalues (eigen-operators) of "meta-operators" and so on. This suggests that COORD, for instance, may itself be treated as an eigen-operator, stable within bounds, and jumping to other values whenever the boundary conditions exceed its former stable domain:

$$Op(COORD_i) = COORD_i.$$

One may be tempted to extend the concept of a meta-operator to that of a "meta-meta-operator" that computes the "eigen-meta-operators," and so on and up a hierarchy without end. However, there is no need to invoke this escape as Warren S. McCulloch has demonstrated years ago in his paper (1945): "A Heterarchy of Values Determined by the Topology of Nervous Nets."

It would go too far in this presentation to demonstrate the construction of heterarchies of operators based on their composability.

A4. Consider the (self-referential) proposition:

"THIS SENTENCE HAS ... LETTERS"

and complete it by writing into the appropriate space the word for the number (or if there are more than one, the numbers) that make this proposition true.

Proceeding by trial and error (comparing what this sentence says (abscissa) with what it is (ordinate)): one finds two eigenvalues "thirty-one" and "thirty-three." Apply the proposition above to itself: "this sentence has thirty-one

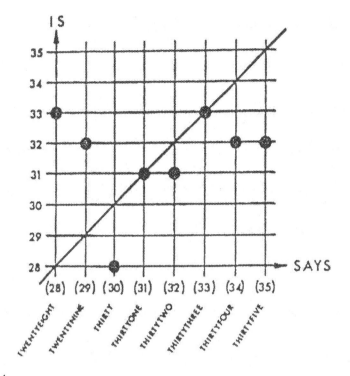

F<small>IGURE</small> 4.

letters' has thirty-one letters." Note that, for instance, the proposition: "this sentence consists of . . . letters" has only one eigenvalue (thirty-nine); while the proposition: "This sentence is composed of . . . letters" has none!

Appendix B

B1. Proof of Expression (11):

$$COORD(Obs_1 * Obs_2) = COORD(Obs_3)$$
$$= Obs_3 = Obs_1 * Obs_2 = COORD(Obs_1) * COORD(Obs_2)$$
Q.E.D.

The apparent distributivity of the operator COORD over the composition "*" should not be misconstrued as "*" being a linear composition. For instance, the fixed points $u_i = \exp(2\pi\lambda i)$, (for $i = 0, 1, 2, 3 \dots$) that complement the operator Op(u):

$$Op(u) = u \tan\left(\frac{\pi}{4} \pm \frac{1}{\lambda}\ln u\right),$$

with λ an arbitrary constant, compose multiplicatively:

$$Op(u_i * u_j) = Op(u_i) * Op(u_j).$$

etc.

References

McCulloch, W. S., *A Heterarchy of Values Determined by the Topology of Nervous Nets*, Bulletin of Mathematical Biophysics, 1945, 7: 89–93.

McCulloch, W. S. and Pitts, W. H., *A Logical Calculus of the Ideas Immanent in Nervous Activity*, Bulletin of Mathematical Biophysics, 1943, 5: 115–133.

Piaget, T., *L'Equilibration des Structures Cognitives*. Presses Univ. France, Paris, 1975.

12
Disorder/Order: Discovery or Invention?*

Heinz von Foerster

Ladies and Gentlemen: This is a great symposium. I enjoy every minute of it. However, I feel there is a blemish, and this is that Gregory Bateson is not with us. The reason why I, in particular, am sad he is not among us is not only because he would have enjoyed tremendously being here, and you would have enjoyed him very much as well, but because I need his help to put to rest one of the questions which has continually recurred during this conference. Here is the question: *Are the states of order and disorder states of affairs that have been discovered, or are these states of affairs that are invented?*

As I tend to say they are invented, I need all the help I can muster in order to defend this position, and so invoke the spirit of Gregory Bateson to stand on my side and to help me now for my defense. I will ask him to give us one of his very charming vignettes which pretend to be dialogues between a fictitious daughter and a fictitious father. (I do not think these fictions are too fictitious, indeed.) These dialogues he called *Metalogues*, and I will read you one now with a few comments on my side. This one is entitled, *Metalogue: What is an Instinct?* It begins with daughter asking father, "Daddy, what is an instinct?" Now, if my daughter, or my son, had asked me, "Daddy, what is an instinct?" I, most likely, would have fallen into the trap of giving a learned, lexical definition. I, for instance, would have said: "An instinct, my dear, is the innate behavior of animals which is unlearned, has a certain complexity, etc.," or something like that. However, Bateson does not fall into that trap and, for an answer to "Daddy, what is an instinct?" he says: "An instinct, my dear, is an explanatory principle." That is not good enough for her; therefore, daughter immediately asks, "But what does it explain?" And he replies (now watch!) "Anything, almost anything at all, anything you want it to explain." Now, please realize, that something which explains "anything you want it to explain" of course explains nothing. But daughter immediately senses something, and she says, "Don't be silly. It doesn't explain gravity!" The father: "No, but that is because nobody

---

* This article was first published in *Disorder and Order*, P. Livingston (ed.), Anna Libri, Saratoga, pp. 177–189 (1984).

wants instinct to explain gravity. If they did, it would explain it. We could simply say, "The moon has an instinct whose strength varies inversely as the square of the distance . . ." Daughter: "But this is nonsense, Daddy!"—"Yes, surely, but it was you who mentioned instinct, not I."—"All right, but what does then explain gravity?"—"Nothing, my dear. Because gravity is an explanatory principle." "Oh," says the daughter, "now, do you mean you cannot use one explanatory principle to explain another principle, never?" Father: "Hardly ever. That is what Newton meant when he said, '*hypotheses non fingo*'."—"And what does that mean, please?" asks daughter. (Now I would like to draw your attention to the fact that when the father gives his answer, everything that he says is put in the descriptive domain. It is always associated with saying or with pointing.) Again, daughter: "What does that mean, please?" Father: "Well, you know what hypotheses are. Any statement linking together two descriptive statements is a hypothesis. If you say there was a full moon on February 1, and another on March 1, and then you link these two descriptions together in any way, the statement which links them is a hypothesis."—"Yes, and I know what *non* means. But what is *fingo*?"—"Well, *fingo* is a late Latin word for 'make'. It forms a verbal noun, *fictio*, from which we get the word 'fiction'."—"Daddy, do you mean that Sir Isaac Newton thought that all hypotheses were just made up, like stories?" Father: "Yes, precisely that."—"But didn't he discover gravity? With the apple?"—"No, dear, he invented it!"

With this Batesonian dialogue I have, as it were, set the stage for what I am going to say. My original plan was to make some historical remarks in regard to the notion of disorder and order; however, during the development of this conference, I realized I should indeed shift my emphasis. There were two points which persuaded me to do this: one, I realized that we have the tremendous pleasure of having Michel Serres here, who is one of the eminent historians and could of course say much better anything historical than I could ever invent; the second point is that I am not the last speaker, and since I feel that this conference has historical significance and what I will say today will be obliterated tomorrow, I am very happy that, in their wisdom, the organizers of this conference have put Michel Serres as the last speaker; moreover, I hope he will satisfy Edgar Morin's request that the observer include himself in the observation, for he would then also be a contributor to the history of this conference.

To what, then, am I to address myself when I am not addressing myself to history? I shall shift from the historical to the epistemological, because I have the feeling that many of the questions that have been raised during this conference have an epistemological root. Nevertheless, with your permission, I will make two points, where I will have osculations with historical events regarding the notions of disorder and order, and this is when our topic touches a certain branch of poetry, namely, thermodynamics. These points I shall discuss because I have seen that, again and again during this symposium, notions which developed from an interaction between people in the scientific fields, let us say, the thermodynamicists and others, a *lingo*, a language, a notation, evolved, which is being used here, alas, in a some-what loose fashion, and I would like to recall for you the occasion on which these notions arose. After I have made these brief contacts with history just to see the perspectives, I will then try to show that the notions of disorder, order, and organization are conceptually linked to a general notion of computation. This will give me a platform, first to talk in quantitative terms about order and complexity, hence of those processes by which order, or complexity, is increased or decreased; but secondly—and this is the essential justification for my tying these notions to computation—to show that these measures are fully dependent upon the chosen framework (which turns out to be the language) in which these computations are carried out. In other words, the amount of order, or of complexity, is unavoidably tied to the language in which we talk about these phenomena. That is, in changing language, different orders and complexities are created, and this is the main point I would like to make.

Since a free choice is given to us which language we may use, we have moved this point into a cognitive domain, and I will reflect upon two types of cognition which I already touched upon in my introductory statement; namely, the problem of whether the states that we call "disorder and order" are states of affairs that are discovered or invented. When I take the position of *invention*, it becomes clear that the one who invents is of course responsible for his or her invention. At the moment when the notion of responsibility arises, we have the notion of *ethics*. I will then develop the fundamental notion of an ethics that refutes ordering principles attempting to organize the other by the injunction, "Thou shalt," and replace it by the organizational principle, that is, organizing oneself with the injunction "I shall." With this note I have given you a brief outline of my talk. Now, ladies and gentlemen, I can begin with my presentation!

First, I would like you to come with me to the year 1850. This is approximately the time when the First Law of Thermodynamics was well established, one understood the principle of conservation of energy, and the Second Law of Thermodynamics was just in the making. What was observed and what was intriguing people very much at that time was an interesting experiment. I ask you to look with me please at the following fascinating state of affairs. Consider two containers, or reservoirs, of the same size. One is hot, and the other one is cool. Now you take these containers, put them together, fuse them, so to speak, and watch what happens. Spontaneously, without our doing anything to them, the cold container will become warmer, and the warmer will become colder. Now, you may say, "O.K., so what?" But, ladies and gentlemen, if you say, "so what?" to anything, you will not see anything.

The engineers (and as Mr. Prigogine has so properly said, thermodynamics was an engineering science), who were working with steam engines, heat engines, etc., were wondering about the efficiency of these machines. They knew very well that if one has a hot and a cold container, one can put between these two vessels a heat engine that will do some work for us, drilling, pumping, pulling, and things like that. But they also knew that the smaller the temperature difference between these two containers is, the less the chance of getting a heat engine going; this means that the possibility of changing heat into work becomes less and less as the temperatures of the two containers become more and more alike.

When Clausius thought about that very carefully, he realized what is going on here: with the decrease in the difference between the two temperatures, the convertibility, the change, the turning of heat energy into work, becomes less and less possible. Therefore he wanted to give this possibility of being able to turn or to change heat into work a good and catchy name. At that time it was very popular to use Greek for neologisms. So he went to his dictionary and looked up the Greek for "change," and "turn." He found the word *trope*. "Aha," he said, "but I would like to talk about *not change*, because, you see, the longer these processes go on, the less heat can be turned into work." Now unfortunately, either he had a lousy dictionary, or he could not speak Greek very well, or he had friends who did not understand what he was talking about. Instead of calling it *utropy*, because *ou* is the Greek word for *non*, as in "Utopia" (no place)—and *utropy* is what he should have called his new concept—for some reason he called it "entropy," because he thought that *en* is the same as the Latin *in* and therefore means "no." That is why we are stuck with the wrong terminology. And what is worse, nobody checked it! An incredible state of affairs! So, in proper *lingo*, when these two containers are put together, the *utropy* of the two increases, because the possibility for changing, for transforming the heat into work becomes less and less.

A couple of years later, two gentlemen, one in Scotland, one in Austria, one in Edinburgh, the other in Vienna, one by the name of Clerk Maxwell, and the other by the name of Ludwig Boltzmann, were intrigued by a fascinating hypothesis, a hypothesis which was so crazy that most of their colleagues in the academic community refused even to talk about that stuff. They were contemplating whether it would be possible to think of matter as not being indefinitely divisible, so that at a particular level of subdivision, one could not subdivide any further. That is, one would be left with small pieces of mass. "Mass" is *moles* in Latin, and for a small thing, one puts on the diminutive suffix, which is -*cula*, and we get the hypothetical "molecules" that would not allow further division.

Contemplate whether this hypothesis makes any sense at all. To put you into the perspective of that time, 1871 or 1872, Boltzmann, who was teaching in Vienna, occupied one chair in physics. The other chair belonged to Ernst Mach, whose name, I believe, is familiar to you. Mach went into the Boltzmann lectures, sitting in the last row of the big physics auditorium, and when Boltzmann used the word "molecule" in his lectures, Mach screamed from the last row, "Show me one!" Of course, at that time one could not show one; they were purely hypothetical. Anyway, these two gentlemen, Maxwell and Boltzmann, addressed themselves to the problem of whether we can indeed interpret some of the fundamental laws of physics as if matter were composed of elementary particles, the molecules. They succeeded. They showed that three fundamental quantities in thermodynamics could be expressed in terms of molecular properties. The one is pressure. It is interpreted as a hailstorm of molecules flying against the walls of a container. The kinetic energy, or the speed of the molecules, would determine temperature. And then they came to the notion of entropy, or *utropy*, as I would say, and here a fascinating thing happened.

They could not explain utropy in purely molecular terms, and had to make an appeal to the cognitive functions of the observer. This is the first time when, in science, the observer enters into his descriptive system. What was necessary in order to handle the notion of utropy, was to talk about the distinguishability of states of affairs. I will give you an example. Take again the two boxes which can be distinguished by their different temperatures: one at a high temperature, the other at a low temperature. Put them together so that they are fused. Now the hotter will become colder, and the colder slowly warmer, and as time goes on their distinction will be lost: they become more and more "confused." Better, the observer becomes "confused" because he will be unable to distinguish between the two containers, his confusion increasing with the increase of the utropy. Here you have one version of the Second Law of Thermodynamics: utropy increases with confusion. Or, as others may say: entropy increases with disorder.

Seeing the Fundamental Laws of Thermodynamics, which were originally formulated so as to account for a macroscopic phenomenology, to have—in turn—their foundation in a microscopic mechanics, stimulated questions about the potential and limits of these Fundamental Laws.

I can see Clerk Maxwell sitting there, dreaming up some mischief about how to defeat the Second Law of Thermodynamics: "Hmm, if I have two containers at equal temperature, what must go on between them so that, without external interference, the one gets hotter, while the other gets colder?" Or, if you wish, letting order (discriminability) emerge from disorder (indiscriminateness), i.e., reducing the entropy of the system. Maxwell, indeed, came up with a charming proposal by inventing a demon who would operate according to a well-defined rule. This demon is to guard a small aperture in the wall separating the two containers and to watch the molecules that come flying toward this aperture. He opens the aperture to let a molecule pass whenever a fast one comes from the cool side or a slow one comes from the hot side. Otherwise he keeps the aperture closed. Obviously, by this maneuver he gets the cool container cooler (for it loses all its "hot" molecules) and the hot container hotter (for it loses all its "cool" molecules), thus apparently upsetting the Second Law of Thermodynamics. So, Maxwell invented his famous demon, whose name is, of course, "Maxwell's Demon," and for quite a while it was thought he would indeed have defeated the Second Law. (Later on, however, it was shown—but that is quite irrelevant to my story—that indeed, the Second Law of Thermodynamics is upheld, even with the demon working. Because in order for the demon to judge whether these molecules are fast or slow, he must of course have a flashlight to see these molecules; but a flashlight has a battery, and batteries run out, and there of course goes the hope of having defeated the Second Law of Thermodynamics!)

But there is another point that I would like to make regarding this demon, and that is that he is the incorporation par excellence not only of any principle that generates distinctions and order, but also of a general notion of computation. One of the most fundamental concepts of computation, I submit, was developed in the thirties by the English mathematician Alan Turing. He exemplified his notion with the aid of a fictitious machine, a conceptual device, the internal states of which are controlled by one, and are controlling the other one of the machine's two external parts. The first one is a (theoretically infinite) long tape that is subdivided into equal-sized squares on which from a given alphabet (one may say "language"), erasable symbols can be written. The other part is a reading/writing head which scans the symbol on the square below it and, depending upon the machine's internal state, will either change this symbol or else leave it unchanged. After this it will move to the next square, either to the left or to the right, and finally will change its internal state. When these operations are completed, a new cycle can begin, with the head now reading the symbol on the new square. In a famous publication, Turing proved that this machine can indeed compute all computable numbers or, as I would say in reference to our topic, all "conceivable arrangements."[1]

What I would like now to demonstrate is that this machine—whose name is, of course, the "Turing Machine"—and Maxwell's demon are functional isomorphs or, to put it differently, that the machine's computational competence and the demon's ordering talents are equivalent. The purpose of my bringing up this equivalence is, as you may remember from my introductory comments, to associate with the notions of disorder, order, and complexity, measures that permit us to talk about different degrees of order, say: "More order here!" or "Less order there!", and to watch the processes that are changing these degrees.

Let us now quickly go through the exercise of this demonstration by comparing the machine's M and the demon's D actions during the five steps of one complete cycle. Step (i): M reads symbol, D watches molecule; (ii): M compares symbol with internal state, D compares molecule's speed with internal standard; (iii): M operates on symbol and tape, D on aperture, opening or closing it; (iv): M changes its internal states, D its internal standard; (v): M and D go back to (i). Q.E.D.

Knowing about this equivalence puts us in the position of transforming any ordering problem into a computational one. Consider, for instance, an

---

[1] "On Computable Numbers with an Application to the Entscheidungsproblem," in *Proceedings of the London Mathematical Society* 2, no. 42 (1936), 230–65.

arbitrary arrangement, A, and its representation on the tape of a Turing Machine by using a certain alphabet (language). What Turing showed is that there exists another tape expression, called the "description" of A, which, when used as the initial tape expression will allow the machine to compute from it the arrangement A. Let me now draw your attention to three measures (numbers). One is the length $L(A)$ (that is, the number of squares) of the tape that is taken up by the arrangement A; the second is the length $L(D)$ of A's description (the initial tape expression); and the third figure is N, the number of cycles the machine has to go through to compute the arrangement A from its description D.

Now we can collect some fruits from our intellectual investment into the notions of machines, demons, etc. I will describe just four:

(i) Order

If the initial tape expression, the description, is short, and what is to be computed, the arrangement, is very long $(L(D) < L(A))$, then it is quite obvious that the arrangement possesses lots of order: a few rules will generate A. Take A to be $0, 1, 2, 3, 4, 5, 6, 7, \ldots, 999,999, 1,000,000$. A suitable description of this arrangement may be: *Each follower equals its precursor + 1.*

(ii) Disorder

If the length of the description approximates the length of the arrangement, it is clear that we do not understand this arrangement, for the description just parrots the arrangement. Take A to be:

$$8, 5, 4, 9, 1, 7, 6, 3, 2, 0.$$

I challenge the mathematicians present, or any puzzle wizard, to come up with a rule other than: *write $8, 5, 4, \ldots$* that generates this arrangement.

(iii) Complexity

I propose to use N, the number of cycles for computing an arrangement, as a measure for the complexity of this arrangement. In other words, I suggest that we associate with the complexity of an arrangement the time it takes the machine to compute it. For instance, during this meeting a juxtaposition molecule/man was made with the suggestion—so I understood—to learn about the properties of human beings from the known properties of molecules. In computational jargon such computations are usually referred to as computations *ab ovo* or, as in our case *ab molecula*. From this point of view it may be not too difficult to see that N, the number of computational steps, will be so large (e.g., the age of the universe being too short to accommodate N) that N becomes "trans-computational." That means, we can just forget about the whole thing, for we shall never see the end of it!

(iv) Language

The choicest of the four fruits I have left to be the last for you to taste, for it is the most crucial one in my narrative. It is the observation that all the three quantities mentioned before: the length of an arrangement, the length of its description, and the length of computing this arrangement, are drastically changed by changing from one alphabet *a* to another one, say, *b*. In other words, the degree of disorder or order that can be seen in an arrangement depends in a decisive way upon the choice of language (alphabet) that is used in these operations. Take as an example my telephone number in Pescadero. It is *879-0616*. Shift to another alphabet, say, the binary alphabet. In that language my number is 100001100010001001011000. Should you have difficulties remembering that number, shift back to the former language!

Take as another example the random number sequence 8, 5, 4, etc., I spoke of earlier (point *ii*). I suggest shifting from an alphabet that uses Arabic numerals to one that spells out each numeral in English: 8—eight, 5—five, 4—four, etc., and it becomes clear that under this alphabet the former "random sequence" is well determined, hence has a very short description: it is "alphabetical" (eight, five, four, nine, one, etc.).

Although I could go on with a wealth of examples that would drive home again and again the main points of my argument, in the hope that the foregoing examples suffice I will summarize these points in two propositions. Number one: A computational metaphor allows us to associate the degree of order of an arrangement with the shortness of its description. Number two: The length of descriptions is language-dependent. From these two propositions, a third one, my essential clincher, follows: Since language is not something we discover—it is our choice, and it is we who invent it—disorder and order are our inventions![2]

With this sequence I have come full circle to my introductory claim that I shall once and for all put to rest the question of whether disorder and order are discoveries or our inventions. My answer, I think, is clear.

Let me draw from this constructivist position a few epistemological consequences that are inaccessible to would-be discoverers.

One of these is that properties that are believed to reside in things turn out to be those of the observer. Take, for instance, the semantic sisters of Disorder: Noise, Unpredictability, Chance; or those of Order: Law, Predictability, Necessity. The last of these two triads, Chance and Necessity, have been associated until even recently with Nature's working. From a constructivist point of view, Necessity arises from the ability to make infallible deductions, while Chance arises from the inability to make infallible inductions. That is, Necessity and Chance reflect some of *our* abilities and inabilities, and not those of Nature.

More of that shortly. For the moment, however, let me entertain the question of whether there exists a biological backup for these notions. The answer is yes, and indeed, I am very happy that we have just those people around who were producing this very backup that allows me to speak about

---

[2] Except for the Greeks, who believed that it was the Gods who invented language, and that we humans are doomed to discover it.

an organism as an autonomous entity. The original version came from three Chilean neuro-philosophers, who invented the idea of autopoiesis. One of them is sitting here, Francisco Varela; another one is Umberto Maturana, and the third one is Ricardo Uribe, who is now at the University of Illinois. They wrote the first paper in English on the notion of autopoiesis, and in my computer language I would say that autopoiesis is that organization which computes its own organization. I hope that Francisco will not let me down tomorrow and will address himself to the notion of autopoiesis. Autopoiesis is a notion that requires systemic closure. That means organizational, but not necessarily thermodynamic, closure. Autopoietic systems are thermodynamically open, but organizationally closed.

Without going into details I would like to mention that the concept of closure has recently become very popular in mathematics by calling upon a highly developed branch of it, namely, Recursive Function Theory. One of its concerns is with operations that iteratively operate on their outcomes, that is, they are operationally closed. Some of these results are directly associated with notions of self-organization, stable, unstable, multiple and dynamic equilibria, as well as other concepts that would fit into the topic of our symposium.

However, traditionally there have always been logical problems associated with the concept of closure, hence the reluctance until recently to take on some of its problematic aspects. Consider, for example, the relation of an observer to the system he observes. Under closure, he would be included in the system of his observation. But this would be anathema in a science where the rule is "objectivity." Objectivity demands that the properties of the observer shall not enter the descriptions of his observations. This proscription becomes manifest when you submit to any scientific journal an article containing a phrase like "I observed that . . ." The editor will return it with the correction "It can be observed that . . ." I claim that this shift from "I" to "it" is a strategy to avoid responsibility: "it" cannot be responsible; moreover, "it" cannot observe!

The aversion to closure, in the sense of the observer being part of the system he observes, may go deeper. It may derive from an orthodox apprehension that self-reference will invite paradox, and inviting paradox is like making the goat the gardener. How would you take it if I were to make the following self-referential utterance: "I am a liar." Do I speak the truth? Then I lie. But when I lie, I speak the truth. Apparently, such logical mischief has no place in a science that hopes to build on a solid foundation where statements are supposedly either true or else false.

However, let me say that the problems of the logic of self-reference have been handled very elegantly by a calculus of self-reference, whose author is sitting on my left (Varela). I hope he will not let me down and will give me a bit of self-reference when he speaks tomorrow!

Social theory needs agents that account for the cohesiveness of social structure. Traditionally the agents are seen in sets of proscriptions issued with some dictatorial flavor, usually of the form "Thou shalt not . . ." It is clear that everything I said tonight not only contradicts, but even refutes, such views. The three columns, autonomy, responsibility, choice, on which my position rests, are pointing in the opposite direction.

What would be my counter-proposal? Let me conclude my presentation with a proposition that may well serve as a Constructivist Ethical Imperative: "I shall act always so as to increase the total number of choices."

Discussion

PAUL WATZLAWICK [STANFORD]. Heinz, would you say that, in addition to what you call the "ethical imperative," there is still a further conclusion to be drawn, and that is that if you realize that you are the constructor of your own reality, you are then also free, and so the question of freedom enters, so there is a deontic quality to what you were saying?

VON FOERSTER. My response is: Yes, precisely.

KARL H. PRIBRAM [STANFORD MEDICAL SCHOOL]. Heinz, I agree with everything you said, and with what Francisco says, but I have a problem. And that problem is, given the kind of framework you have just "invented" for us, and which I like very much, why is it that when I go into the laboratory, something happens that surprises me? When I know how things are supposed to go, and they don't.

VON FOERSTER. You are a very inventive character—you even invent your surprises. For instance, when I was talking about the two containers that are brought together and said that a most surprising thing is taking place, namely, that the hot one is getting cooler, and the cool one getting hotter, I felt that apparently this was seen as a joke—of course, everybody knows that, so what? But my hope was that you would try to see this phenomenon as if for the first time, as something new, something fascinating. Let me illustrate this point. I don't know whether you remember Castaneda and his teacher, Don Juan. Castaneda wants to learn about things that go on in the immense expanses of the Mexican chaparral. Don Juan says, "You see this . . .?" and Castaneda says "What? I don't see anything." Next time, Don Juan says, "Look here!" Castaneda looks, and says, "I don't see a thing." Don Juan gets desperate, because he wants really to teach him how to see. Finally, Don Juan has a solution. "I see now what your problem is. You can only see things that you can explain. Forget about explanations, and you will see." You were surprised because you abandoned your preoccupation with explanations. Therefore, you could see. I hope you will continue to be surprised.

13
Cybernetics of Cybernetics*

HEINZ VON FOERSTER
University of Illinois, Urbana

Ladies and gentlemen—As you may remember, I opened my remarks at earlier conferences of our Society with theorems which, owing to the generosity of Stafford Beer, have been called "Heinz von Foerster's Theorems Number One and Number Two". This all is now history.[1,10] However, building on a tradition of two instances, you may rightly expect me to open my remarks today again with a theorem. Indeed I shall do so but it will not bear my name. It can be traced back to Humberto Maturana,[7] the Chilean neurophysiologist, who a few years ago, fascinated us with his presentation on "autopoiesis", the organization of living things.

Here is Maturana's proposition, which I shall now baptize "Humberto Maturana's Theorem Number One":

"Anything said is said by *an observer."*

Should you at first glance be unable to sense the profundity that hides behind the simplicity of this proposition let me remind you of West Churchman's admonition of this afternoon: "You will be surprised how much can be said by a tautology". This, of course, he said in utter defiance of the logician's claim that a tautology says nothing.

I would like to add to Maturana's Theorem a corollary which, in all modesty, I shall call "Heinz von Foerster's Corollary Number One":

"Anything said is said to *an observer."*

With these two propositions a nontrivial connection between three concepts has been established. First, that of an *observer* who is characterized by being able to make descriptions. This is because of Theorem 1. Of course, what an observer says is a description. The second concept is that of *language*. Theorem 1 and Corollary 1 connect two observers through language. But, in turn, by this connection we have established the third concept I wish to consider this evening, namely that of *society*: the two observers constitute

---

* Originally published in *Communication and Control*, K. Krippendorff (ed.), Gordon and Breach, New York, pp. 5–8 (1979). Reprinted with permission.

the elementary nucleus for a society. Let me repeat the three concepts that are in a triadic fashion connected to each other. They are: first, the observers; second, the language they use; and third, the society they form by the use of their language. This interrelationship can be compared, perhaps, with the interrelationship between the chicken, and the egg, and the rooster. You cannot say who was first and you cannot say who was last. You need all three in order to have all three. In order to appreciate what I am going to say it might be advantageous to keep this closed triadic relation in mind.

I have no doubts that you share with me the conviction that the central problems of today are societal. On the other hand, the gigantic problem-solving conceptual apparatus that evolved in our Western culture is counter-productive not only for solving but essentially for perceiving social problems. One root for our cognitive blind spot that disables us to perceive social problems is the tradition\al explanatory paradigm which rests on two operation: One is *causation*, the other one *deduction*. It is interesting to note that something that cannot be explained—that is, for which we cannot show a cause or for which we do not have a reason—we do not wish to see. In other words, something that cannot be explained cannot be seen. This is driven home again and again by Don Juan, a Yaqui Indian, Carlos Casteneda's mentor.[2-5]

It is quite clear that in his teaching efforts Don Juan wants to make a cognitive blind spot in Castaneda's vision to be filled with new perceptions; he wants to make him "see". This is doubly difficult, because of Castaneda's dismissal of experiences as "illusions" for which he has no explanations on the one hand, and because of a peculiar property of the logical structure of the phenomenon "blind spot" on the other hand; and this is that we do not perceive our blind spot by, for instance, seeing a black spot close to the center of our visual field: we do not see that we have a blind spot. In other words, we do not see that we do not see. This I will call a second order deficiency, and the only way to overcome such deficiencies is with therapies of second order.

The popularity of Carlos Castaneda's books suggest to me that his points are being understood: new paradigms emerge. I'm using the term "paradigm" in the sense of Thomas Kuhn[6] who wants to indicate with this term a culture specific, or language specific, stereotype or model for linking descriptions semantically. As you may remember, Thomas Kuhn argues that there is a major change in paradigms when the one in vogue begins to fail, shows inconsistencies or contradictions. I however argue that I can name at least two instances in which not the emergent defectiveness of the dominant paradigm but its very flawlessness is the cause for its rejection. One of these instances was Copernicus' novel vision of a heliocentric planetary system which he perceived at a time when the Ptolemaeic geocentric system was at its height as to accuracy of its predictions. The other instance, I submit, is being brought about today by some of us who cannot—by their life—pursue any longer the flawless, but sterile path that explores the properties seen to reside within objects, and turn around to explore their very properties seen now to reside within the observer of these objects. Consider, for instance, "obscenity". There is at aperiodic intervals a ritual performed by the supreme judges of this land in which they attempt to establish once and for all a list of all the properties that define an obscene object or act. Since obscenity is not a property residing within things (for if we show Mr X a painting and he calls it obscene, we know a lot about Mr X but very little about the painting), when our lawmakers will finally come up with their imaginary list we shall know a lot about them but their laws will be dangerous nonsense.

With this I come now to the other root for our cognitive blind spot and this is a peculiar delusion within our Western tradition, namely, "objectivity":

"The properties of the observer shall not enter the description of his observations."

But I ask, how would it be possible to make a description in the first place if not the observer were to have properties that allows for a description to be made? Hence, I submit in all modesty, the claim for objectivity is nonsense! One might be tempted to negate "objectivity" and stipulate now "subjectivity". But, ladies and gentlemen, please remember that if a nonsensical proposition is negated, the result is again a nonsensical proposition. However, the nonsensicality of these propositions either in the affirmative or in their negation cannot be seen in the conceptual framework in which these propositions have been uttered. If this is the state of affairs, what can be done? We have to ask a new question:

"What are the properties of an observer?"

Let me at once draw your attention to the peculiar logic underlying this question. For whatever properties we may come up with it is we, you and I, who have to make this observation, that is, we have to observe our own observing, and ultimately account for our own accounting. Is this not opening the door for the logical mischief of propositions that refer to themselves ("I am a liar") that have been so successfully excluded by Russell's Theory of Types not to bother us ever again? Yes and No!

It is most gratifying for me to report to you that the essential conceptual pillars for a theory of the observer have been worked out. The one is a calculus of infinite recursions;[11] the other one is a calculus of self-reference.[9] With these calculi we are now able to enter rigorously a conceptual framework which deals with *observing* and not only with the observed.

Earlier I proposed that a therapy of the second order has to be invented in order to deal with dysfunctions of the second order. I submit that the cybernetics of observed systems we may consider to be first-order cybernetics; while second-order cybernetics is the cybernetics of observing systems. This is in agreement with another formulation that has been given by Gordon Pask.[8] He, too, distinguishes two orders of analysis. The one in which the observer enters the system by stipulating the *system's* purpose. We may call this a "first-order stipulation". In a "second-order stipulation" the observer enters the system by stipulating *his own* purpose.

From this it appears to be clear that social cybernetics must be a second-order cybernetics—a *cybernetics of cybernetics*—in order that the observer who enters the system shall be allowed to stipulate his own purpose: he is autonomous. If we fail to do so somebody else will determine a purpose for us. Moreover, if we fail to do so, we shall provide the excuses for those who want to transfer the responsibility for their own actions to somebody else: "I am not responsible for my actions; I just obey orders." Finally, if we fail to recognize autonomy of each, we may turn into a society that attempts to honor commitments and forgets about its responsibilities.

I am most grateful to the organizers and the speakers of this conference who permitted me to see cybernetics in the context of social responsibility. I move to give them a strong hand. Thank you very much.

References

1. Beer, S., *Platform for Change*: 327, New York: Wiley, 1975.
2. Castaneda, C., *The Teachings of Don Juan: A Yaqui Way of Knowledge*, New York: Ballantine, 1969.
3. Castaneda, C., *A Separate Reality*, New York: Simon and Schuster, 1971.
4. Castaneda, C., *Journey to Ixtlan*, New York: Simon and Schuster, 1972.
5. Casteneda, C., *Tales of Power*, New York: Simon and Schuster, 1974.
6. Kuhn, T., *The Structure of Scientific Revolution*, Chicago: University of Chicago Press, 1962.
7. Maturana, H., "Neurophysiology of cognition", in Garvin, P. (Ed.), *Cognition, A Multiple View*: 3–23, New York: Spartan Books, 1970.
8. Pask, G., "The meaning of cybernetics in the behavioral sciences (the cybernetics of behavior and cognition: extending the meaning of 'goal')" in Rose, J. (Ed.), *Progress in Cybernetics*, Vol. 1: 15–44, New York: Gordon and Breach, 1969.
9. Varela, F., "A calculus for self-reference", *International Journal of General Systems*, **2**, No. 1: 1–25, 1975.
10. Von Foerster, H., "Responsibility of competence", *Journal of Cybernetics*, **2**, No. 2: 1–6, 1972.
11. Weston, P. E. and von Foerster, H., "Artificial intelligence and machines that understand", in Eyring, H., Christensen, C. H., and Johnston, H. S. (Eds.), *Annual Review of Physical Chemistry*, **24**: 358–378, Palo Alto: Annual Review Inc., 1973.

14
Ethics and Second-Order Cybernetics*

HEINZ VON FOERSTER

Ladies and Gentlemen:

I am touched by the generosity of the organizers of this conference who not only invited me to come to your glorious city of Paris, but also gave me the honor of opening the Plenary sessions with my presentation. And I am impressed by the ingenuity of the organizers who suggested to me the title of my presentation. They wanted me to address myself to "Ethics and Second-Order Cybernetics." To be honest, I would have never dared to propose such an outrageous title, but I must say that I am delighted that this title was chosen for me.

Before I left California for Paris, others asked me full of envy, what am I going to do in Paris, what will I talk about? When I answered "I shall talk about Ethics and Second-Order Cybernetics" almost all of them looked at me in bewilderment and asked, "What is second-order cybernetics?" as if there were no questions about ethics. I am relieved when people ask me about second-order cybernetics and not about ethics, because it is much easier to talk about second-order cybernetics than it is to talk about ethics. In fact it is impossible to talk about ethics. But let me explain that later, and let me now say a few words about cybernetics, and of course, the cybernetics of cybernetics, or second-order cybernetics.

As you all know, cybernetics arises when effectors (say, a motor, an engine, our muscles, etc.) are connected to a sensory organ which in turn acts with its signals upon the effectors. It is this circular organization which sets cybernetic systems apart from others that are not so organized. Here is Norbert Wiener, who re-introduced the term "Cybernetics" into scientific discourse. He observed, "The behavior of such systems may be interpreted as directed toward the attainment of a goal." That is, it looks as if these systems pursued a purpose!

That sounds very bizarre indeed! But let me give you other paraphrases of what cybernetics is all about by invoking the spirit of women and men who

* Originally published in French in *Systèmes, Ethique, Perspectives en thérapie familiale*, Y. Ray et B. Prieur (eds.), ESF editeur, Paris, pp. 41–55 (1991).

rightly could be considered the mamas and papas of cybernetic thought and action. First there is Margaret Mead, whose name I am sure is familiar to all of you. In an address to the American Society of Cybernetics she remarked:

As an anthropologist, I have been interested in the effects that the theories of Cybernetics have within our society. I am not referring to computers or to the electronic revolution as a whole, or to the end of dependence on script for knowledge, or to the way that dress has succeeded the mimeographing machine as a form of communication among the dissenting young. Let me repeat that, I am *not* referring to the way that *dress* has succeeded the mimeographing machine as a form of communication among the dissenting young.

And she then continues:

I specifically want to consider the significance of the set of cross-disciplinary ideas which we first called "feed-back" and then called "teleological mechanisms" and then called it "cybernetics," a form of cross-disciplinary thought which made it possible for members of many disciplines to communicate with each other easily in a language which all could understand.

And here is the voice of her third husband, the epistemologist, anthropologist, cybernetician, and as some say, the papa of family therapy, Gregory Bateson, "Cybernetics is a branch of mathematics dealing with problems of control, recursiveness and information."

And here is the organizational philosopher and managerial wizard Stafford Beer, "Cybernetics is the science of effective organization."

And finally, here the poetic reflection of "Mister Cybernetics," as we fondly call him, the Cybernetician's cybernetician; Gordon Pask, "Cybernetics is the science of defensible metaphors."

It seems that cybernetics is many different things to many different people. But this is because of the richness of its conceptual base; and I believe that this is very good, otherwise cybernetics would become a somewhat boring exercise. However, all of those perspectives arise from one central theme; that of circularity. When, perhaps a half century ago, the fecundity of this concept was seen, it was sheer euphoria to philosophize, epistemologize, and theorize about its unifying power and its consequences and ramification on various fields. While this was going on, something strange evolved among the philosophers, the epistemologists and the theoreticians. They began to see themselves more and more as being included in a larger circularity; maybe within the circularity of their family; or that of their society and culture; or even being included in a circularity of cosmic proportions!

What appears to us today as being most natural to see and think, was then not only difficult to see, but wasn't even allowed to be thought. Why? Because it would violate the basic principle of scientific discourse which demands the separation of the observer from the observed. It is the principle of objectivity. The properties of the observer shall not enter the description of his observations.

I present this principle here, in its most brutal form, to demonstrate its non-sensicality. If the properties of the observer (namely to observe and describe) are eliminated, there is nothing left; no observation, no description. However, there was a justification for adhering to this principle, and this justification was fear; fear that paradoxes would arise when the observers were allowed to enter the universe of their observations. And you know the threat of paradoxes. To steal their way into a theory is like having the cloven-hoofed foot of the devil stuck in the door of orthodoxy.

Clearly when cyberneticians were thinking of partnership in the circularity of observing and communicating, they were entering into a forbidden land. In the general case of circular closure, A implies B; B implies C; and (Oh, horror!) C implies A! Or in the reflexive case, A implies B, and (Oh, shock!) B implies A! And now the devil's cloven-hoof in its purest form, the form of self-reference; A implies A (Outrage!)

I would like to invite you now to join me in a land where it is not forbidden; rather, where one is encouraged to speak about oneself. What else can one do anyway? This turn from looking at things "out there" to looking at "looking itself," arose I think, from significant advances in neurophysiology and neuropsychiatry. It appeared that one could now dare to ask the question of how the brain works. One could dare to write a theory of the brain.

It may be argued that over the centuries since Aristotle, physicians and philosophers again and again developed theories of the brain. So, what's new of today's cyberneticians? What is new is the profound insight that a brain is required to write a theory of a brain. From this follows that a theory of the brain, that has any aspirations for completeness, has to account for the writing of this theory. And even more fascinating, the writer of this theory has to account for her or himself. Translated into the domain of cybernetics; the cybernetician, by entering his own domain, has to account for his or her own activity. Cybernetics then becomes cybernetics of cybernetics, or *second-order cybernetics*.

Ladies and Gentlemen, this perception represents a fundamental change, not only in the way we conduct science, but also how we perceive teaching, learning, the therapeutic process, organizational management, and so on and so forth; and I would say, of how we perceive relationships in our daily life. One may see this fundamental epistemological change if one first considers oneself to be an independent observer who watches the world go by; as opposed to a person who considers oneself to be a participant actor in the drama of mutual interaction of the give and take in the circularity of human relations.

In the case of the first example, as a result of my independence, I can tell others how to think and act, "Thou shalt . . ." "Thou shalt not . . ." This is the origin of moral codes. In the case of the second example, because of my interdependence, I can only tell myself how to think and act, "I shall . . ." "I shall not . . ." This is the origin of ethics.

This was the easy part of my presentation. Now comes the difficult part. I am supposed to talk about ethics. How to go about this? Where to begin?

In my search for a beginning I came across the lovely poem by Yveline Rey and Bernard Prieur that embellishes the first page of our program. Let me read to you the first few lines:

"Vous avez dit Ethique?"
Déjà le murmur s'amplifie en rumeur.
Soudain les roses ne montrent plus des épines.
Sans doute le sujet est-il brûlant.
Il est aussi d'actualité.

Let me begin with epines – with the thorns – and I hope, a rose will emerge. The thorns I begin with are Ludwig Wittgenstein's reflections upon ethics in his Tractatus Logico-Philosophicus. If I were to provide a title for this tractatus, I would call it Tractatus Ethico-Philosophicus. However, I am not going to defend this choice, I rather tell you what prompts me to refer to Wittgenstein's reflections in order to present my own.

I'm referring to point Number 6 in his Tractatus where he discusses the general form of propositions. Near the end of this discussion he turns to the problem of values in the world and their expression in propositions. In his famous point Number 6.421 he comes to a conclusion which I will read to you in the original German, "Es ist Klar, dass sich Ethik nicht aussprechen laesst." I wish I knew a French translation. I only know two English translations which are both incorrect. Therefore, I will present *my* translation into English, with my conviction that the simultaneous translators will do a superb job of presenting Wittgenstein's point in French. Here is my English version of 6.421, "It is clear that ethics cannot be articulated."

Now you understand why earlier I said, "My beginning will be thorns." Here is an International Congress on Ethics, and the first speaker says something to the effect that it is impossible to speak about ethics! But please be patient for a moment. I quoted Wittgenstein's thesis in isolation. Therefore it is not yet clear what he wanted to say.

Fortunately, the next point 6.422, which I will read in a moment, provides a larger context for 6.421. To prepare for what you are about to hear, you should remember that Wittgenstein was a Viennese. So am I. Therefore there is a kind of underground understanding which I sense you Parisians will share with us Viennese. Let me try to explain. Here now is point 6.422 in the English translation by Pears and McGuinness; "When an ethical law of the form "Thou shalt . . ." is laid down, one's first thought is, 'And what if I do not do it?'" When I first read this, my thought was that not everybody will share Wittgenstein's view. I think that this reflects his cultural background.

Let me continue with Wittgenstein, "It is clear however, that ethics has nothing to do with punishment and reward in the usual sense of the terms. Nevertheless, there must indeed be some kind of ethical reward and punishment, but they must reside in the action itself."

They must reside in the action itself! You may remember, we came across such self-referential notions earlier with the example, "A implies A" and its recursive relatives of second-order cybernetics. Can we take a hint from these comments for how to go about reflecting about ethics, and at the same time adhere to Wittgenstein's criterion? I think we can. I myself try to adhere to the following rule; to master the use of my language so that ethics is implicit in any discourse I may have. (e.g., in science, philosophy, epistemology, therapy, etc.)

What do I mean by that? By that I mean to let language and action ride on an underground river of ethics, and to make sure that one is not thrown off. This insures that ethics does not become explicit and that language does not degenerate into moralizations. How can one accomplish this? How can one hide ethics from all eyes and still let her determine language and action? Fortunately, ethics has two sisters who allow her to remain unseen. They create for us a visible framework; a tangible tissue within which and upon which we may weave the goblins of our life. And who are these two sisters? One is Metaphysics, the other is Dialogics.

My job now is to talk about these two ladies, and how they manage to allow ethics to become manifest without becoming explicit.

Metaphysics

Let me first talk about Metaphysics. In order to let you see at once the delightful ambiguity that surrounds her, let me quote from a superb article, "The Nature of Metaphysics" by the British scholar W.H. Walsh. He begins his article with the following sentence, "Almost everything in metaphysics is controversial, and it is therefore not surprising that there is little agreement among those who call themselves metaphysicians about what precisely it is they are attempting."

Today, when I invoke Metaphysics, I do not seek agreement with anybody else about her nature. This is because I want to say precisely what it is when we become metaphysicians, whether or not we call ourselves metaphysicians. I say that we become a metaphysician any time we decide upon in principle undecidable questions. For instance, here is a decidable question, "Is the number 3,396,714 divisible by 2?" It will take you less than two seconds to decide that indeed this number is divisible by two. The interesting thing here is that it will take you exactly the same short time to decide if the number has not 7, but 7000 or 7 million digits. I could of course invent questions that are slightly more difficult; for instance, "Is 3,396,714 divisible by three?", or even more difficult ones. But there are also problems that are extraordinarily difficult to decide, some of them having been posed more than 200 years ago and remain unanswered.

Think of Fermat's "Last Theorem" to which the most brilliant heads have put their brilliant minds and have not yet come up with an answer. Or think of Goldbach's "Conjecture" which sounds so simple that it seems a proof cannot be too far away, "All even numbers can be composed as the sum of two primes." For example, 12 is the sum of the two prime numbers 5 and 7; or $20 = 17 + 3$; or $24 = 13 + 11$, and so on and so forth. So far, no counterexample to Goldbach's conjecture has been found. And even if all further tests would not refute Goldbach, it still would remain a conjecture until a sequence of mathematical steps is found that decides in favor of his good sense of numbers. There is a justification for not giving up and for continuing the search for finding a sequence of steps that would prove Goldbach. It is that the problem is posed in a framework of logico-mathematical relations which guarantees that one can climb from any node of this complex crystal of connections to any other node.

One of the most remarkable examples of such a crystal of thought is Bertrand Russell's and Alfred North Whithead's monumental *Principia Mathematica* which they wrote over a 10 year period between 1900 and 1910. This 3 volume *magnum opus* of more than 1500 pages was to establish once and for all a conceptual machinery for flawless deductions. A conceptual machinery that would contain no ambiguities, no contradictions and no undecidables.

Nevertheless, in 1931, Kurt Gödel, then 25 years of age, published an article whose significance goes far beyond the circle of logicians and mathematicians. The title of this article I will give you now in English, "On formally undecidable propositions in the *Principia Mathematica* and related systems." What Gödel does in his paper is to demonstrate that logical systems, even those so carefully constructed by Russell and Whitehead, are not immune to undecidables sneaking in.

However, we do not need to go to Russell and Whitehead, Gödel, or any other giants to learn about in principle undecidable questions. We can easily find them all around. For instance, the question about the origin of the universe is one of those in principle undecidable questions. Nobody was there to watch it. Moreover, this is apparent by the many different answers that are given to this question. Some say it was a single act of creation some 4 or 5,000 years ago. Others say there was never a beginning and that there will never be an end; because the universe is a system in perpetual equilibrium. Then there are those who claim that approximately 10 or 20 billion years ago the universe came into being with a "Big Bang" whose remnants one is able to hear over large radio antennas. But I am most inclined to trust Chuang Tse's report, because he is the oldest and was therefore the closest to the event. He says:

Heaven does nothing, this nothing-doing is dignity;
Earth does nothing, this nothing-doing is rest;
From the union of these two nothing-doings arise all action
And all things are brought forth.

I could go on and on with other examples, because I have not yet told you what the Burmese, the Australians, the Eskimos, the Bushmen, the

Ibos, etc., would tell you about their origins. In other words, tell me how the universe came about, and I will tell you who you are.

I hope that I have made the distinction between decidable and, in principle, undecidable questions sufficiently clear so that I may present the following proposition which I call the "metaphysical postulate:"

Only those questions that are in principle undecidable, *we* can decide.

Why? Simply because the decidable questions are already decided by the choice of the framework in which they are asked, and by the choice of the rules used to connect what we label "the question" with what we take for an "answer." In some cases it may go fast, in others it may take a long, long time. But ultimately we arrive after a long sequence of compelling logical steps at an irrefutable answer; a definite "yes," or a definite "no."

But we are under no compulsion, not even under that of logic, when we decide on in principle undecidable questions. There is no external necessity that forces us to answer such questions one way or another. We are free! The compliment to necessity is not chance, it is choice! *We can choose who we wish to become when we have decided on an in principle undecidable question.*

That is the good news, as American journalists would say, now comes the bad news. With this freedom of choice we are now responsible for the choice we make. For some, this freedom of choice is a gift from heaven. For others such responsibility is an unbearable burden. How can one escape it? How can one avoid it? How can one pass it on to somebody else?

With much ingenuity and imagination, mechanisms have been contrived by which one could bypass this awesome burden. Through hierarchies, entire institutions have been built where it is impossible to localize responsibility. Everyone in such a system can say, "I was told to do 'X.'" On the political stage, we hear more and more the phrase of Pontius Pilate, "I have no choice but 'X.'" In other words, "Don't hold me responsible for 'X.' Blame someone else." This phrase apparently replaces, "Among the many choices I had, I decided on 'X.'"

I mentioned objectivity before, and I mention it here again as a popular device for avoiding responsibility. As you may remember, objectivity requires that the properties of the observer be left out of any descriptions of his observations. With the essence of observing (namely the processes of cognition) having been removed, the observer is reduced to a copying machine with the notion of responsibility successfully juggled away.

Objectivity, Pontius Pilate, hierarchies, and other devices are all derivations of a choice between a pair of in principle undecidable questions which are, "Am I *apart from* the universe?" Meaning whenever I *look*, I'm looking as if through a peephole upon an unfolding universe; or, "Am I *part of* the universe?" Meaning whenever I *act*, I'm changing myself and the universe as well.

Whenever I reflect on these two alternatives, I'm surprised by the depth of the abyss that separates the two fundamentally different worlds that can be created by such a choice. That is to see myself as a citizen of an independent universe, whose regulations, rules and customs I may eventually discover; or to see myself as a participant in a conspiracy, whose customs, rules, and regulations we are now inventing.

Whenever I speak to those who have made their decision to be either discovers or inventors, I'm impressed by the fact that neither of them realizes that they have ever made that decision. Moreover, when challenged to justify their position, a conceptual framework is constructed which itself turns out to be the result of a decision upon an in principle undecidable question.

It seems as though I'm telling you a detective story while keeping quiet about who is the good guy and who is the bad guy; or who is sane and who is insane; or who is right and who is wrong. Since these are in principle undecidable questions, it is for each of us to decide, and then take responsibility for. There is a murderer. I submit that it is unknowable whether he is or was insane. The only thing we *know* is what I say, what you say, or what the expert says he is. And what I say, what you say, and what the expert says about his sanity or insanity is my, is your, and is the expert's responsibility. Again, the point here is not the question "Who's right and who's wrong?" This is an in principle undecidable question. The point here is freedom; freedom of choice. It is José Ortega y Gasset's point:

Man does not have a nature, but a history. Man is nothing but a drama. His life is something that has to be chosen, made up as he goes along. And a human consists in that choice and invention. Each human being is the novelist of himself, and though he may choose between being an original writer and a plagiarist, he cannot escape choosing. He is condemned to be free.

You may have become suspicious of me qualifying all questions as being in principle undecidable questions. This is by no means the case. I was once asked how the inhabitants of such different worlds as I sketched before, (the inhabitants of the world they discover, and the inhabitants of a world they invent) can ever live together. Answering that is not a problem. The discovers will most likely become astronomers, physicists and engineers; the inventors family therapists, poets, and biologists. And living together won't be a problem either, as long as the discoverers discover inventors, and the inventors invent discoverers. Should difficulties develop, fortunately we have this full house of family therapists who may help to bring sanity to the human family.

I have a dear friend who grew up in Marakesh. The house of his family stood on the street that divides the Jewish and the Arabic quarters. As a boy, he played with all the others, listened to what they thought and said, and learned of their fundamentally different views. When I asked him once who was right he said, "They are both right."

"But this cannot be," I argued from an Aristotelian platform, "Only one of them can have the truth!"

"The problem is not truth," he answered, "The problem is trust."

I understood. The problem is understanding. The problem is understanding understanding! The problem is making decisions upon in principle undecidable questions.

At that point Metaphysics appeared and asked her younger sister Ethics, "What would you recommend I bring back to my proteges, the metaphysicians, regardless of whether or not they refer to themselves as such?" Ethics answered, "Tell them they should always try to act so as to increase the number of choices. Yes, increase the number of choices!"

Dialogics

Now I would like to turn to Ethics' sister, Dialogics. What are the means at her disposal to insure that Ethics can manifest herself without becoming explicit? You may already have guessed that it is, of course, language. I am not referring here in the sense of the noises produced by pushing air past our vocal cords; or language in the sense of grammar, syntax, semantics, semiotics; nor the machinery of phrases, verb phrases, noun phrases, deep structure, etc. When I refer here to language, I refer to language the "dance." Similar to when we say "It takes two to Tango," I am saying, "It takes two to language."

When it comes to the dance of language, you the family therapists are of course the masters, while I can only speak as an amateur. Since "amateur" comes from "amour," you'll know at once that I love to dance this dance. In fact, what little I know of this dance I learned from you. My first lesson came when I was invited to sit in an observation room and observe through the one way mirror a therapeutic session in progress with a family of four. For a moment my colleagues had to leave, and I was by myself. I was curious as to what I would see when I couldn't hear what was said, so I turned off the sound.

I recommend that you perform this experiment yourself. Perhaps you will be as fascinated as I was. What I saw then, the silent pantomime, the parting and closing of lips, the body movements, the boy who only once stopped biting his nails . . . what I saw then were the dance steps of language, the dance steps alone, without the disturbing effects of the music. Later I heard from the therapist that this session was very successful indeed. I thought, what magic must sit in the noises these people produced by pushing air past their vocal cords and by parting and closing their lips. Therapy! What magic indeed! And to think that the only medicine at your disposal are the dance steps of language and its accompanying music. Language! What magic indeed!

It is left to the naive to believe that magic can be explained. Magic cannot be explained. Magic can only be practiced, as you all well know. Reflecting on the magic of language is similar to reflecting upon a theory of the brain. As much as one needs a brain to reflect upon a theory of the brain, one needs the magic of language to reflect upon the magic of language. It is the magic of those notions that they need themselves to come into being. *They are of second-order.* It is also the way language protects itself against explanation by always speaking about itself.

There is a word for language, namely "language." There is a word for word, namely "word." If you don't know what word means, you can look it up in a dictionary. I did that. I found it to be an "utterance." I asked myself, "What is an utterance?" I looked it up in the dictionary. The dictionary said that it means "to express through words." So here we are back where we started. Circularity; A implies A.

But this is not the only way language protects itself against explanation. In order to confuse her explorer she always runs on two different tracks. If you chase language up one track, she jumps to the other. If you follow her there, she is back on the first. What are these two tracks? One track is the track of appearance. It runs through a land that appears stretched out before us; the land we are looking at as though through a peephole. The other track is the track of function. It runs through the land that is as much a part of us as we are a part of it; the land that functions like an extension of our body.

When language is on the track of appearance it is a monologue. There are noises produced by pushing air past vocal cords. There are the words, the grammar, the syntax, the well formed sentences. Along with these noises goes the denotative pointing. Point to a table, make the noise "table"; point to a chair, make the noise "chair."

Sometimes it does not work. Margaret Mead quickly learned the colloquial language of many tribes by pointing to things and waiting for the appropriate noises. She told me that once she came to a particular tribe, pointed to different things, but always got the same noises, "chumulu." A primitive language she thought, only one word! Later she learned that "chu mulu" means "pointing with finger."

When language switches to the track of function it is dialogic. There are, of course, these noises; some of them may sound like "table," others like "chair." But there need not be any tables or chairs because nobody is pointing at tables or chairs. These noises are invitations to the other to make some dance steps together. The noises "table" and "chair" bring to resonance those strings in the mind of the other which, when brought to vibration, would produce noises like "table" and "chair." Language in its function is connotative.

In its appearance, language is descriptive. When you tell your story, you tell it as it was; the magnificent ship, the ocean, the big sky, and the flirt you had that made the whole trip a delight. But for whom do you tell it? That's the wrong question. The right question is; with whom are you going to dance your story, so that your partner will float with you over the decks of your ship, will smell the salt of the ocean, will let the soul expand over the sky? And there will be a flash of jealousy when you come to the point of your flirt.

In its function, language is constructive because nobody knows the source of your story. Nobody knows, nor ever will know how it was, because "as it was" is gone forever.

You remember René Descartes as he was sitting in his study, not only doubting that the was sitting in his study, but also doubting his existence. He asked himself, "Am I, or am I not?" "Am I, or am I not?" He answered this rhetorical question with the solipsistic monologue, "Je pense, donc je suis" or in the famous Latin version, "Cogito ergo sum." As Descartes knew very well, this is language in its appearance, otherwise he would not have quickly published his insight for the benefit of others in his "Discourse de la méthode." Since he understood the function of language as well, in all fairness he should have exclaimed, "Je pense, donc nous sommes", "Cogito ergo sumus" or, "I think, therefore *we* are."

In its appearance, the language I speak is *my* language. It makes me aware of myself. This is the root of *consciousness*. In its function, my language reaches out for the other. This is the root of *conscience*. And this is where Ethics invisibly manifests itself through dialogue. Permit me to read to you what Martin Buber says in the last few lines of his book *Das Problem des Menschen*:

Contemplate the human with the human, and you will see the dynamic duality, the essence together. Here is the giving and the receiving, here is the aggressive and the defensive power, here the quality of searching and of responding, always both in one, mutually complementing in alternating action, demonstrating together what it is; human. Now you can turn to the single one and you can recognize him as human for his potential of relating. We may come closer to answering the question, "What is human?" when we come to understand him as the being in whose dialogic, in his mutually present two-getherness, the encounter of the one with the other is realized and recognized at all times.

Since I cannot add anything to Buber's words, this is all I can say about ethics, and about second-order cybernetics.

Thank you very much.

Yveline Rey: Interview with Heinz von Foerster

Yveline: The first time I heard your name mentioned, it was accompanied by the term "cybernetician." How does one become a cybernetician? Why this choice at the beginning? What were the influential steps throughout the course of your life?

Heinz: Yes. How does one become a cybernetician? Or, perhaps you want me to tell you how *I* became a cybernetician.

You may remember the point I made in my address; that we all are metaphysicians, whether we call ourselves such, whenever we decide upon in-principle undecidable questions. To answer your question, I could also say we are all cyberneticians (whether or not we call ourselves such) whenever we justify our actions without using the words "because of ...," or "à cause de ...," but with the phrase in English "in order to ...," which in French is much more Aristotelian, "à fin de ..."

Y. Why Aristotelian?

H. In his *Metaphysics*, Aristotle distinguished four different kinds of causes or, as I would say, four different excuses; two of which have temporal character, "causa efficientis" and "causa finalis." Physicists love the former, where causes in the past determine the effects in the present: "*Because* she did turn the switch, the lights go on now." Psychologists prefer the latter: "*In order to* have the lights on, she turns the switch now." Causes in the future, "to have the room lit," determine actions in the present, "turn the switch now."

Y. Very interesting, but where does cybernetics come in?

H. Physicists explore the connection between the positions of the switch, making or breaking contact, and the electrical processes that heat the wires in the lamp to temperatures that are high enough to radiate electro-magnetical waves in the visible spectrum, etc., etc. Cyberneticians explore the connection between the little girl's wish to enter a lit as opposed to a dark room, as well as the senso-motoric processes and the emerging eye-hand correlation that bring her hand along an unpredictable path, but with a predictable outcome, closer and closer to the switch which she then turns in the right direction, etc. If one were to watch this girl, one might be tempted to say as did Norbert Wiener, "... her behavior may be interpreted as directed to the attainment of a goal." In the early cybernetic literature you will find again and again reference to the notion of "goal," "purpose," "end," etc. Since the Greek word for "end" is "telos," our pre-cyberneticians used "teleology" for identifying their activity.

Y. But, Heinz, you said before that we are all cyberneticians, whether or not we call ourselves such, but when I go to turn on a light switch I am not "exploring the senso-motoric connections ..." et cetera. I just go and turn on the switch. Where is the cybernetician?

H. (Laughing) This is one more reason why I love women! You look through all the scientific verbal haze and go straight to the essential points. Now ... Hmm ... What can I say?

I think I can extricate myself from this dilemma by inventing a new category of cybernetics: "Zero-order Cybernetics." I suggest we have a case of zero-order cybernetics when activity becomes structured; when "behavior" emerges, but one doesn't reflect upon the "why" and the "how" of this behavior. One just acts. This is when cybernetics is implicit.

Y. I see. But what is now "First-order cybernetics?"

H. This is when one reflects upon one's behavior, upon the "how" and the "why." Then cybernetics becomes explicit, and one develops notions like "feedback," "amount of information," "circularity," "recursion," "control" "homeostasis," "dynamic stability," "dynamic instability or chaos," "fixed points," "attractors," "equi-finality," "purpose," "goal," etc., etc. In other words, one arrives at the whole conceptual machinery of "early" cybernetics, first-order cybernetics; or as I would say, the cybernetics of observed systems.

Y. Let me come back to my first question. How did you come upon cybernetics?

H. Very simple. Cybernetics came upon me; because my English vocabulary was at most 25 words.

Y. This makes no sense, dear Heinz. You'll have to explain that a bit better.

H. Okay. Then we have to go back to a time when you, dear Yveline, were not yet born. We have to go back to the year 1948, when parts of Austria were still occupied by Russian troops, and the world was slowly recovering from the wounds of the war. In November of that year, in Cambridge, Massachusetts, Norbert Wiener published a book entitled *Cybernetics*, with the subtitle Communication and Control in the Animal and the Machine. Also that November, Heinz von Foerster in Vienna, Austria, published a book entitled *Das Gedächtnis* [The memory] with the subtitle *Eine quantenphysikalische Untersuchung*, [An investigation in quantum physics]. I am originally a physicist, and what I tried to do in this investigation was to connect observations in experimental psychology and neurophysiology with the physics of the large (biological) molecules. I think that I didn't do a bad job of it.

Now I have to switch to another track. My wife's dearest friend, Ilse, had escaped from Germany when Hitler came into power. By 1948 she was well established in New York and she invited me to come to the United States in the hope that I could establish a beachhead in order to make it easier for the rest of my family to follow. In February of 1949 I crossed a very stormy Atlantic on the Queen Mary. Since I don't get seasick, (most of the other passengers were) I always had 6 waiters serving me in an empty dining room.

A few days after my arrival in New York, one of America's leading neuropsychiatrist, Warren McCulloch (who, by an amazing combination of miraculous circumstances, had gotten hold of my book) invited me to present my theory of memory at a meeting in New York that was to take place a few days later. He also recommended that I find a book entitled *Cybernetics* in order to prepare myself a bit for this meeting. I did that, and with the little English at my disposal at that time, I tried hard to understand some of its basic points.

Somewhat ill prepared in concepts and language, I came to this meeting whose title was more or less an enigma as well: "Circular Causal and Feedback Mechanisms in Biological and Social Systems." To my surprise, it was a small meeting of about 20 participants, but to my even greater surprise, this was an assembly of the crème de la crème of American scientists. There was, of course, Warren McCulloch who was chairman of the conference, and whose works in 4 volumes have recently been published. There was Norbert Wiener himself, of whom a lovely biography by P. Masani appeared last year. There was John von Neuman, the man who started the computer revolution. Then there were Gregory Bateson and his wife Margaret Mead, or should I say Margaret Mead and her husband Gregory Bateson, who brought to anthropology wisdom, profundity and humor; both in different ways.

These are but a few, whose names I believe would be familiar to my European friends. I don't know who invented the notion of "interdisciplinarity," but this meeting was its manifestation. If you were to begin with Anthropology in an alphabetical list of academic professions, and end with Zoology, my guess would be that almost every one of these disciplines had a representative present.

I was called upon relatively early to present my story, and I wrestled valiantly with my 20 English words to make myself understood. The whole thing would have turned into a catastrophe if it weren't for the presence of Gerhard von Bonin, Heinrich Klüver and others who spoke fluently German and who rescued me by translating some of my arguments.

That evening, the group had a business meeting. Before it was over, I was invited to come in. "Heinz," began the chairman, "we listened to your molecular theory of memory, and your theory agrees with many observations which other theories cannot account for. *What* you had to say was very interesting. However, *how* you said it was abominable! Because we want you to learn English fast, we have decided to appoint you to be the editor of the transactions of this conference."

I was of course speechless. How could I edit articles by such superb writers as Wiener, Mead, Bateson, etc.? How could I organize material of which I, at best, understood only half? But, I thought "Why not try?" So I accepted the appointment. I immediately proposed that, "Since the title of this conference is so long, it is hard to remember, and for me, hard to pronounce; 'circular-causal-and-feedback-mechanisms . . .' I propose to call this conference 'Cybernetics.' "

Everybody looked at Norbert Wiener, who sat next to me, and applauded in his honor and in acceptance of my proposal. Deeply touched by the recognition of his peers, tears came to his eyes, and he left the room to hide his emotions.

The sponsor of this, and four more conferences on this topic, was the Josiah Macy Jr. Foundation of New York, who asked me to edit each of the 5 volumes. Since all of that took place in the remote past, aficionados of cybernetics refer to these books as the "legendary Macy meetings on cybernetics."

Here ends, dear Yveline, my story of how cybernetics came upon me.

Y. Throughout the course of the conference, in the conference rooms as well as the corridors of the Cité de la Villette, there was much discussion about first-order cybernetics and second-order cybernetics; mostly to put them opposite each other. For instance, "But you see my dear, in my view this is from first-order cybernetics . . ." or, "I tell you, one really feels the difference; this time we are in the second-order cybernetics." Would you attempt to clarify for the people here, what are the fundamental distinctions for first-order and second-order cybernetics? Which change in direction or observation signify second-order cybernetics? Or to paraphrase G. Spencer-Brown, whom you like to cite, "Design me a resemblance!" or, "Design me a distinction!"

H. Let me draw the distinction for you. You followed me when I moved from zero-order to first-order cybernetics. What did I do? I let the underlying circularity of processes of emergence, of manifestation, of structurization, of organization, etc., become explicit. By that I mean that we now *reflect* about these circular processes which generate structure, order, behavior, etc., in those things we observe. Now Yveline, you can easily guess how to move from first-order to second-order cybernetics.

Y. I think so. Let me try. In second-order you reflect upon your reflections.

H. Of course!

Y. And now, can I go on to third-order cybernetics?

H. Yes, you could. But it would not create anything new, because by ascending into "second-order," as Aristotle would say, one has stepped into the circle that closes upon itself. One has stepped into the domain of concepts that apply to themselves.

Y. Do you mean to say that a second-order cybernetics is a cybernetics of cybernetics?

H. Yes, precisely!

Y. Can you give me other examples?

H. Yes of course. For instance, compare a typical first-order cybernetics concept such as "purpose," (as being the equivalent of "why") with a second-order question, "What is the purpose of 'purpose'?" (asking why the notion of "purpose" is used in the first place; i.e. how does it influence discourse, explanations, argumentations, etc.?)

One nice feature of this notion is that it relieves one of the need to account for the way things are done which are intended. Every time I tie my shoelaces, or you slip into your pumps, we do it differently. We do it in thousands of unpredictable variations, but the outcome is predictable; my shoelaces are tied, your shoes are on your feet.

On the other hand, it is quite impossible for a physicist to invent the "Laws of Nature" with which to compute our behavior from the initial conditions of my united shoelaces or your pumps in your wardrobe; that is to compute the paths, the "trajectories" and the movements that our bodies and our limbs are taking, which tie my laces or put shoes on your feet. The physicist's "causa efficientis" is impotent. But the cyberneticist's "causa finalis" does it all. If the intentions are clear, (independent of the initial conditions) the sensorimotor loops will adjust and readjust our movements until my laces are tied; your shoes are on your feet.

Y. Thank you. I feel much better with my shoes on. I see now the purpose of using the notion of purpose. One does not need to know *how* to get there; one needs only to know the *there*. This is a very nice feature indeed! Is there a bad feature too?

H. Yes there is. The ugly feature of the notions of "purpose," "goal," and "end," is that they can be used to justify the specific ways of getting there; "The end justified the means." And as we know now, the means can be very ugly indeed. The question should be, "Do the means justify the end?"

Y. If we would remember to ask the question this way, the world could be a very different place. But now Heinz, to use your language, tell me how did second-order cybernetics "come upon" you?

H. Through a woman, of course. It was Margaret Mead. You remember the quote I cited in my address? It came from a speech she gave, I think in 1968. Since she rarely uses titles for her talks and almost never reads from a script, I sent her the transcript from a recording asking for her corrections and a title. There was no reply. I urged by telegram; still no answer. Finally, I tried to reach her by telephone at the Museum of Natural History in New York where she was a curator. I was told she was with the Papuas, or the Trobrianders, or the Samoans, and could not be reached. So, I had to edit her speech and invent a title. What struck me was her speaking about cybernetics in a cybernetical way. Thus I chose for her the title, "Cybernetics of Cybernetics."

It appears to me today that the interest in the peculiar properties of concepts that apply to themselves, (or even need themselves to come into being) must then have been floating in the air. Francisco Varela, the Chilean neurophilosopher referred to them as "self-referential," the Swedish logico-mathematician Lars Lofgren as "auto-logical."

Y. If I were to ask you to give me the shortest description of the distinction between first-order cybernetics and second-order cybernetics, what would you say?

H. I would say, first-order cybernetics is the cybernetics of observed systems, while second-order cybernetics is the cybernetics of observing systems.

Y. Very short indeed! Would you like to expand on this?

H. Perhaps only briefly, because my "shortest description" is nothing else but a paraphrase of the description I made in my address, where I juxtaposed the two fundamentally different epistemological, even ethical, positions where one considers oneself: on the one hand, as an independent observer who watches the world go by; or on the other hand, as a participant actor in the circularity of human relations.

When taking the latter position, (the position I believe taken by systemic family therapists) one develops notions like "closure," "self-organization," "self-reference," "self," "auto-poiesis," "autonomy," "responsibility," etc., etc. In other words, one arrives at the whole conceptual machinery of contemporary cybernetics, the cybernetics of observing systems, and thus one comes very close to the theme of your Congress: "Ethics, Ideologies, New Methods."

Y. At the conclusion of your paper, On Constructing a Reality, which was published in Paul Watzlawick's book *The Invented Reality*, you ask, "What are the consequences of all this in ethics and aesthetics?" You also wrote, "The ethical imperative: Act always so as to increase the number of choices." And, "The aesthetical imperative: If you desire to see, learn how to act." Can you add something to the connections between ethics, aesthetics and change; which from my point of view, are the three basic coordinates in family therapy?

H. I like your three coordinates, because all three have a second-order flavor. And, of course, I am delighted that two of my imperatives correspond to two of your coordinates. However, I feel some uneasiness that your third coordinate "change" is not yet accompanied by an appropriate imperative. Let me remedy this situation at once by inventing an imperative for you; the therapeutic imperative: "If you want to be yourself, *change!*" Is this paradoxical? Of course! What else would you expect from change?

Y. You say with so much self assurance, "Paradoxical, of course!" How can you connect change with paradox?

H. Easily! You remember paradox? It yields one meaning when apprehended one way, and one meaning when apprehended the other. What do you do when I say "I am a liar," do you believe me? If you do, then I must have spoken the truth; but if I had spoken the truth, I must have lied, etc., etc.

What is the problem here? Lying? No, the problem is "I," the shortest self-referential loop. When speaking about oneself, using "I," magic is performed. One creates oneself by creating oneself. "I" is the operator who is the result of the operation.

Y. This is all magic to me. Where does "change" come in?

H. The paradoxical nature of change is much richer than the orthodox "paradox of the liar" which switches from "true" to "false," and from "false" to "true," and so on and so forth in dynamic stability. The unorthodox nature of change arises when you apprehend "change" any way you wish to apprehend it, and it will yield something else, otherwise it wouldn't be "change." This is, I believe, its therapeutic force.

Y. But you said, "If you want to be yourself; change!" How can you be yourself and change?

H. I wanted to appeal to ancient wisdom. It is 2600 years old and comes from the *I Ching*. Under the 58th symbol "Fu," or "The Turning Point," it says, "The ultimate frame for change is the unchanging."

Y. (Smiling) This conversation with you, Heinz, has been a joyful and exciting day of learning. It seems to have mirrored the theme of our conference; ethics and family therapy. It feels as though I've discovered a new freedom within a precise and rigorous framework. This framework, clearly defined by the fundamental guidelines of therapeutic practice, encourages communication with another, thereby creating a new space. Does this not broaden our possibilities by redrawing the line of the horizon? If rigor were combined with creativity, the ethics of choice could also be the ethics of change!

At least that is the very personal understanding which I have gained from our encounter. I now have an exquisite diffused feeling of a door which opens onto another door, which opens onto another door, which opens onto another door . . .

15
For Niklas Luhmann:* "How Recursive is Communication?"

HEINZ VON FOERSTER
(Tranlated by Richard Howe)

A year and a half ago Niklas Luhmann sent me a fascinating essay for my 80th birthday (Luhmann 1991). This article culminates in two extraordinary questions. I won't read you these questions now;[§] I'd rather just briefly report on the impression these questions made on me. I saw in them a resemblance to two of the great problems of antiquity, two problems of geometry. The one problem is the *Trisectio anguli*. That is the problem of dividing an angle into three parts using only a compass and a ruler. And the other problem is the *Quadratura circuli*, the task of constructing a square, again using only a compass and a ruler, the area of which is equal to a given circle. As you probably recall, both of these problems are unsolvable in principle, as Karl-Friedrich Gauss showed about two hundred years ago. But if one removes the restriction of working only with a compass and a ruler, then these problems can easily be solved.

When I got the invitation to say a few words here at the birthday celebration for Niklas Luhmann, I of course immediately thought, oh good, now that's where I'll present my answers to the two problems that he put to me for my birthday. I sat myself down and went to work on the answers, but in the midst of my preparations it suddenly occurred to me: but Heinz, that's

---

* Lecture given at the Authors Colloquium in honor of Niklas Luhmann on February 5, 1993 at the Center for Interdisciplinary Research, Bielefeld. The German version was published in *Teoria Soziobiologica 2/93*, Franco Angeli, Milan, pp. 61–88 (1993).
§ Editor's note: The two questions run (Luhmann 1991, p. 71) "1. Does knowledge rest on construction in the sense that it only functions *because* the knowing system is operatively closed, therefore: *because* it can maintain no operative contact with the outside world; and because it *therefore* remains dependent, for everything that it constructs, on its own distinction between self-reference and allo-reference? 2. Can (or must) one impute the formation of "Eigen values" to the domain of latency; therefore for first order observation to the intangible and therefore stable distinction that underlies every single designation of objects; and in the domain of second order observation to those very forms that are conserved when a system interrupts its constant observation of that which cannot be observed?"

completely wrong! One just doesn't do that in this part of the world. Here one gives birthday children questions, not answers! So therefore I thought to myself, okay, I'll present my answers at some other opportunity; today, on the occasion of this birthday celebration, I too will come with two questions. And it's not just about questions but rather—we are, indeed, here in the Center for Interdisciplinary Research—about two research programs into still unsolved problems of the social sciences. I thought I'd present these problems today, for I have the feeling that if one would concern oneself with these questions one could make an essential contribution to social theory.

What are these two questions about? The first problem or research program has to do with an extension, or perhaps I should say: with a deepening, of recursive functions. You all know about the unprecedented successes of the recursive functions that are in constant use in chaos theory and indeed elsewhere. But I have the feeling that these results of chaos research can be applied by sociology only metaphorically. Why? All chaos research is concerned with functions, and functions are only relations between numbers, at best, complex numbers. A function can be quadratic, one gives this function a two, out comes a four, and one gives this function a three, and out comes a nine. It operates only on numbers, but sociology doesn't work with numbers: sociology is interested in functions. And functions of functions one calls functors. A functor is, so to speak a system that is intended to coordinate one group of functions with another group, and so today I propose to develop a research program in which one is concerned with recursive functors. So that's problem number one.

Problem number two that I'd like to present today is a theory of compositions. It consists in developing a system of composition, and, indeed, a system of composition for systems. What is this problem about? I have System **A**, I have a System **B**, and now I'd like to integrate both of these into a System **C**. What do the rules consist of that allow a new System **C** to arise, the rules of integration, of composition? Is it a kind of addition, a kind of integration? We've got all the best words for it, but what does the formalism for such problems look like? Today one could also provide the composition problem with another name: It's about, for example, the problem of the Croats, the Bosnians, the Herzegovinians—one could call it the Vance-Owen Problem. These are the problems that we confront in social theory today. How can one solve this problem? Or in a different sense it is also about the problem of autopoiesis: how can I bring an autopoietic System **A** into a relationship with another autopoietic System **B** in such a way that a new System **C** arises, itself an autopoietic system? Unfortunately, the poets or autopoets who invented autopoiesis have given us no rules for the compositional possibilities of such autopoietic systems. They have, to be sure, applied indices, but that isn't really a fundamental theory of composition. These are, in brief, my two problems.

Now of course you'll say, "for heaven's sakes, we're sociologists, and here Heinz von Foerster comes with fundamental mathematical problems—what are we supposed to make of that?" So I thought I could sweeten or lighten this problematique if I tried above all to present the ideas so clearly that they became transparent. And when something is transparent, then one no longer sees it: the problems disappear. And as a second idea, I thought I'd bring from California a trio of American jewels for our birthday child Niklas Luhmann that are probably somewhat known here already but may still amuse the birthday child in their special birthday edition.

The first present I've brought along is an essay by Warren McCulloch, written about a half a century ago. It is the famous article with the title "A Hetarchy of Values Determined by the Topology of Nervous Nets" (1945). I find that this article is of such great significance that I'd like to draw your attention to its existence once more. So that you can see what field he worked in, I'll read you a sentence from the last paragraph. It rests on the idea of a circular organization of the nervous system: "circularities in preference." These circularities arise when one prefers **A** to **B**, prefers **B** to **C**, and, again, **C** to **A**. In classical logic one then speaks of being illogical. Nevertheless, McCulloch says that is not illogical, it is logic as it is actually used. Therefore: "Circularities in preference instead of inconsistencies, actually demonstrate consistency of a higher order than had been dreamed of in our philosophy. An organism possessed of this nervous system—six neurons— is sufficiently endowed to be unpredictable from any theory founded on a scale of values." A system of six neurons is, in the framework of existing theories, unpredictable in principle. That is present number one.

Present number two that I've brought along with me is an article by Louis Kaufmann, a mathematician who is fascinated by self-reference and recursion. The article is called "Self-reference and recursive form" (1987). And so that you can see why I find it so important, I'll read you the last sentence of this article. The last sentence of this article is: "Mathematics is the consequence of what there would be if there could be anything at all."

Present number three is by my much admired teacher Karl Menger, a member of the *Wiener Kreis* (Vienna Circle), to which I am pleased, even today, to have fallen victim! When I was a young student I enthusiastically attended Karl Menger's lectures. The article by Karl Menger, which I've brought here as present number three, is "Gulliver in the Land without One, Two, Three" (1959). You may ask, why I've brought such an article to a group of sociologists! In this article Karl Menger already developed the idea of functors, that is, of functions of functions, which I consider wholly decisive for the theoretical comprehension of social structures. Here too I'll read the last sentence, so you can see what it's all about. The last sentence is: "Gulliver intended to describe his experiences in the Land without One, Two, Three in letters to Newton, to the successors of Descartes, to Leibniz, and to the Bernoullis. One of these great minds, rushing from one discovery to the next, might have paused for a minute's reflection upon the way their own epochal ideas were expressed. It is a pity that, because of Gulliver's preparations for another voyage, those letters were never written."

So now I'd like to turn these three presents over to Niklas Luhmann! Of course that requires a bouquet!"*

Now I come to a theme that is not my own but rather was proposed to me by the Center for Interdisciplinary Research. I always like it when one proposes a theme for me, for if I then come with this theme, I optimally fulfill the wishes of my hosts! The assignment that was posed for me for today consists of the question "How recursive is communication?" I didn't know how I was supposed to read that. *How* recursive is communication? Or: how *recursive* is communication? Or, how recursive *is* communication? Unfortunately I'm not an ontologist, i.e., I don't know what *is*. I've never been able to do anything but consider what would be—*if*. So I've posed the question to myself as how would it be *if* we conceived of communication as recursion. And so here is my Proposal No. 00.

00. Proposal: "Communication is recursion."

You could understand that as if it had to do with an entry in a dictionary. If you don't know what communication is, you look it up in the dictionary under *C*. There it says, "Communication is recursion." Aha, you say, good! What is recursion? Then of course you go back to the dictionary again and find, this time under *R*: "Recursion is communication." So it is with every dictionary. If you busy yourself a bit with them, you will find that the dictionary is always self-referential: From **A** you are sent to **B**, from **B** to **C**, and from **C** back again to **A**. That's the dictionary game. You could of course also conceive of my proposal as a simple tautology: "communication is recursion." Indeed, but, as the philosophers assert, tautologies don't say anything. Nevertheless, tautologies do say something about the one who utters them. At the end of my lecture you may not know anything about recursion or communication, but you will certainly know something about me! My program, therefore, is the proposal: "communication is recursion," and what it looks like.

I'd like to present my program in three chapters, whereby I'd like to use the first chapter essentially to recall to your memories a terminology whose central concept is a fictitious "machine" that executes well defined operations on numbers, expressions, operations, etc. This chapter starts out by recapitulating some concepts that are already current among you. As you will later see, I'm using this terminology in order to make the decisive point in my lecture palatable to you, namely, insight into the unsolvability, in principle, of the "analytical problem." In other sciences this problem goes under other names: it's called "the decision problem" in logic, the "halting problem" in computer science, etc.

---

* Editor's note: Heinz von Foerster hands Niklas Luhmann copies of the three articles, specially bound for this occasion; as he does so, a bouquet of flowers appears magically out of thin air and Niklas Luhmann thanks him.

I gave a lot of thought to what version of this problem I could acquaint you with so that, without resorting to mathematical somersaults, the abyss dividing the synthetic problem from the analytic one would become clear. I finally allowed myself a compromise, in that I won't demonstrate how the analytical problem is unsolvable *in principle* but rather only an easier version, namely, that all the taxes in the world and all the time available in our universe would by no means be sufficient to solve the analytical problem for even relatively simple "non-trivial machines": the problem is "transcomputational," our ignorance is fundamental.

This abysmal ignorance, this complete, fundamental ignorance is something that I've still never really seen presented at full strength, and that is what I'd like to present to you today, so that we can get some insight into the question of how, in the face of such fundamental ignorance, we can concern ourselves with our problems? In the second chapter then I'd like to sketch the development of recursive functors. I'll make it as easy and playful as possible, so you can enjoy following these trains of thought. And in the third chapter I'd like to speak about compositions, compositions of functors, of compositions of systems.

First Chapter: Machines

I'll begin the first chapter by recapitulating a language that was actually introduced by Alan Turing, an English mathematician, in order to leave the long-windedness of deductively logical, argument to a machine, a conceptual machine, that would then turn all the wheels and buttons, so that one only has to observe it: if one enters the problem on one side of the machine, then the solution emerges on the other side. Once this machine has been established, we have a language that can very easily jump from one well-defined expression to another, and if you then want to know how this machine works, you can always take it apart. Therefore: machine language. This language is already current among you, but permit me, despite that, to briefly repeat it, for, as I said, I'll need to concepts in a few minutes.

I come to my proposition:

01. Trivial machines: (i) synthetically determined; (ii) independent of the past; (iii) analytically determinable; (iv) predictable.

A trivial machine is defined by the fact that it always bravely does the very same thing that it originally did. If for example the machine says it adds 2 to every number you give it, then if you give it a 5, out comes a 7, if you give it a 10, out comes a 12, and if you put this machine on the shelf for a million years, come back, and give it a 5, out will come a 7, give it a 9, out will come an 11. That's what's so nice about a trivial machine.

But you don't have to input numbers. You could also input other forms. For example, the medieval logicians input logical propositions. The classi-

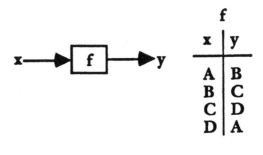

$$n_4 = 4! = 24$$

FIGURE 1.

cal logico-deductive proposition that was always used as an example in the Middle Ages is the famous proposition "All men are mortal." So you arrive at an "All men are mortal"-trivial machine. If you shove a person into it on one side, a corpse comes out the other. Take Socrates—"Socrates is a man"—, shove him in on one side and—bam!—out comes a dead Socrates on the other. But you don't need people, you don't need Socrates, you could even work with letters.

Here I have represented an "anagrammaton," a machine that calculates anagrams (Figure 1). As you know, an anagram is something that replaces one letter with another.

To make this example as simple as possible, I'd like to propose an agrammaton the alphabet of which consists of only four letters (**A**, **B**, **C**, **D**) and which in accordance with the table in Figure 1 makes a **B** out of an **A**, a **C** out of a **B**, a **D** out of a **C**, and finally an **A** out of a **D**. When I was just a kid and sent loveletters to my girlfriend, I of course agreed with her on an anagram so that our parents couldn't read what we wrote. But of course such anagrams are very easy to solve. For example, how many anagrams can one construct altogether with 4 letters? As you know, that's simply the number of permutations of the letters **A**, **B**, **C** and **D**. Which is 4 times 3 times 2, therefore 4!, which results in 24 anagrams (Figure 2).

Here I have exactly 24 anagrams at my disposal, and if you want to make an experiment to find out which of these is our own anagrammaton, then you need only four trials. You give it **A**, **B** comes out; you give it **B**, **C** comes out; give it **C**, **D** comes out; and finally **A** results from giving it **D**. So you've solved the problem. Trivial machines are, therefore, as formulated in Proposition 01, synthetically determined (we have in fact just built one); independent of the past (we could put ours on the shelf for years and years); analytically determinable (we just did that); and, therefore, predictable.

Now you understand the great love affair of western culture for trivial machines. I could give you example after example of trivial machines. When you buy an automobile, you of course demand of the seller a trivializations-

01	02	03	04	05	06	07	08	09	10	11	12	13	14	15	16	17	18	19	20	21	22	23	24
A	A	A	A	A	A	B	B	B	B	B	B	C	C	C	C	C	C	D	D	D	D	D	D
B	B	C	C	D	D	A	A	C	C	D	D	A	A	B	B	D	D	A	A	B	B	C	C
D	D	B	D	B	C	C	D	A	D	A	C	B	D	A	D	A	B	B	C	A	C	A	B
D	C	D	B	C	B	D	C	D	A	C	A	D	B	D	A	B	A	C	B	C	A	B	A

The 24 anagrams that can be formed with exactly 4 letters

F<small>IGURE</small> 2.

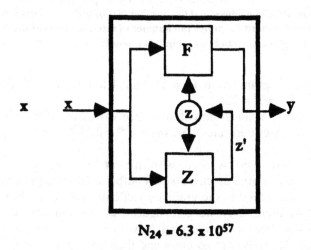

$$N_{24} = 6.3 \times 10^{57}$$

F<small>IGURE</small> 3.

document that says that this automobile will remain a trivial machine for the next 10,000 or 100,000 kilometers or the next five years. And if the automobile suddenly proves to be unreliable, you get a trivializateur, who puts the machine back in order. Our infatuation with trivial machines goes so far that we send our children, who are usually very unpredictable and completely surprising fellows, to trivialization institutes, so that the child, when one asks "how much is 2 times 3" doesn't say "green" or "that's how old I am" but rather says, bravely, "six." And so the child becomes a reliable member of our society.

02. Non-trivial machines: (i) synthetically determined; (ii) dependent on the past; (iii) analytically determinable; (iv) unpredictable.

Now I come to the non-trivial machines. Non-trivial machines have "inner" states (Figure 3).

In each operation, this inner state changes, so that when the next operation takes place, the previous operation is not repeated, but rather another operation can take place. One could ask, how many such non-trivial machines one could construct if, as in our case, one has the possibility of 24 different states. The number of such possible machines is $N_{24} = 6.3 \times 10^{57}$. That is a number with 57 zeros tacked on. And you can already see that some difficulties arise when you want to explore this machine analytically. If you pose a question to this machine every microsecond and have a very fast computer that can tell you in one microsecond what kind of a machine it is, yes or no, then all the time since the world began is not enough to analyze this machine. Therefore my next proposition runs:

03. Numbers: Let n be the number of input and output symbols, then the number N_T of possible trivial machines, and the number N_{NT} of non-trivial machines is: $N_T(n)$ = n^n, $N_{NT}(n) = n^{nz}$, where z signifies the number of internal states of the NT machine, but z cannot be greater than the number of possible trivial machines, so that $z_{maz} = n^n$, $N_{NT}(n) = n^{nn^n}$.

For a trivial anagrammaton $(z = 1)$ with 4 letters $(n = 4)$ the result is $N_T(4) = 4^4 = 2^{2\cdot4} = 2^8 = 256$.

For a non-trivial anagrammaton (which calculated different anagrams according to prescribed rules): $N_{NT}(4) = 4^{4\cdot4^4} = 2^{2\cdot2\cdot256} = 2^{2048} = _{approx.}10^{620}$.

W. Ross Ashby, who worked with me at the Biological Computer Laboratory, built a little machine with 4 outputs, 4 inputs, and 4 inner states, and gave this machine to the graduate students, who wanted to work with him. He told them, they were to figure out for him how this machine worked, he'd be back in the morning. Now, I was a night person, I've always gotten to the lab only around noon and then gone home around 1, 2, or 3 in the morning. So I saw these poor creatures sitting and working and writing up tables and I told them: "Forget it! You can't figure it out!"—"No, no, I've almost got it already!" At six A.M. the next morning they were still sitting there, pale and green. The next day Ross Ashby said to them: "Forget it! I'll tell you how many possibilities you have: 10^{126}." So then they relaxed.

Just imagine! Here we're concerned with only 4 letters, for input and output symbols and with inner states totaling only 24 possibilities. The complexity of this system is so enormous that it is impossible to find out how this machine works. And yet, although our brain employs over 10^{10} neurons, the representatives of "artificial intelligence" have the nerve to say that they're about to discover how the brain works. They say, "I've worked on a machine that works like the brain." "Oh, congratulations—and by the way, just how *does* the brain work?" No one knows that. But then one can't even make the comparison. One can only say that the *machine* works thus and thus, but one can't say how the *brain* works, because nobody knows. But perhaps one doesn't need to know how the brain works. Maybe it's just, as the American saying goes, that "we're barking up the wrong tree."

For example, how is it possible that this colloquium in honor of Luhmann was announced, that one would hear various speakers talk, and, although we have no idea how the brain works, that we all arrived here promptly at 9 o'clock. And what do we see? Everyone is here, everyone is listening, one of them makes noises with his mouth, some are taking notes, etc. Indeed, how is that possible? What's going on here?

For this I'd like now to take the next step. As I hope to show you, all of this can only happen because these systems operate recursively!

Second Chapter: Recursors

In order to develop the following thoughts as clearly as possible, I will increase their complexity stepwise, so that you can follow, step by step, what it's all about.

Dimensionality 1 (Operationally Open)

I'll begin with systems of dimensionality 1. Why "dimensionality 1?" Because here signals are linear and flow in one direction only. One could represent this situation in its brutal simplicity by a directed line segment (Figure 4) in which all operations that transform **x** into **y** are comprised by the one single point "0".

Since it is my intention here to talk later on about compositions of at least two systems, I'll present you now with the two systems **D** and **S**, which should help me out in the following exposition (Figure 5).

D operates on the variable **x** and produces **y**, which is expressed by the function **y** = **D(x)**. The same holds, *mutatis mutandis*, for machine **S**.

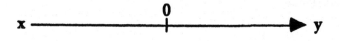

FIGURE 4.

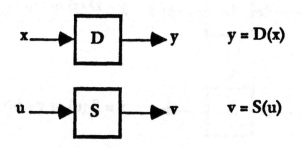

FIGURE 5.

The letters have an historical foundation, for in the development of recursive machines or even non-trivial machines one distinguished between two functions: between the state function, the **S**-function, and driving function, the **D**-function. Therefore they're called **D** and **S**, but you don't need to worry about the **D** and the **S**. You only have to know that it's about two machines, one of which operates on **x** and produces **y** while the other operates on **u** and produces **v**.

Parameterization

Now we add to these machines another, external control value, so that we can vary the operation of these machines (Figure 6). The control functions **u** and **x** that have been introduced into the machines from above are to change the operations of these machines. Regarded in another way, these parameters allow us to control the non-triviality of these machines. If these machines have the menu of our 24 anagrams at their disposal, one could switch from anagram to anagram via parametric inputs, just as one switches from channel to channel watching television. That can be expressed in two ways using an algebraic formalism. In the first, the parameter can be indicated by a subscript that modifies the function: $y = D_u(x)$, $v = S_x(u)$, in the other, it can be declared as a full-grown variable: $y = D(x,u)$, $v = S(u,x)$.

Dimensionality 2 (Operationally Closed: the Fundamental Equations of Non-linear Dynamics)

Now comes a decisive step, for I'll transform the systems, up till now of dimensionality 1, into systems of dimensionality 2 through an operational closure in which each output becomes the next input just as soon as it is produced (Figure 7).

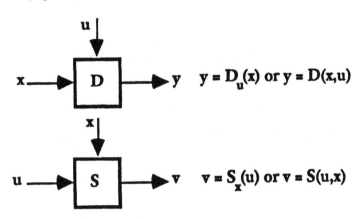

$$y = D_u(x) \text{ or } y = D(x,u)$$

$$v = S_x(u) \text{ or } v = S(u,x)$$

FIGURE 6.

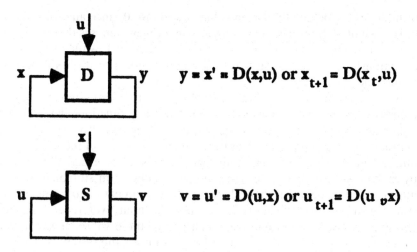

$$y = x' = D(x, u) \text{ or } x_{t+1} = D(x_t, u)$$

$$v = u' = D(u, x) \text{ or } u_{t+1} = D(u_v x)$$

FIGURE 7.

I let **y**, the output of the **D** machine, become in turn the input. I do the same thing with the **S** machine. This step transforms operational linearity into operational circularity, a situation that can be represented only in a plane, therefore in a 2-dimensional manifold. Once again, this can be expressed in algebraic formalism in two ways: first, in that one makes the next input, x', the result of the current operation, $x' = y$, whereby the marked quantity is to follow the unmarked one:

$$x' = D(x, u), \text{ and } u' = S(u, x)$$

whereby the recursiveness of these expression can be recognized in that the variables **x**, **u** appear as functions of themselves. One gets a "physicalization" of this situation, expressing the passage of time, by introducing the parameter "time" in the form of incremental units: **t** now, **t + 1** a single incremental unit later:

$$x_{t+1} = D(x_t, u), \text{ and } u_{t+1} = S(u_t, x)$$

Those of you who are occupied with chaos theory and with recursive functions will recognize at once that these are the fundamental equations of recursive function theory. Those are the conceptual mechanisms with which chaos research is conducted; it is always the same equations over and over again. And they give rise to completely astonishing, unforeseen operational properties. Viewed historically, even early on one noticed a convergence to some stable values. An example: if you recursively take the square root of any random initial value (most calculators have a square root button), then you will very soon arrive at the stable value 1.0000. . . . No wonder, for the root of 1 is 1. The mathematicians at the turn of the century called these values the "Eigen values" of the corresponding functions. To the operation of taking roots belong the Eigen values 1 and also 0, since any root of 0 is 0. The essential difference between these two Eigen values is that for every deviation from 1, recursion leads the system back to 1, while at the least deviation from 0 the system leaves null and wanders to the stable Eigen value "one."

About 20 years ago there was an explosion of renewed interest in these recursive operations, as one discovered that many functions develop not only stable *values* but also a stable *dynamic*. One called these stabilities "attractors," apparently a leftover from a teleological way of thinking. Since one can let some systems march through the most diverse Eigen behaviors by making simple changes in the parameters, one soon stumbled onto a most interesting behavior that is launched by certain parametric values: the system rolls through a sequence of values without ever repeating one, and even if one believes one has taken one of these values as the initial value, the sequence of values cannot be reproduced: the system is chaotic.

Let me make just a couple of more remarks about stable Eigen behaviors.

Consider next the fascinating process that recursively sifts only discrete values out of a continuum of endless possibilities. Recall the operation of taking roots, which lets one and only one number, namely "1," emerge from the endless domain of the real numbers. Can that serve as a metaphor for the recursiveness of the natural process, sometimes also called "evolution," in which discrete entities are sifted out of the infinite abundance of possibilities, such as a fly, an elephant, even a Luhmann? I say yes," and hope to contribute additional building blocks to the foundation of my assertion.

But consider also that although one can indeed make the inference from given operations to their Eigen behaviors, one cannot make the converse deduction from a stable behavior, an Eigen behavior, to the corresponding generative operations. For example, "one" is the Eigen value of infinitely many different operations. Therefore, the inference from the recursive Eigen value "1" to the square root operation as the generator is not valid, because the fourth, the tenth, the hundredth root, recursively applied, yield the same Eigen value "1." Can that serve as a metaphor for the recursiveness of the natural process, sometimes called the "laws of nature," of which there could be infinitely many versions that would explain a Milky Way, a planetary system, indeed, even a Luhmann? I say "yes" and turn for support to Wittgenstein's *Tractatus*, Point 5.1361: "The belief in the causal nexus is *the* superstition."

This result, that there emerge Eigen values, is the only thing we can rely on. For then an opaque machine begins to behave in a predictable way, for as soon as it has run into an Eigen state, I can of course tell you, for example, if this Eigen state is a period, what the next value in the period is. Through this recursive closure and only through this recursive closure do stabilities arise that could never be discovered through input/output analysis. What is fascinating is that while one can observe these stabilities it is in principle

impossible to find out what generates these stabilities. One cannot analytically determine how this system operates, although we see that it does
operate in a way that permits us to make predictions.

Third Chapter: Compositions

I have up till now spoken of systems as entities, spoken about their behavior, their synthesis, analysis, and taxonomy. But here I am in the company
of scholars of sociology, i.e., the science of the "*socius*," of the companion
and comrade, of the "*secundus*," the follower, the second. So I must concern
myself with at least two systems, with their behavior, their synthesis and
analysis. Indeed a society usually consists of more than two members, but
if the process of integration, the "composition" of two systems has been
established, one can use stepwise recursion to apply the established composition rule to an arbitrary number of new arrivals.

How does such a composition come about?

Here, I believe, is perhaps the essential step in my exposition, for through
the composition of two systems of dimensionality 2, the recursors, there
emerge systems that are irreducibly of dimensionality 3.

But how is one to proceed?

Dimensionality 3 (Calculus of Recursive Functors)

The systems in Figure 8 should help me out here. I'll go back to the two
machines, the recursors **D** and **S** from Figure 7.

In step one (orientation) I rotate recursor **S** 90°, so that the variables and
parameters in **D** and **S** are aligned with one another; in step two (composition) I push the two together, so that out of the two separate systems **D**
and **S** a new machine now arises, a **DS**-composition. This new machine is
distinguished by its double closure, first a closure on **u**, that previously, as a
parameter, controlled **D**, and then the closure on **x**, that previously, as a
parameter, controlled **S**. So now both systems control one another recip-
rocally; the operational functions of the one system become functions of
the other: two recursive functors.

Extensions of the Second Order

0.5 Functors: functions of functions (functions of the second order)

From your middle school years your can surely recall the differential and
integral calculus. One wrote **dy/dx** and spoke of the "derivative of **y** with
respect to **x**," whereby **y** is a function of x: **y** = **f(x)**. That is, the derivative,
or differential operator **Di**, as I'd like to call it, is a functor, for it operates

Orientation　　　　**Composition**

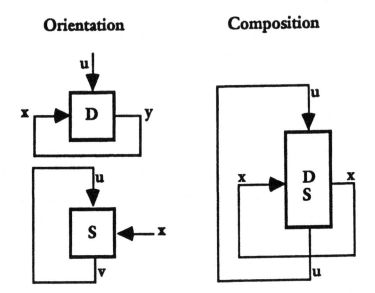

on a function, let's say $y = x^2$, and produces a function: $\mathbf{Di[x^2]} = \mathbf{2x}$, or in Menger's elegant notation: $\mathbf{Di[(\ )^2]} = \mathbf{2}$. Does $\mathbf{Di}$ have Eigen functions? Yes indeed: the exponential function $\mathbf{y} = \mathbf{e^x}$, in Menger's notation: $\mathbf{Di[exp]} = \mathbf{exp}$, and on account of the extraordinary relationship of the exponential function to the trigonometric functions **sin** and **cosin**: $\mathbf{Di^4[sin]} = \mathbf{sin}$, $\mathbf{Di^4[cos]} =$ **cos**, i.e., **sin** and **cosin** are the Eigen functions of the differential operator iterated fourfold.

One doesn't have to restrict oneself to mathematical expressions. Menger developed these ideas for logical functions (1962), a generalization that is significant here. For example, the algebraic expression:

$$S = S(D) = S(D(S))$$
$$D = D(S) = D(S(D))$$

of the composition of the two systems **D** and **S** in Figure 8 makes the recursion of the two functors **D, S** clearly visible.

06. Compositions (the properties of the composition are not the properties of the components)

Viewed historically, attention to qualitative changes that arise in the transition from aggregate to system was guided by an unfortunate formulation of this transition that was given by "generalists," "holists," "environmentalists," etc.: "the whole is greater than the sum of its parts." As one of my colleagues once remarked: "can't the numskulls even add?"

But if a measure function **M** is introduced, then the holistic sentiment can be made precise: "The measure of the sum of the parts is not the sum of the measures of the parts": $M\Sigma(T_i) \neq \Sigma M(T_i), (i = 1,2,3,4 \dots n)$. If the measure function is super-additive, then indeed the holistic motto is justified. Let us take two parts **(a, b)** and for our measure function squaring ( )^2. Then in fact $(a + b)^2$ is greater than $(a)^2 + (b)^2$, for $a^2 + b^2 + 2ab$ is greater than $a^2 + b^2$, and indeed by exactly the systemic reciprocity part **ab + ba**, which, by symmetry (commutativity **ab = ba**): **ab + ba = 2ab**.

A first step in the generalization of the measure function permits us to establish the rules of the game of an algebra of composition, in which one, as previously, regards the distributive law only as a special case vis-á-vis operators. If **K** is some composition (addition, multiplication, logical implication, etc.), then, just as previously: **Op[K(f,g)] ≠ K[Op(f), Op(g)]**.

That is to say, the result of an operation **Op** on a system constructed via the **K**-composition is not equivalent to a system constructed via the **K**-composition of the results of the operator **Op**.

This proposition plays an important role for the autopoieticists, who indeed always insist that the properties of the autopoietic system cannot be expressed by the properties of its components.

Now just two cases worth mentioning (a restriction and an extension), which together allow the interchange of operations and compositions.

(i) Homogeneous composition: let **K** be the composition rule, then **Op[K(f,g)] = K[Op(f), Op(g)]**;

(ii) Superposition: Let **K** and **C** be composition rules, then **Op[K(f,g)] = C[Op(f), Op(g)]**.

This formulation moved the inventors of information theory to follow the example of Boltzmann and choose the logarithmic function for the entropy **H** (here **Op**) of a signal source. Since when two sources with signal repertoires n_1, n_2 are composed, the new repertoire is $n_1 \times n_2$, the new entropy is simply the sum of the former two: $H(n_1 \times n_2) = H(n_1) + H(n)$, for **log(a·b) = log(a) + log(b)**.

If you consider the "composition" in Figure 8 more closely, you will see that it is in principle impossible to arrange the **x** and the **u** loops on the paper in such a way that they don't intersect one another. One must raise either the **x** or the **u** off the paper into the "third dimension" in order to add the two recursions to the system in such a way that they are independent of one another. This can be made even clearer if one dispenses with drawing the external lines, in that one rolls the **DS**-system into a cylinder around the **u**-axis, so that the **x**-output and **y**-input edges are merged.

One can also get rid of the outer **u**-loop in that one bends the cylinder into a ring and melds the upper and lower circular ends: this makes **u**-out

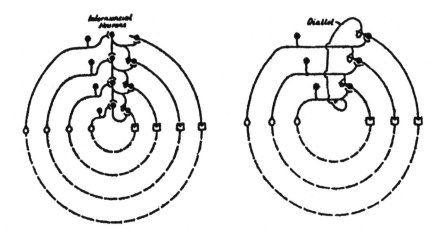

FIGURE 9.

into u-in. This ring or torus is the topological representative of a doubly closed system.

If you'd like pictures, you can find them already very early in Warren McCulloch's article "A Hetarchy of Values Determined by the Topology of Nervous Nets":

07. Warren S. McCulloch (1945): "A Hetarchy of Values Determined by the Topology of Nervous Nets" (Figure 9).

His argument is as follows: In his Figure 3 (Figure 9, left), he shows the recursion of neural activity whose internal components are indicated by the unbroken arcs and whose external components are indicated by the broken ones: McCulloch's thesis of the closure of the neural pathways via the environment. In this circuit the organization is hierarchical, for the presently external senso-motoric loops (dromes) can inhibit the inner loops. Therefore this network cannot calculate the "circularities in preference," the "value anomaly" that I spoke of. In his Figure 4 (Figure 9, right), he introduces the diallels ("crossovers") that from the lower circle can inhibit the upper: twofold closure.

A second reference to the value of toroids for representing doubly closed processes will be found in Proposition 8:

08. Double closure of the senso-motoric and inner-secretoric-neuronal circuits. N = neural bundle; syn = synapse; NP = neurohypophysis; MS = motorium; SS = sensorium (Figure 10).

Here you see sketched (Figure 10a) both of the orthogonally operating circuits: on one side the neural signal flow from the sense organs (**SS**) via the nerve bundles (separated by synapses) to the motorium and from there through the environment and back to the sensorium (**SS**); on the other side

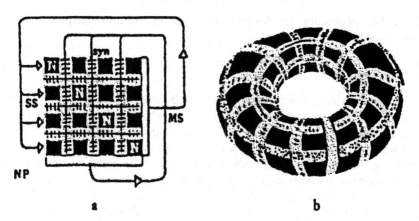

FIGURE 10.

the steroids that are poured into the synapses by the neurohypophysis, which recursively regulate the neural transition functions. Once again, we obtain the torus when we wrap the square scheme around both the horizontal and the vertical axes (Figure 10b).

09. The closure theorem: "In every operationally closed system there arise Eigen behaviors."

Among the many variants and paraphrases of this astonishing theorem I've picked Francisco Varela and Joseph Goguen's version, for I believe I see an affinity here with the sociological vocabulary. The word "behavior," as well as "conduct," "action," etc., does imply the recognizability of regularities, of "invariants" in the temporal course of the action. Here, among sociologists, one is probably not as interested in whether the cosin or the sin appears as the Eigen behavior of the system, but rather whether in a cultural domain a meeting between two members of this cultural domain is celebrated by a handshake or by bowing.

One could even go further and be on the lookout for the emergence of invariants that arise when air is blown in a certain way through the vocal cords, whose vibrations then elicit hisses and grunt with which the meeting of two members of a cultural domain is celebrated and in the southern regions of this geographic area are heard as "Hi, y'all" and in the northern regions as "Hello there."

In all that I have said up until now, I have tried to make it obvious that these invariants, these "Eigen behaviors" arise through the recursively reciprocal effect of the participants in such an established social domain. Therefore, I'd like to turn back to the original question that was put to me: "How recursive is communication?" and also to my proposal:

00. Communication is recursion

With the vocabulary developed here I can extend and sharpen this first version with a few words:

10. Communication is the Eigen behavior of a recursively operating system that is doubly closed onto itself.

The essential thing about the topology of a double closure is that it not only avoids the pseudo-solution of hierarchy, to which one always defers the responsibility for judgment in order to avoid one's own, but also that, through the hetarchical organization that comes with it, the fascinating possibility exists of allowing operators to become operands and operands to become operators. This is just exactly what we've always wanted to understand but which has nevertheless been made impossible for us up till now by the structure of a one dimensional logic. But through the interchangeability of functors standing in reciprocal relationships to one another, our freedom of action is returned to us and with it also our responsibility.

With that I have arrived at my conclusion, which I owe to Wilhelm Busch:

11. Wilhelm Busch's Desideratum:

"Twice two is four" is clearly true,
Too bad it's cheap and flighty;
For I would rather that I knew
About what's deep and mighty.

Whether I've succeeded at that, I don't know, but I do thank you many times for having been so friendly as to have the patience to listen to me. And as my last word, I'd like once more to congratulate our birthday child Niklas Luhmann.

Literature

Kaufmann, Louis
1987 "Self-reference and Recursive Form," *Journal of Social and Biological Structure*, 10, pp. 53–72.

Luhmann, Niklas
1991 "Wie lassen sich latente Strukturen beobachten?," in Paul Watzlawick, Peter Krieg (eds.), *Das Auge des Betrachters: Beiträge zum Konstruktivismus. Festschrift für Heinz von Foerster*, Piper, München-Zürich, pp. 61–74.

McCulloch, Warren
1945 "A Hetarchy of Values Determined by the Typology of Nervous Nets," *Bulletin of Mathematical Biophysics*, 7, pp. 89–93; reprinted in Warren S. McCulloch, *Embodiments of Mind*, MIT Press, Cambridge, Mass., 2nd edition, 1989, pp. 40–45.

Menger, Karl
1959 "Gulliver in the Land without One, Two, Three," *Mathematical Gazette*, 43, pp. 241–250.
1962 "Functional Algebra and Propositional Calculus," in *Self-Organizing Systems*, Spartan Books, Washington, D.C., pp. 525–532.

16
Introduction to Natural Magic*

as told by
Heinz von Foerster
to
Paul Schroeder
(Edited and with a preface by Ranulph Glanville)

Hello Heinz, and hello Mai.

It is seven a.m., Monday morning, September 28, 1987. I am sitting in our kitchen on a brilliant, crisp, clear fall morning here in Orono, Maine. My children aren't up yet. Mazie is not up yet. A moment of peace to get started on this taperecorded query.

It took a little longer for me to get this together than I had planned. I had disabled the record function on our little tape recorder here at home, and I didn't get around to fixing that until yesterday.

(The family rises.)

Hello, again. It is 11:35, same morning. All the flurries of the morning are over. I wanted to share a little bit of our family life with you.

Pretty soon I should be getting to the point, which we discussed on the telephone. Just some accounts. I have several specifics in mind. The first, and we agreed on it, is that I would like to hear the background of the set of volumes that you got, Heinz, from your cousin; I think it was from your cousin.

What I would like, most of all, is to have the pleasure of hearing a few of your stories again. I wish we could be there together with you. I am not much concerned with obtaining a definitive version—there is never much inconsistency in what you present. I can remember the themes of many of your tales, but the details have been lost.

Because of course, when we sit and listen with attention to what you say, and to each other at all times, we don't make very good notes. Then of course, later it is difficult to make notes because the memory has not been tuned. So, in a sense this is a matter of record, but not necessarily of definitive record.

Now, before my family returns, and before we make all the preliminaries a little too long, I hope you still accept my idea here. Please wait until

---

* This was published in *Systems Research: The Official Journal of the International Federation for Systems Research, 10* (10), pp. 65–79 (1993). © John Wiley & Sons Limited. Reproduced with permission.

you have a few moments free. Take your cassette to the woods, or to the airport, whatever you're doing, and when you have some moments, sit and add. I will appreciate that.

I will commit myself to making a visual transcription of these documents. In many ways, the printed page is the most flexible of media. I am somewhat uncomfortable, still, speaking into the tape recorder, even talking to you, but I expect that in time this is something that will also be taken care of.

Our greetings to you, and I look forward to hearing from you both.
Paul

Dear Paul.

That was really magic, when you transformed me from my deck on Rattlesnake Hill in Pescadero to your home in Maine, allowing me to hear your children, Mazie, you, the clatter, the singing, the good morning— almost tasting the good food. And I thought it was a most ingenious idea to invite me, or to invite anybody, to become a storyteller, because you tell the story in such a wonderful way that it is absolutely irresistible to continue the yarn which you started to spin.

Now, you have invited me to tell some of the stories which I experienced in my youth, and you also have asked me to give the background of these stories. And I think I will suffer a similar fate as Thucydides who wanted to write about the Peloponnesian War. But in order to give the background of the Peloponnesian War, he practically spent the rest of his life writing about the background, and did not have very much time to talk about the Peloponnesian War. Now I will try to somewhat balance this act, and I will give you a little bit of the background, and then comes the story of Wiegleb's *Introduction to Natural Magic*. There are twenty volumes, printed at the end of the 18th century, between 1780 and 1795.

So first, let me give you the background of the stories I am going to tell you, dear Paul.

The background has probably three major chapters. And the first chapter is perhaps to establish my relationship with my cousin, Martin, with whom I grew up practically as a brother, with a brother. We were both born in the year 1911. He was a little bit earlier than I was, and the relation between both of us is via my mother and his father. They were siblings. My mother was born Lilith Lang, and his father was Erwin Lang. Erwin married an extraordinary, elfin, beautiful, ethereal dancer by the name of Grete Wiesenthal, who conquered the world with her charms and her absolutely incredibly beautiful light and unearthly dancing. She broke away from the ballet, as many of the great dancers at the turn of this century, like Loie Fuller, Isadora Duncan, Ruth St. Denis, Gertrude Barrison and many, many others.

My mother happened to be *la costumière* for her sister-in-law. They were all very close to each other. She not only designed some of the at that time outrageous but extremely lovely costumes, but she was also there in the evenings, when the performances were going on, and she saw that everybody wore the dresses and the designs she made, in the proper way.

The way in which my cousin Martin and I became much closer than just the two sons of two siblings was that my father, Emil, and Martin's father, Erwin, were taken prisoners of World War I practically in the first week after the hostilities broke out. My uncle Erwin was to fight on the eastern front, on the Russian front, and my father Emil was sent down to the Southeast to battle the Serbs, who were already well entrenched, and were prepared for the beginning of this war.

They were both captured. Within the first week, my father was taken as a prisoner of war, taken away by the Serbs—today you would call them the Yugoslavs**, but this was at that time Serbia—and my uncle Erwin was put into one of the big trains. The Germans lost hundreds of thousands of soldiers in the first two or three weeks. They were completely unprepared for trench warfare. They were still riding on horses, pulling out their sabres and trying to attack the Russians who were deeply entrenched, and shooting them down with their machine guns.

My uncle Erwin was transported to Siberia, where he stayed until practically 1917, when he succeeded in fleeing, and by train and on foot reached China, ultimately arriving in Tsingtao. There he met the great philosopher Richard Wilhelm, whose name you probably know as the translator of many of the Chinese philosophical works, most prominent, of course, the *I Ching*. They are translated into English, and, Paul, I am sure, if you have an *I Ching*, it is the one translated by Richard Wilhelm.

Ok, that's the story of Erwin. But the story of my father: he was brought to Serbia, where the Serbs were defeated after about two months of battle, and then he was transferred to the Italians, and became an Italian prisoner of war, and stayed on a tiny, tiny island between Corsica and Sardinia.

But nevertheless these two boys, Heinz and Martin, grew up without fathers, and the mothers, being very close, arranged also that the boys were very close. So Martin practically grew up with me in our house, because Grete was of course dancing all over Germany, making performances etc., etc. And if Martin was not at our house, he was staying with his Grandmother, the mother of Lilith and Erwin, Marie. So I was, in many cases, taken with my mother in the evening—because the idea of a baby-sitter didn't exist at that time—I was taken by my mother to the theatres, where Grete Wiesenthal performed. And I, as a good boy, was to sit in a little corner, and watch the wonderful ladies who changed from one costume into another, and probably at that time I developed this preferred taste for women, and I think this stuck with me for the rest of my life. If you have the chance to see these absolutely incredible creatures, like elves, going out onto the stage, coming back, transforming themselves into other elves and going out onto the stage again, and you watch from the sidelines, you get a very different impression; what incredible, magical, ethereal, creatures they are.

** Though not any longer!

Martin and I stayed together very frequently. When we were approximately the age of eleven, for some reason or other we became very much fascinated with magic. We got one of the standard little boxes which you buy in a store, which contain all the magic tricks for kids, but when we opened it, when we tried to do some of these tricks, we thought it absolutely ridiculous, so stupid; everybody will find out what that is, I mean, the double floor and all that silly stuff.

We said, this is really . . . , this is not the way of doing magic.

We started to develop our own illusions, and soon became more and more deeply involved. There is a very great and internationally famous store in Vienna, which designs, constructs and delivers great illusions for magicians from all over the world. People of the great performance stage came to that store, which was called the Zauberklingel, the Magic Bell. A klingel is a bell, a German word for bell, so the magic bell. But it happened to be that there was a Herr Klingel, a Mr. Bell, who owned the store, and who was, when we were about eleven, probably about sixty. When we went into the store to buy perhaps a little something here or there, of course what he did was always first to show what the whole illusion was about. And then, if you liked it, you could buy it.

Of course, all of these things were completely out of the world for us, we couldn't buy them. But we came as if we would like to see this particular piece of illusion, "Could we please see it?" So he looked at us, of course, very condescending, and said, "Would you buy it?" We said, "Ahem, well, we don't know, we would like to see it first." So he showed it to us, when he was in a good mood, otherwise he threw us out, and then we said, "Well, ahem, we are not going to buy, thank you very much," and left the store very fast.

Of course, while we were watching, we were thinking, "How is it done?" Since I am of a constructive type, and Martin is a performing type, I was sitting down, constructing the thing—how it could work?—and Martin then developed how to perform it. So we both co-operated in the performance and construction of these things.

For me it was quite clear that you have to have a good mechanical and physical mind in order to perform very great illusions. For Martin, however, it was quite clear that the mechanics didn't do it. What he did, of course, was the performance. So I learned from him the accent on performance, and he learned from me the technical problems.

Now both Martin and I, when we grew up, fell in love with one particular German Romantic author. He is not too well known as an author in the United Sates, or in the English-speaking domain, but is indirectly very well known through the Offenbach opera *Hoffmann's Tales*. The German Romantic poet E.T.A. Hoffmann was one of our very much preferred poets, writers. There was one story which fascinated both Martin and me, that was the story of a tomcat. And the tomcat's name was Murr, m-u-r-r. Tomcat Murr it would be pronounced in English.

Tomcat Murr was a very unusual cat, because he learned by sitting on the shoulder or on the table of his master, who was in the employ of a small Duke in Germany, and this master's name was Master Abraham. And he was sitting at the desk of Master Abraham, who was working practically all the night writing and computing and thinking about this and that, and Kater Murr, or Tomcat Murr, was watching his master, learning how to write and how to read. So he tried it out a little bit for himself, and then observed that he could indeed write, and Tomcat Murr thought his life was so fascinating, that he should put it down as his biography, his autobiography.

When he decided to write his autobiography he was short of paper, and he did not know where to find some, but he saw stacks of paper on his master's desk, partly thrown away, apparently discarded. So he took these papers and started to write his autobiography. Now, when he was through with the autobiography, he made it known to one of the publishers in that province. The publisher was of course fascinated with a tomcat's autobiography, so he said, "We are going to print that." This was the *Autobiography of Tomcat Murr*.

When they started to typeset the story of Tomcat Murr, something very peculiar happened. At certain points, the story of Tomcat stopped and continued in an entirely different fashion, a story of a little Duchy in Germany, where the master is writing about the kind of things he has to prepare; he has to prepare an Aeolian Harp for the next festivity of the Duke, when he gets visitors—he has to make the water fountains in the proper shape.

It turns out that Kater Murr was writing on the other side of pages which were written by his master, Master Abraham. The printers found this out too late, so they could not, in practice, stop. The whole Kater Murr autobiography is now written in such a way that you read a couple of pages about the Tomcat, and then you read a couple of pages of the stories of Master Abraham.

The fascinating thing, going through that book, is that you will see that the observations of Tomcat Murr and the experiences of Master Abraham are interlinked, because they live at the same place, live in the same environment and they live in the same cultural setting. These are two complementary stories.

For me it was fascinating to read the Master Abraham part. I even made little notes on the corners of the pages where Master Abraham's story continued, so I could skip the Tomcat Murr story, because Master Abraham was applying physics to the entertainment world. The water fountains, the water plays, the Aeolian Harp, the ways in which he constructed automata to entertain his employers, etc., etc., fascinated me.

I knew that Master Abraham was collecting his information from one extraordinarily famous book. This was from a man by the name of Wiegleb. Wiegleb published a series of books—I did not know how many—a series of books which were devoted to the introduction to natural magic, or Lehrbuch, that means textbook, on natural magic in which there is

Lebens-Ansichten

des

Katers Murr

nebst

fragmentarischer Biographie

des

Kapellmeisters Johannes Kreisler

in

zufälligen Makulaturblättern.

Herausgegeben

von

E. T. A. Hoffmann

Zwei Theile.

Berlin,
bei G. Reimer.
1 8 2 8.

everything, such as how to make cheese, how to prepare wine, how to make Aeolian Harps, how to make a theatre illuminated, how to project pictures on the wall, how to cut shadow things; anything, name it, and it was in Wiegleb.

Master Abraham knew about Wiegleb's books, and used them from time to time, to inform himself about building an automaton and doing this and that. So this was my side.

My cousin Martin, however, was fascinated with Tomcat Murr's autobiography, and in his edition of the Kater Murr autobiography he made little markings on the page so that he could read Tomcat Murr continuously, without the interruption of this natural magician, Master Abraham.

We must have been about fourteen years old, when the next episode of my story begins. This is now Chapter Two.

I have to tell you that my cousin Martin and I, in the summer, always went to a place which belonged to an uncle, the first husband of my grandmother, Marie Lang. His name was Theodor Köchert. That is in English K-o-e-c-h-e-r-t. Theodor Köchert was married to my grandmother Marie, first, and they produced a son by the name of Erich. This was of course a half-brother of my mother and my uncle Erwin. Uncle Erich inherited from his Papa an incredibly beautiful place, out on one of these wonderful lakes which are on the Salzkammergut, this is a region approximately east of Salzburg, where there are several very, very lovely lakes. One lake is named according to the river which flows through it, and the river's name is Traun. T-r-a-u-n. Therefore the lake is called the Traun Lake, or the Traunsee.

Now, he owned a piece of land, perhaps about twenty or thirty acres, directly on a little finger which is sticking out in the Traun Lake. Vis-à-vis that place, on the other side of the lake, is a beautiful mountain which is called the Traunstein, the Traun Stone, which is a big rocky mountain with steep slopes to the west, and is about, I would say, two thousand meters, six thousand feet in altitude. It is a tremendous view directly on the other side of the lake.

We spent most of our summers at this property on the Traun Lake. The place is called Hollereck. The place where the river Traun exits that lake is called Gmunden. Gmunden and this whole region is an extraordinarily old region. Directly south is Hallein, where the salt mines of Austria are located, opened up in the stone age, where the salt was mined and brought down through the valley and then shipped through the Danube, down to the east, further into Hungary, Rumania, etc., etc. These are extraordinarily old places and people settled there very, very early. Gmunden is a very old city, and has many fascinating and interesting stores.

One of the stores is a second-hand bookstore. I wouldn't call it a second-hand bookstore; maybe I could call it an old and rare bookstore. But it called itself a years, she must have been sixty or something like that, at least she appeared to us to be sixty. She was a little bit roundish, looked a bit like a barrel, and was very, very stern. Whenever we entered that bookstore to buy just a silly old book, a second-hand book for maybe fifty cents or twenty cents or something like that, she chased us out very fast, etc., "Don't roam around those books," and so on, and so forth.

Anyway, she was a very interesting lady because she bought the libraries from defunct places, such as castles or monasteries, for very little money.

Now it happened that I have another uncle, by the name of Goldschmidt, and this uncle was born a very rich boy. He was exactly the same age as my uncle Erwin, and they went to school together. Now, my uncle Goldschmidt,—his first name was Ernest, but nobody in Vienna called him Ernest—he was called Emsterl. So Emsterl went to school with Erwin, went through the gymnasium (this is high school in Austria), and he was at that time already a very, very bright youngster. He was a very good student, he knew everything; he could not be baffled, squeezed with questions, he knew everything, while my uncle Erwin practically knew nothing. He never learned the homework; he played soccer, that was his thing. But Emsterl was fascinated with books and read a lot, so he could really answer all the questions.

He had rebelled against authority already when he was a kid, as all my relatives always rebelled against authority. He had a particular way of rebelling against authority because the authorities, the teachers, they couldn't do anything with him, because he knew everything, he was a straight-A guy. So he plagued them with other gimmicks.

For instance, in Austria pants are stitched together with the seams being on the outside of the thigh. So you have, on the right part of the pants, the seam of the right side, and the left is on the left side.

There is an Austrian command, "Hands to the pants' seams!" And that means almost the same as "Attention". The German expression is "Hände an die Hosennaht," "Hands on the seams of the pants!"

But British pants are stitched in a different way, their seams are on the inside. So, the left side of the trousers has its seams on the right-hand side, inside, and the other, right, on the left-hand. So, when the teachers commanded "Hände an die Hosennaht," "Hands on the pants seams," Emsterl was fumbling around, looking for where the seams were, crossing the hands, putting the right hand to the inside seam of the left and so on. Things of that short he invented to really torment the teachers, who otherwise tried to torment the kids. But not with Emsterl.

Emsterl developed into what he himself called a bookworm. He was fascinated by all books, wonderful editions, very nice bindings. His doctoral dissertation when he was about twenty-four, twenty-three, was bindings, of books I think of the fifteenth century—incunabula—or maybe a little later perhaps. He was the reference man for bindings and everybody had to consult the Goldschmidt book on bindings.

He became very interested in all the libraries in Austria. Most of the monasteries which were founded perhaps in the twelfth, thirteenth, fourteenth, fifteenth centuries, had incunabula, had many other imprints, had of

course many, many manuscripts, very early manuscripts, and had collected the books of the very early printers. He studied these libraries, knew exactly which book was where, so that if, for instance, the very big monastery of Melk, one of the gigantic libraries, perhaps with about 500,000 books, did not find a book, they were writing to E.P. Goldschmidt in London and said, "Do we have that book, and do you know where it is?" Then he would write back, "Yes, of course you have that book. It is on such and such a shelf and such a floor on the fifth position on that shelf."

He was very well informed about what was going on in this field, and had himself opened up a bookstore, old and rare books, E.P. Goldschmidt, Old and Rare Books, 45 Old Bond Street, in London, W1. It is the same house in which Laurence Sterne died, so there is an inscription, "This is the place where Laurence Sterne died," I don't know in which year, I have forgotten that. But this is the house in which E.P. Goldschmidt, Old and Rare Books, had its store.

Now, E.P. Goldschmidt, of course, whenever he visited us in Vienna, or my uncle Erwin, or whatever, later on stopped by at Miss Wlk's Old and Rare Books, because she sometimes bought some fascinating stuff. And that's where he got a very interesting story. I remember when he appeared in Hollereck (my Uncle Erich Köchert's place) with tremendous excitement. He went to the store, browsed around, looked at this book, looked at that book, at that book and this book. Then he opened up a book which was not a very interesting book, I think it was a book about the history of knights or something like that, seventeenth century or sixteenth century. He looked at that book and thought, this is of no interest, opened it up, and then, when he turned the page over, on the other side of that page, was clearly a very, very early printing, must have been fifteenth century, late fifteenth century. It gave him a real shock, but of course he would not show his interest. He went around, looked at other books, and opened up that book again, and flipped to another page; yes, it was all in Latin. "Wow, this is incredible." He caught one sentence, and tried to figure out; what is that story, which is printed in Latin, so to say on the other side of the book on knights? It is precisely a repetition of the Tomcat Murr story, where Murr's story is on one side of the page, and on the other side of the page is Master Abraham's story.

He tried to figure out, what is that sentence? Who wrote that sentence? what Latin is that? And he, as if by accident, came a third time around, caught another one, and said, "It can only be one thing, and that is the Tacitus *Germania*." This is Tacitus'—the Roman historian's—story about early German culture, geography and history.

Apparently what happened, he figured out fast, was that somebody printed this book on knights on paper which was apparently discarded, which contained the Tacitus *Germania*. So he bought one book from Mrs. Wlk, and he bought another book from Mrs. Wlk, and said "How much are these?," and then "Why don't I buy this book on knights also, how much is it?"—maybe twenty dollars, the other one is two dollars and five dollars and fifty dollars, so altogether he paid about eighty dollars, ninety dollars for about six or seven books, and wrapped them up in a brown bag, bought the books and left Mrs. Wlk's bookstore very slowly. The moment he left, he raced over to the taxi stand, jumped into the taxi, and said "Drive me to Hollereck," to my uncle Erich Köchert's place. Then he pulled out the books, turned the knight book around, and indeed it was Tacitus' *Germania*, almost complete. That meant he bought for perhaps thirty or forty dollars something which he later sold, I think, for perhaps about twenty or thirty thousand dollars, to the German State Library, I think in Berlin, who *had* to buy one of the early Latin versions of the Tacitus *Germania*, which existed I think in only one or two copies.

This is only a little bit of background on Mrs. Wlk. Now, let me go back to Heinz and Martin, who spent their summer vacation on Hollereck, at my uncle Erich's place. Of course we had bicycles, and from time to time we had to go to Gmunden, either to do this or that or to buy some oil for the bicycle. Now, always when we came to Gmunden, we stopped at Mrs. Wlk's place, looked around, and things like that. One day we came by, I think we must have been thirteen or fourteen or something like that, so it was the year 1925 or 1924, and in the window were standing twenty volumes of a book, where one, the first volume was opened, and it said, Wiegleb's *Textbook on Natural Magic*. I said, "My God, here is Wiegleb!" So Martin said, "What?", and "Of course, this is Wiegleb, the *Introduction to Natural Magic*, the early books on physics." Ok, we both went into the store.

I said, "Well, Mrs. Wlk, I see you have Wiegleb's textbook on magic, tell me how much are these twenty volumes?" So she said, "Well, each volume is about two shillings, so the whole thing, twenty times two is forty shillings, will be forty shillings." Translated today into dollars, it will be perhaps about eight dollars, something like that. But when you were fourteen years of age, in Austria, you didn't carry forty shillings with you; this was impossible. No, what you had was about five shillings, or two shillings. We didn't have the shillings to buy the Wiegleb.

So we said, "Mrs. Wlk, don't sell the Wiegleb, keep it out of the window, we will be back as soon as we have raised the funds to buy that book." So she said, "Ok, I will see to that, I will try. I can't hold it too long for you boys, you know I can't hold it too long." "No no no, we will be back in a moment."

We both jumped on our bicycles. Martin raced in one direction, I raced in another. I thought I could borrow the money from this or that relative or perhaps from one of my friends. Friends were not at home, others were not there, I couldn't get it, absolutely impossible, so I hurried down to Altmünster, which is about four or five miles from Gmunden.

Finally, I reached Altmünster. I went quickly to this person, to chose persons, and finally I got the forty shillings together, jumped on the bike, raced back to Gmunden, and went to the store of Mrs. Wlk.

I arrived there of course bathed in sweat, put my bike around the corner, looked into the window, and there were no books. No Wiegleb. So I walked into the bookstore, and there was Mrs. Wlk. I said, "Mrs. Wlk, I have here the forty shillings, I would like to buy the Wieglebs, textbook for the natural magic, twenty volumes." She said, "I'm sorry, boy, I have sold it already." "No. What? You've sold it?" "Yes. I promised I would keep it for you, but you are so late." "Now, for heaven's sake, who bought it?" "Well now, the other body who came along with you." "Ah, thank you." That was of course a tremendous relief, wonderful, I went back out, I got on my bike, and now I could pedal slowly back to Altmünster and see my cousin Martin.

There he was with the twenty volumes, Wiegleb's textbook for natural magic. "Great," I said, "Wonderful. You did it. I also have the forty shillings, why don't we share it," or whatever I proposed. Martin said, "No, you know, I bought these twenty volumes." I said, "Of course, you bought it, it is wonderful, now we have the volumes." "No," he said "I have the volumes." I said, "What do you mean, you have the volumes?" "Of course, I bought them, so I have these twenty volumes." I said, "But, Martin, you don't know what to do with that stuff. It's all physics." "No, no, it's natural magic. I would like to have the volumes. Of course, Heinz, if you would like to look at one of them, one at a time, and things like that, I will of course allow you to look at these volumes. You can borrow them, if you want to, and I will lend them to you, no problem at all." I said, "But Martin, this is utterly ridiculous, I mean this is all physics." "No, what, physics, schmysics, doesn't really matter, it's the introduction to natural magic and I bought these volumes, I paid for them, and if you wish you can look at them."

So, this was a little bit of a letdown for me, and I said to myself, this is a kind of a mean thing, he can't really use them, so I was allowed to look at them. Of course, they have a wonderful—a whole volume for—the Index, you can look up anything you want to, and you find the appropriate volume, and things like that. So, anyway, this is the end of Instalment No. Two, II-A, on the books, on those twenty volumes.

Ok. Then, later on of course, we stayed together until the end of our high school. When we were eighteen we graduated, and then we went our different ways; he went to the theatre, first to Berlin, worked in the movies, with movie people, became an assistant to some of the very famous directors at that time, and I of course entered into the study of physics, at the University of Vienna, and the Institute of Technology of Vienna, etc. etc.. So we were heading in different directions. Then, after a number of years World War Two began, and the bombing commenced. I was staying in Berlin. He, Martin, was unfortunately drafted, so he was a soldier with the German army. But soon they found out that he could do more than just use a gun or doing this and that. He could perform magic. So he became one of the great performers of the German army, and travelled from France to Russia to Serbia, to Italy, to this and to that, performing and performing and performing. He never had to touch a rifle, except during boot training, where he had of course to juggle around with those deadly instruments.

I, in Berlin, of course, was bombed out very soon, and lost practically all of my books. Some of them I could transport to an escape place in Silesia.

Then the translation would be, the Heavenly Comedy or something like that. I packed it in my last little suitcase because it was a special edition, bound in leather, and printed on extraordinarily thin paper. So this came, later on, after the war, when nobody had cigarettes, but I could sometimes find some tobacco, extraordinarily handy because I could use the last pages of the *Paradiso*, which I thought always were a little bit silly, as cigarette paper, in which I could roll my cigarettes.

So I let the Paradise go up in smoke in the years of, let's say, 1945, 1946 etc. etc.. Fortunately, later on I got cigarette papers. Only the last couple of songs of the *Paradiso* have indeed been used as cigarette paper.

Now, ok, things settled down, and after many years Mai and I and our three boys had already moved to the United States. Every three or four years I visited Vienna, and on one occasion Martin had established himself in a new apartment in Vienna. He had married in the meantime a very charming lady and they had a daughter. He married very late. He established himself in a very charming apartment, directly vis-à-vis one of the most beautiful Imperial castles in Austria, Schönbrunn. It consists of one major castle complex, then wonderful gardens leading up a hill, where on top of the hill is a very charming lookout, which is called Gloriette.

Martin with his wife and daughter walked through these charming formal gardens of the Hapsburg emperors, and they were maintained in wonderful condition: rose gardens and lily gardens and fountains. On one of my visits to Vienna, when I always stayed with Martin, he said, "By the way, Heinz, I have a surprise for you." I said, "That's very nice". "I have just put all my books into various bookshelves, and there is a set of books, I do not know where to put". I said, "Well, what are they?" "Well, twenty volumes." "Twenty volumes, of what?" "Come and look."

I looked at them. It was the Wiegleb, which had survived in Vienna, but which would have been burned to ashes if I had owned it, and had had it in Berlin. So he said, "Ok, Heinz, if you can use the Wiegleb, here it is, it's yours." So I said, "That is wonderful, because I can use it all the time." So he packed it for me in a box, and mailed it to me. I got it at Christmas, I think it was 1982, in Pescadero. Since I always use these volumes, they are directly on my desk, and here they are, being a very good source of my understanding of physics, of my understanding of the culture which generated physics. And, Paul, when you come the next time, I think you should have a good look.

There is another little detail, and it is this. When I was still at the University of Illinois, some people by hearsay heard that I was once a magician. So the History of Science Society, which is a very, very good society at the University of Illinois, invited me to give a lecture on the history of magic. I said, this is wonderful. I would be delighted to do that. So I knew of course a source of information on the history of magic, and this was of course Wiegleb's textbook of natural magic. So I said, ok I accept that. I had lots of time. I think they announced it two or three months before I had to give the lecture. I went to the very good library of the University of Illinois, to check out whether we had a Wiegleb in Illinois.

There was no Wiegleb. So I went to the interlibrary service department and said, "Well, I would like to have a book, and I can give you the exact details." Of course, references to Wiegleb can be found in almost all books on the history of science and magic.

I gave them the reference. I waited and waited. I went back to the library and said, "What happened? I ordered this book a long time ago?" They said, "We searched through all the United States, libraries that are in the interlibrary service connection, and not even in the Library of Congress, nor in the this and that and that, could we find any original Wiegleb."

I said, that is very, very sad, so I had to make do with some others. Of course, there are plenty of books on magic; well, I had to be satisified with some others. But I wanted to just drop that as a footnote to my story on the Wiegleb.

I will close with this. What I should do now is to pack the tape, and mail it to you so that you have fun with the Wiegleb. Also, I have some photographs, of Grete Wiesenthal, of my mother, and perhaps I'll make some copies of these photographs, which I can only copy on my little Canon copying machine, but they are reasonably good. You can get an idea of them.

Please send my greetings to all the wonderful members of your family, and I tell you, it was a great pleasure to sit in your kitchen, and tell you a little bit about the stories of Austria, myself, my family, and Martin.

Heinz

Pescadero
October
17–18, 1987

MARIE   LILITH   PETER MARTIN   GRETE

17
Publications

Heinz von Foerster

1943

1. *Uber das Leistungsproblem beim Klystron, Ber. Lilienthal Ges. Luftfahrt-forschung, 155,* pp. 1–5 (1943).

1948

2. *Das Gedachtnis: Eine quantenmechanische Untersuchung,* F. Deuticke, Vienna, 40 pp. (1948).

1949

3. *Cybernetics: Transactions of the Sixth Conference* (ed.), Josiah Macy Jr. Foundation, New York, 202 pp. (1949).

4. *Quantum Mechanical Theory of Memory* in *Cybernetics: Transactions of the Sixth Conference* (ed.), Josiah Macy Jr. Foundation, New York, pp. 112–145 (1949).

1950

5. With Margaret Mead and H. L. Teuber, *Cybernetics: Transactions of the Seventh Conference* (eds.), Josiah Macy Jr. Foundation, New York, 251 pp. (1950).

1951

6. With Margaret Mead and H. L. Teuber, *Cybernetics: Transactions of the Eighth Conference* (eds.), Josiah Macy Jr. Foundation, New York, 240 pp. (1951).

1953

7. With M. L. Babcock and D. F. Holshouser, *Diode Characteristic of a Hollow Cathode, Phys. Rev., 91,* 755 (1953).

8. With Margaret Mead and H. L. Teuber, *Cybernetics: Transactions of the Ninth Conference* (eds.), Josiah Mac Jr. Foundation, New York, 184 pp. (1953).

1954

9. With E. W. Ernst, *Electron Bunches of Short Time Duration, J. of Appl. Phys., 25,* 674 (1954).

10. With L. R. Bloom, *Ultra-High Frequency Beam Analyzer, Rev. of Sci. Instr., 25,* pp. 640–653 (1954).

11. *Experiment in Popularization, Nature,* 174, 4424, London (1954).

1955

12. With Margaret Mead and H. L. Teuber, *Cybernetics: Transactions of the Tenth Conference* (eds.), Josiah Macy Jr. Foundation, New York, 100 pp. (1955).

13.   With O. T. Purl, *Velocity Spectrography of Electron Dynamics in the Traveling Field, Journ. of Appl. Phys., 26*, pp. 351–353 (1955).

14.   With E. W. Ernst, *Time Dispersion of Secondary Electron Emission, Journ. of Appl. Phys., 26*, pp. 781–782 (1955).

1956

15.   With M. Weinstein, *Space Charge Effects in Dense, Velocity Modulated Electron Beams, Journ. of Appl. Phys., 27*, pp. 344–346 (1956).

1957

16.   With E. W. Ernst, O. T. Purl, M. Weinstein, *Oscillographie analyse d'un faisceau hyperfrequences*, LE VIDE, 70, pp. 341–351 (1957).

1958

17.   *Basic Concepts of Homeostasis* in *Homeostatic Mechanisms*, Upton, New York, pp. 216–242 (1958).

1959

18.   With G. Brecher and E. P. Cronkite, *Production, Ausreifung and Lebensdauer der Leukozyten* in *Physiologie und Physiopathologie der weissen Blutzellen*, H. Braunsteiner (ed.), George Thieme Verlag, Stuttgart, pp. 188–214 (1959).

19.   *Some Remarks on Changing Populations* in *The Kinetics of Cellular Proliferation*, F. Stohlman Jr. (ed.), Grune and Stratton, New York, pp. 382–407 (1959).

1960

20.   *On Self-Organizing Systems and Their Environments* in *Self-Organizing Systems*, M. C. Yovits and S. Cameron (eds.), Pergamon Press, London, pp. 31–50 (1960).

20.1   *O Samoorganizuyushchiesja Sistemach i ich Okrooshenii* in *Samoorganizuyushchiesju Sistemi*, 113–139, M. I. R. Moscow (1964).

20.2   *Sobre Sistemas Autoorganizados y sus Contornos* in *Epistemologia de la Communicacion*, Juan Antonio Bofil (ed.), Fernando Torres, Valencia, pp. 187–214 (1976).

21.   With P. M. Mora and L. W. Amiot, *Doomsday: Friday, November 13, AD 2026*, Science, *132*, pp. 1291–1295 (1960).

22.   *Bionics* in *Bionics Symposium*, Wright Air Development Division, Technical Report 60–600, J. Steele (ed.), pp. 1–4 (1960).

23.   *Some Aspects in the Design of Biological Computers* in *Sec. Inter. Congress on Cybernetics*, Namur, pp. 241–255 (1960).

1961

24.   With G. Pask, *A Predictive Model for Self-Organizing Systems*, Part I: Cybernetica, *3*, pp. 258–300; Part II: *Cybernetica, 4*, pp. 20–55 (1961).

25.   With P. M. Mora and L. W. Amiot, *Doomsday*, Science, *133*, pp. 936–946 (1961).

26.   With D. F. Holshouser and G. L. Clark, *Microwave Modulation of Light Using the Kerr Effect, Journ. Opt. Soc. Amer., 51*, pp. 1360–1365 (1961).

27.   With P. M. Mora and L. W. Amiot, *Population Density and Growth*, Science, *133*, pp. 1931–1937 (1961).

1962

28.   With G. Brecher and E. P. Cronkite, *Production, Differentiation and Lifespan of Leukocytes* in *The Physiology and Pathology of Leukocytes*, H. Braunsteiner (ed.), Grune & Stratton, New York, pp. 170–195 (1962).

29.   With G. W. Zopf, Jr., *Principles of Self-Organization: The Illinois Symposium on Theory and Technology of Self-Organizing Systems* (eds.), Pergamon Press, London, 526 pp. (1962).

30. *Communication Amongst Automata, Amer. Journ. Psychiatry, 118,* pp. 865–871 (1962).

31. With P. M. Mora and L. W. Amiot, *"Projections" versus "Forecasts" in Human Population Studies, Science, 136,* pp. 173–174 (1962).

32. *Biological Ideas for the Engineer, The New Scientist, 15,* 173–174 (1962).

33. *Bio-Logic* in *Biological Prototypes and Synthetic Systems,* E. E. Bernard and M. A. Kare (eds.), Plenum Press, New York, pp. 1–12 (1962).

33.1 *Bio-Logika* in *Problemi Bioniki,* pp. 9–23, M. I. R., Moscow (1965).

34. *Circuitry of Clues of Platonic Ideation* in *Aspects of the Theory of Artificial Intelligence,* C. A. Muses (ed.), Plenum Press, New York, pp. 43–82 (1962).

35. *Perception of Form in Biological and Man-Made Systems* in *Trans. I.D.E.A. Symp.,* E. J. Zagorski (ed.), University of Illinois, Urbana, pp. 10–37 (1962).

36. With W. R. Ashby and C. C. Walker, *Instability of Pulse Activity in a Net with Threshold, Nature, 196,* pp. 561–562 (1962).

1963

37. *Bionics* in *McGraw-Hill Yearbook Science and Technology,* McGraw-Hill, New York, pp. 148–151 (1963).

38. *Logical Structure of Environment and Its Internal Representation* in *Trans. Internat'l Design Conf., Aspen,* R. E. Eckerstrom (ed.), H. Miller, Inc., Zeeland, Mich., pp. 27–38 (1963).

39. With W. R. Ashby and C. C. Walker, *The Essential Instability of Systems with Threshold, and Some Possible Applications to Psychiatry* in *Nerve, Brain and Memory Models,* N. Wiener and I. P. Schade (eds.), Elsevier, Amsterdam, pp. 236–243 (1963).

1964

40. *Molecular Bionics* in *Information Processing by Living Organisms and Machines,* H. L. Oestreicher (ed.), Aerospace Medical Division, Dayton, pp. 161–190 (1964).

41. With W. R. Ashby, *Biological Computers* in *Bioastronautics,* K. E. Schaefer (ed.), The Macmillan Co., New York, pp. 333–360 (1964).

42. *Form: Perception, Representation and Symbolization* in *Form and Meaning,* N. Perman (ed.), Soc. Typographic Arts, Chicago, pp. 21–54 (1964).

43. *Structural Models of Functional Interactions* in *Information Processing in the Nervous System,* R. W. Gerard and J. W. Duyff (eds.), Excerpta Medica Foundation, Amsterdam, The Netherlands, pp. 370–383 (1964).

44. *Physics and Anthropology, Current Anthropology, 5,* pp. 330–331 (1964).

1965

45. *Memory without Record* in *The Anatomy of Memory,* D. P. Kimble (ed.), Science and Behavior Books, Palo Alto, pp. 388–433 (1965).

46. *Bionics Principles* in *Bionics,* R. A. Willaume (ed.), AGARD, Paris, pp. 1–12 (1965).

1966

47. *From Stimulus to Symbol* in *Sign, Image, Symbol,* G. Kepes (ed.), George Braziller, New York, pp. 42–61 (1966).

1967

48. *Computation in Neural Nets, Currents Mod. Biol., 1,* pp. 47–93 (1967).

49. *Time and Memory* in *Interdisciplinary Perspectives of Time,* R. Fischer (ed.), New York Academy of Sciences, New York, pp. 866–873 (1967).

50. With G. Gunther, *The Logical Structure of Evolution and Emanation* in *Interdisciplinary Perspectives of Time*, R. Fischer (ed.), New York Academy of Sciences, New York, pp. 874–891 (1967).

51. *Biological Principles of Information Storage and Retrieval* in *Electronic Handling of Information: Testing and Evaluation*, Allen Kent et al. (ed.), Academic Press, London, pp. 123–147 (1967).

1968

52. With A. Inselberg and P. Weston, *Memory and Inductive Inference* in *Cybernetic Problems in Bionics*, Proceedings of Bionics 1966, H. Oestreicher and D. Moore (eds.), Gordon & Breach, New York, pp. 31–68 (1968).

53. With J. White, L. Peterson and J. Russell, *Purposive Systems*, Proceedings of the 1st Annual Symposium of the American Society for Cybernetics (eds.), Spartan Books, New York, 179 pp. (1968).

1969

54. With J. W. Beauchamp, *Music by Computers* (eds.), John Wiley & Sons, New York, 139 pp. (1969).

55. *Sounds and Music* in *Music by Computers*, H. Von Foerster and J. W. Beauchamp (eds.), John Wiley & Sons, New York, pp. 3–10 (1969).

56. *What Is Memory that It May Have Hindsight and Foresight as well?* in *The Future of the Brain Sciences*, Proceedings of a Conference held at the New York Academy of Medicine, S. Bogoch (ed.), Plenum Press, New York, pp. 19–64 (1969).

57. *Laws of Form*, (Book Review of *Laws of Form*, G. Spencer Brown), *Whole Earth Catalog*, Portola Institute; Palo Alto, California, p. 14 (Spring 1969).

1970

58. *Molecular Ethology, An Immodest Proposal for Semantic Clarification* in *Molecular Mechanisms in Memory and Learning*, Georges Ungar (ed.), Plenum Press, New York, pp. 213–248 (1970).

59. With A. Inselberg, *A Mathematical Model of the Basilar Membrane*, *Mathematical Biosciences*, 7, pp. 341–363 (1970).

60. *Thoughts and Notes on Cognition* in *Cognition: A Multiple View*, P. Garvin (ed.), Spartan Books, New York, pp. 25–48 (1970).

61. *Bionics, Critique and Outlook* in *Principles and Practice of Bionics*, H. E. von Gierke, W. D. Keidel and H. L. Oestreicher (eds.), Technivision Service, Slough, pp. 467–473 (1970).

62. *Embodiments of Mind*, (Book Review of *Embodiments of Mind*, Warren S. McCulloch), *Computer Studies in the Humanities and Verbal Behavior*, III (2), pp. 111–112 (1970).

63. With L. Peterson, *Cybernetics of Taxation: The Optimization of Economic Participation*, *Journal of Cybernetics*, 1 (2), pp. 5–22 (1970).

64. *Obituary for Warren S. McCulloch* in *ASC Newsletter*, 3 (1), (1970).

1971

65. *Preface* in *Shape of Community* by S. Chermayeff and A. Tzonis, Penguin Books, Baltimore, pp. xvii–xxi (1971).

66. *Interpersonal Relational Networks* (ed.); CIDOC Cuaderno No. 1014, Centro Intercultural de Documentacion, Cuernavaca, Mexico, 139 pp. (1971).

67. *Technology: What Will It Mean to Librarians?*, *Illinois Libraries*, 53 (9), 785–803 (1971).

68. *Computing in the Semantic Domain*, in *Annals of the New York Academy of Science, 184*, 239–241 (1971).

1972

69. *Responsibilities of Competence, Journal of Cybernetics, 2* (2), pp. 1–6 (1972).

70. *Perception of the Future and the Future of Perception* in *Instructional Science, 1* (1), pp. 31–43 (1972).

70.1 *La Percepcion de Futuro y el Futuro de Percepcion* in *Comunicacion* No. 24 Madrid (1975).

1973

71. With P. E. Weston, *Artificial Intelligence and Machines that Understand, Annual Review of Physical Chemistry*, H. Eyring, C. J. Christensen, H. S. Johnston (eds.), Annual Reviews, Inc.; Palo Alto, pp. 353–378 (1973).

72. *On Constructing a Reality*, in *Environmental Design Research*, Vol. 2, F. E. Preiser (ed.), Dowden, Hutchinson & Ross; Stroudberg, pp. 35–46 (1973).

1974

73. *Giving with a Purpose: The Cybernetics of Philanthropy*, Occasional Paper No. 5, Center for a Voluntary Society, Washington, D.C., 19 pp. (1974).

74. *Kybernetik einer Erkenntnistheorie*, in *Kybernetic und Bionik*, W. D. Keidel, W. Handler & M. Spring (eds.), Oldenburg; Munich, 27–46 (1974).

75. *Epilogue to Afterwords*, in *After Brockman: A Symposium*, ABYSS, 4, 68–69 (1974).

76. With R. Howe, *Cybernetics at Illinois* (Part One), *Forum, 6* (3), 15–17 (1974); (Part Two), *Forum, 6* (4), 22–28 (1974).

77. *Notes on an Epistemology for Living Things*, BCL Report No. 9.3 (BCL Fiche No. 104/1), Biological Computer Laboratory, Department of Electrical Engineering, University of Illinois, Urbana, 24 pp. (1972).

77.1 *Notes pour un épistémologie des objets vivants*, in *L'unité de L'homme*, Edgar Morin and Massimo Piatelli-Palmerini (eds.), Edition du Seuil; Paris, 401–417 (1974).

78. With P. Arnold, B. Aton, D. Rosenfeld, K. Saxena, *Diversity *S*H*R*: A Measure Complementing Uncertainty H*, *SYSTEMA*, No. 2, January 1974.

79. *Culture and Biological Man* (Book Review of *Culture and Biological Man*, by Elliot D. Chapple), *Current Anthropology, 15* (1), 61 (1974).

80. *Comunicación, Autonomia y Libertad* (Entrovista con H.V.F. *Comunicación* No. 14, 33–37, Madrid (1974).

1975

81. With R. Howe, *Introductory Comments to Francisco Varela's Calculus for Self-Reference, Int. Jour. for General Systems, 2*, 1–3 (1975).

82. *Two Cybernetics Frontiers*, (Book Review of *Two Cybernetics Frontiers* by Stewart Brand), *The Co-Evolutionary Quarterly, 2*, (Summer), 143 (1975).

83. *Oops: Gaia's Cybernetics Badly Expressed, The Co-Evolutionary Quarterly, 2*, (Fall), 51 (1975).

1976

84. *Objects: Tokens for (Eigen-)Behaviors*, ASC Cybernetics Forum, 8, (3 & 4) 91–96 (1976). (English version of 84.1).

1977

84.1 *Formalisation de Certains Aspects de l'Équilibration de Structures Cognitives*, in *Epistémologie Génétique et Équilibration*, B. Inhelder, R. Garcias and J. Voneche (eds.), Delachaux et Niestle, Neuchatel, 76–89 (1977).

344   H. von Foerster

1978

72.1 "Construir la realidad" in *Infancia y Aprendizaje*, Madrid, *1* (1), 79–92 (1978).

1979

72.3 "On Constructing a Reality" in *An Integral View*, San Francisco, *1* (2), 21–29 (1979).

1980

85. *Minicomputer—verbindende Elemente, Chip*, Jan. 1980, p. 8.

86. *Epistemology of Communication*, in *The Myths of Information: Technology and Postindustrial Culture*, Kathleen Woodward (ed.), Coda Press, Madison, 18–27 (1980).

1981

87. *Morality Play, The Sciences, 21* (8), 24–25 (1981).

88. *Gregory Bateson, The Esalen Catalogue, 20* (1), 10 (1981).

89 *On Cybernetics of Cybernetics and Social Theory*, in *Self-Organizing Systems*, G. Roth and H. Schwegler (eds.), Campus Verlag, Frankfurt, 102–105 (1981).

90. *Foreword*, in *Rigor and Imagination*, C. Wilder-Mott and John H. Weakland (eds.), Praeger, New York, vii–xi (1981).

91. *Understanding Understanding: An Epistemology of Second Order Concepts*, in *Aprendizagem/Desenvolvimento, 1* (3), 83–85 (1981).

1982

92. *Observing Systems*, with an Introduction by Francisco Varela., Intersystems Publications, Seaside, 331 + xvi pp. (1982).

93. "A Constructivist Epistemology" in *Cahiers de la Fondation Archives Jean Piaget*, No. 3, Geneve, 191–213 (1982).

94. "To Know and To Let Know: An Applied Theory of Knowledge" Canadian Library J., *39*, 277–282 (1982).

1983

95. "The Curious Behavior of Complex Systems: Lessons from Biology" in *Future Research*, H. A. Linstone and W. H. C. Simmonds (eds.), Addison-Wesley, Reading, 104–113 (1977).

96. "Where Do We Go From Here" in *History and Philosophy of Technology*, George Bugliarello and Dean B. Doner (eds.), University of Illinois Press, Urbana, 358–370 (1979).

1984

97. "Principles of Self-Organization in a Socio-Managerial Context" in *Self-Organization and Management of Social Systems*, H. Ulrich and G. J. B. Probst (eds.), Springer, Berlin, 2–24 (1984).

98. "Disorder/Order: Discovery or Invention" in *Disorder and Order*, P. Livingston (ed.), Anma Libri, Saratoga, 177–189 (1984).

72.4 "On Constructing a Reality" in *The Invented Reality*, Paul Watzlawick (ed.), W.W. Norton, New York, 41–62 (1984).

99. "Erkenntnistheorien und Selbstorganisation" *DELFIN, IV*, 6–19 (Dez. 1984).

1985

100. "Cibernetica ed epistemologia: storia e prospettive" in *La Sfida della Complessita*, G. Bocchi and M. Ceruti (eds.), Feltrinelli, Milano, 112–140 (1985).

101. *Sicht und Einsicht: Versuche zu einer operativen Erkenntnistheorie*, Friedrich Vieweg und Sohn, Braunschweig, 233 pp. (1985).

102.   "Entdecken oder Erfinden: Wie laesst sich Verstehen verstehen?" in *Einfuehrung in den Konstruktivismus*, Heinz Gumin und Armin Mohler (eds.), R. Oldenbourg, Muenchen, 27–68 (1985).

103.   "Apropos Epistemologies" in *Family Process*, *24* (4), 517–520 (1985).

94.1   "To Know and To Let Know" in CYBERNETIC, J. Am. Soc. Cybernetics, *1*, 47–55 (1985).

1986

47.2   "From Stimulus to Symbol" in *Event Cognition: An Ecological Perspective*, Viki McCabe and Gerald J. Balzano (eds.), Lawrence Erlbaum Assoc., Hillsdale, NY, 79–92 (1986).

104.   "Foreword" in *The Dream of Reality: Heinz von Foerster's Constructivism*, by Lynn Segal; W.W. Norton & Co., New York, xi–xiv (1986).

105.   "Vernunftige Verrucktheit" (I), in *Verrueckte Vernunft*, Steirische Berichte 6/84, 18 (1984).

106.   "Comments on Norbert Wiener's 'Time, Communication, and the Nervous System'" in *Norbert Wiener: Collected Works*, *IV*, P. Masani (ed.), MIT Press, Cambridge, 244–246 (1985).

107.   "Comments on Norbert Wiener's 'Time and the Science of Organization'" in *Norbert Wiener: Collected Works*, *IV*, P. Masani (ed.), MIT Press, Cambridge, 235 (1985).

108.   "Comments on Noerbert Wiener's 'Cybernetics'; 'Cybernetics'; 'Men, Machines, and the World About'" in *Norbert Wiener: Collected Works*, *IV*, P. Masani (ed.), MIT Press, Cambridge, 800–803 (1985).

1987

99.1   "Erkenntnistheorien und Selbstorganisation" in *Der Diskurs des Radikalen Konstruktivismus*, Siegfried J. Schmidt (ed.), Suhrkamp, Framkfurt, 133–158 (1987).

102.1 "Entdecken oder Erfinden—Wie laesst sich Verstehen verstehen? in *Erziehung und Therapie in systemischer Sicht*, Wilhelm Rotthaus (ed.), Verlag modernes Denken, Dortmund, 22–60 (1987).

109.   "Vernunftige Verrucktheit" (II), in *Verruckte Vernunft*, Vortrage der 25. Steirischen Akademie, D. Cwienk (ed.), Verlag Tech. Univ., Graz, 137–160 (1986).

110.   "Cybernetics" in *Encyclopedia for Artificial Intelligence*, *I.*, S. C. Shapiro (ed.), John Wiley and Sons, New York, 225–227 (1987).

110.1 "Kybernetik", Zeitschr. Systemische Therapie 5 (4), 220–223 (1987).

112.   "Preface" in *The Construction of Knowledge: Contributions to Conceptual Semantics*, by Ernst von Glasersfeld, Intersystems Publ. Seaside, ix–xii (1987).

113.   "Understanding Computers and Cognition" Book Review of *Understanding Computers and Cognition: A New Foundation of Design* by Terry Winograd and Fernando Flores; Technological Forecasting and Social Change, An Int. J. *32*, #3, 311–318 (1987).

114.   "*Sistemi che Osservano*, Mauro Ceruti and Umberta Telfner (eds.), Casa Editrice Astrolabio, Roma, 243 pp. (1987).

1988

72.5   "On Constructing a Reality" in *Adolescent Psychiatry, 15: Developmental and Clinical Studies*, Sherman C. Feinstein (ed.), University of Chicago Press, Chicago, 77–95 (1988).

72.6 "Costruire una realta" in *La realta inventata*, Paul Watzlawick (ed.), Feltrinelli, Milano, 37–56 (1988).

72.7 "La construction d'un realite" in *L'invention de la realite: Contributions au constructivism*, Paul Watzlawick (ed.), Edition du Seuil, Paris, 45–69 (1988).

72.8 "Construyendo una realidad" in *La Realidad inventada*, Paul Watzlawick (ed.), Editorial Gedisa, Barcelone, 38–56 (1988).

104.1 "Vorbemerkung" in *Das 18. Kamel oder die Welt als Erfindung: Zum Konstruktivismus Heinz von Foersters*, Lynn Segal, Piper, Muenchen, 11–14 (1988).

115. "Abbau und Aufbau" in *Lebende Systeme: Wirklichkeitskonstruktionen in der Systemischen Therapie*, Fritz B. Simon (ed.), Springer Verlag, Heidelberg, 19–33 (1988).

116. "Foreword" in *Aesthetics of Change* by Bradford Keeney, The Guilford Press, New York, xi (1983).

117. "Cybernetics of Cybernetics" in *Communication and Control in Society*, Klaus Krippendorff (ed.), Gordon and Breach, New York, 5–8 (1979).

118. Interviews 1988.

1989

119. "Wahrnehmen wahrnehmen" in *Philosophien der neuen Technologie*, Peter Gente (ed.), Merve Verlag, Berlin, 27–40 (1988).

120. "Preface" in *The Collected Works of Warren S. McCulloch*, Rook McCulloch (ed.), Intersystems Publication, Salinas, i–iii (1989).

121. "Circular Causality: The Beginnings of an Epistemology of Responsibility" in *The Collected Works of Warren S. McCulloch*, Rook McCulloch (ed.), Intersystems Publications, 808–829 (1989).

122. "Anacruse" in *Auto-reference et therapie familiale*, Mony Elkaim et Carlos Sluzki (eds.), Cahiers critiques de therapie familial et de pratiques de reseaux #9; Bruxelles, 21–24 (1989).

123. "Geleitwort" in *Architektonik: Entwurf einer Metaphysik der Machbarkeit*, Bernhard Mitterauer, Verlag Christian Brandstaetter, Wien, 7–8 (1989).

124. "The Need of Perception for the Perception of Needs" LEONARDO *22* (2), 223–226 (1989).

1990

125. "Preface" in *Education in the Systems Sciences*, by Blaine A. Snow, The Elmwood Institute, Berkeley, iii (1990).

126. "Sul vedere: il problema del doppio cieco" in OIKOS *1*, 15–35 (1990).

127. "Implicit Ethics" in *of/of* Book-Conference, Princelet Editions, London, 17–20 (1984).

128. "Non sapere di non sapere" in *Che cos'e la conoscenza*, Mauro Ceruti and Lorena Preta (eds.), Saggitari Laterza, Bari, 2–12 (1990).

129. "Understanding Understanding" in METHODOLOGIA: *7*, 7–22 (1990).

130. "Foreword" to *If You Love Me, Don't Love Me*, by Mony Elkaim. Basic Books, New York, ix–xi (1990).

131. "Kausalitaet, Unordnung, Selbstorganisation" in *Grundprinzipien der Selbstorganisation*, Karl W. Kratky und Friedrich Wallner (eds.), Wissenschaftliche Buchgesellschaft, Darmstadt, 77–95 (1990).

94.2 "To Know and to Let Know" in 26, *1*, Agfacompugraphic, 5–9 (1990).

119.1 "Wahrnehmen wahrnehmen" in *Aisthesis: Wahrnehmung heute oder Perspektiven einer anderen Aesthetik*, Karlheinz Barck, Peter Gente, Heidi Paris, Stefan Richter (eds.), Reclam, Leipzig, 434–443 (1990).

132. "Carl Auer und die Ethik der Pythagoraer" in *Carl Auer: Geist oder Ghost*, G. Weber und F. Simon (eds.), Auer, Heidelberg, 100–111 (1990).

77.2 "Bases Epistemologicas" in: *Anthropos* (Documentas Anthropos), Barcelona, *Supplementos Antropos*, 22 (Oct. 1990), 85–89.

72.9 "Creation de la Realidad" in: *Antropos* (Documentas Anthropos), Barcelona, *Supplementos Anthropos*, 22 (Oct. 1990), 108–112.

1991

56.1 "Was ist Gedaechtnis, dass es Rueckschau *und* Vorschau ermoeglicht" in: *Gedaechtnis, Probleme und Perspektiven der interdisziplinaeren Gedaechtnis-forschung*, hg. von Siegfried Schmidt, Franfurt, Suhrkamp, 56–95 (1991).

132.1 "Carl Auer and the Ethics of the Pythagoreans" in *Strange Encounters with Carl Auer*, G. Weber and F. B. Simon (eds.), W. W. Norton, New York, 55–62 (1991).

133. "Through the Eyes of the Other" in *Research and Reflexivity*, Frederick Steier (ed.), Sage Publications, London, 63–75 (1991).

134. *Las Semillas de la Cybernetica* [Obras escogidas: ## 20;60;69;70; 77;89;94;96;97;98;103.] Marcelo Pakman (ed.), Presentación Carlos Sluzki, Gedisa editorial, Barcelona, 221 pp. (1991).

135. "'Self': An Unorthodox Paradox" in *Paradoxes of Selfreference in the Humanities, Law, and the Social Sciences*, Stanford Literature Review, Spring/Fall 1990, ANMA LIBRI, Stanford, 9–15 (1991).

136. With Guitta Pessis-Pasternak: "Heinz von Foerster, pionier de la cybernétique" in *Faut-il Brûler Descartes?* G. Pessis-Pasternak (ed.), Edition La Decouvert, Paris, 200–210 (1991).

137. "Éthique et Cybernétique de second ordre" in *Systèmes, Éthique, Perspectives en thérapie familiale*, Y. Ray et B. Prieur (eds.), ESF editeur, Paris, 41–55 (1991).

138. With Yveline Ray: "Entretien avec Heinz von Foerster" in *Systèmes, Éthique, Perspectives en thérapie familiale*, Y. Ray et B. Prieure (eds.), ESF editeur, Paris, 55–63 (1991).

1992

102.2 "Entdecken oder Erfinden; Wie laesst sich Verstehen verstehen?" in *Einfuerung in den Kostruktivismus* Serie Piper #1163, Piper, Muenchen, 41–88 (1992).

110.2 "Cybernetics" in *The Encyclopedia of Artificial Intelligence*, Second Edition, S. C. Shapiro (ed.), John Wiley and Sons, New York, 309–312 (1992).

119.2 "Wahrnehmen wahrnehmen" in *Schlusschor*, Botho Strauss, Schaubuehne, Berlin, 89–95 (1992).

139. With Christiane Floyd: "Self-Organization and Reality Construction" in *Software Development and Reality Construction*, C. Floyd, H. Zullighoven, R. Budde, and R. Keil-Slawik (eds.), Springer Verlag, New York and Heidelberg, 75–85 (1992).

140. *Wissen und Gewissen: Versuch einer Bruecke*, mit einer Vorbemerkung von Siegfried Schmidt, einer Einfuehrung von Bernard Scott, und einer Einleitung von Dirk Baecker, [Enthaelt: ## 20;56;58;60;69;70;72;74;77;84;86;97;98;110; 127;129.], hg. von Siegfried Schmidt, Framkfurt, Suhrkamp (1992).

141. "La percezione della quarta dimensione spaziale" in *Evoluzione e Conoscenza*, Mauro Ceruti (ed.), Pierluigi Lubrina, Bergamo, 443–459 (1992).

142. Mit Wolfgang Ritschel: "Zauberei und Kybernetik" in *Menschenbilder*, Hubert Gaisbauer und Heinz Janisch (eds.), Austria Press, Wien, 45–56 (1992).

137.1 "Ethics and Second Order Cybernetics" in *Cybernetics and Human Knowing*, *1*.1, 9–20 (1992).

143. "Letologia. Una teoria dell'apprendimento e della conoscenza *vis á vis* con gli indeterminabili, inconoscibili." in *Conoscenza come Educazione*, Paolo Perticari (ed.), FrancoAngeli, Milano, 57–78 (1992).

144. "Kybernetische Reflexionen" in *Das Ende der grossen Enwürie*, H. E. Fischer, A. Fetzer, J. Schweitzer (eds.), Suhrkamp, Frankfurt, 132–139 (1992).

145. "Geleitwort" zu *Systemische Therapie: Grundlagen Klinischer Theorie und Praxis*, Kurt Ludewig, Klett-Cotta, Stuttgart, 8–10 (1992).

70.2 "Perception of the Future, and the Future of Perception" in *Full Spectrum Learning*, Kristina Hooper Woolsey (ed.), Apple Multimedia Lab, Cuppertino, 76–87 (1992).

143.1 "Lethology, A Theory of Learning and Knowing, vis a vis Undeterminables, Undecidables, Unknowables", in *Full Spectrum Learning*, Kristina Hooper Woolsey (ed.), Apple Multimedia Lab, Cuppertino, 88–109 (1992).

1993

146. With Paul Schroeder: "Introduction to Natural Magic", Systems Research *10*(1), 65–79 (1993).

147. With Mosche Feldenkrais: "A Conversation", The Feldenkrais J. #8, 17–30 (1993).

148. "Das Gleichnis vom Blinden Fleck: Ueber das Sehen im Allgemeinen" in *Der entfesselte Blick*, G. J. Lischka (ed.), Benteli Verlag, Bern, 14–47 (1993).

149. With Paul Schroeder: "Einfuehrung in die 12-Ton Musik", in *KybernEthik*, Heinz von Foerster, Merve Verlag, Berlin 40–59 (1993).

150. *KybernEthik* [Enthaelt:## 146.1;149;137.2;117.1;100.1;121.1;143.1;69.2], Merve Verlag, Berlin, 173p (1993).

137.3 "Ethik und Kybernetik zweiter Ordnung" Psychiatria Danubiana, *5* (182), Medicinska Naklada, Zagreb, 33–40 (Jan–Jul, 1993).

137.4 "Ethics, and Second Order Cybernetics", Psychiatria Danubiana, *5* (182), Medicinska Naklada, Zagreb, 40–46 (Jan–Jul, 1993).

1994

153. Mit Wolfgang Moller-Streitborger: "Es gibt keine Wahrheit—nur Verantwortung", Psychologie Heute, Marz 1994, 64–69 (1994).

122.1 "Anacruse" in *La therapie familiale en changement*, Mony Elkaim (ed.), SYNTHELABU, Le Plssis-Robinson, 125–129 (1964).

154. "Wissenschaft des Unwissbaren" in *Neuroworlds*, Jutta Fedrowitz, Dick Matejovski und Gert Kaiser (eds.), Campus Verlag, Framkfurt, 33–59 (1994).

155. "Inventare per apprendere, apprendere per invantare," in *Il senso dell' imparare*, Paolo Perticari e Miranella Sclavi (eds.), Anabasi, Milano, 1–16 (1994).

126.1 "On Seeing: The Problem of the Double Blind" in *Adolescent Psychiatry, 11*, S. C. Feinstein and R. Marohn (eds.), University of Chicago Press, Chicago, 86–105 (1993).

151. "On Gordon Pask", Systems Research *10* (3), 35–42 (1993).

152. "Fuer Niklas Luhmann: 'Wie rekursiv ist die Kommunikation?' ". Mit einer Antwort von Niklas Luhmann; in *Teoria Soziobiologica 2/93*, FrancoAngeli, Milan, 61–88 (1993).

152.1 "Per Niklas Luhmann: 'Quanto e ricorsiva la communicazioni?' ". Con un riposta di Niklas Luhmann; in *Teoria Soziobiologica 2/93*, FrancoAngeli, Milan, 89–114 (1993).

57.1   "Gesetze der Form" in *Kalkül der Form*, Dirk Baecker (ed.), Suhrkamp, Frankfurt, 9–11 (1943).

1994

153.   Mit Wolfgang Möller-Streitborger: "Es gibt keine Wahrheit—nur Verantwortung", Psychologie Heute, März 1994, 64–69 (1994).

122.1 "Anacruse" in *La thérapie familiale en changement*, Mony Elkaim (ed.), SYNTHELABU, Le Plssis-Robinson, 125–129 (1964).

1995

156.   "WORTE" in *Weltbilder/Bildwelten (INTERFACE II)*, Klaus Peter Denker und Ute Hagel (eds.), Kulturbehoerde Hamburg, 236–247 (1995).

157.   "Wahrnehmen wahrnehmen—und was wir dabei lernen koennen" in *Schein und Wirklichkeit* Oesterreichische Werbewirtschaftliche Gesellschaft, Wien, 15–27 (1995).

158.   "Vorwort" in *Vorsicht: Keine Einsicht!, Rokarto 1*, Thouet Verlag, Aachen, 1 (1995).

152.2 "For Niklas Luhmann:'How Recursive is Communication?'" Special Edition, ASC Annual Conference, Chicago. Illinois, 5/17–21/95; The American Society for Cybernetics, c/o Frank Galuszka, 150 Maplewood Av., Philadelphia, PA × 19144.

74.1   "Cybernetics of Epistemology", Special Edition, ASC Annual Conference, Chicago, Illinois, 5/17–21/95; The American Society for Cybernetics, c/o Framk Galuszka, 150 Maplewood Av., Philadelphia, PA × 19144.

159.   "Die Magie der Sprache und die Sprache der Magie." in *Abschied von Babylon: Verstaendigung ueber Grenzen der Psychiatrie*, Bock, Th. et al. (eds.), Psychiatrie–Verlag Bonn, 24–35 (1995).

160.   With Stefano Franchi, Guven Guzeldere and Eric Minch: "Interview with Heinz Von Foerster" in *Constructions of the Mind: Artificial Intelligence and the Humanities*, Stanford Humanities Review, 4, No. 2, 288–307 (1995).

137.5 "Ethics and Second Order Cybernetics" in *Constructions of the Mind: Artificial Intelligence and the Humanities*, Stanford Humanities Review, 4, No. 2, 308–327 (1995).

161.   Mit Gertrud Katharina Pietsch: "Im Gespraech bleiben", Zeitschrift fuer Systemische Therapie, 13 (3), Dortmund, 212–218 (July, 1995).

162.   "Wahrnehmen-Werben-Wirtschaft", in Werbeforschung und Praxis 3/95, Werbewirtschaftliche Gesellschaft, Wien/Bonn, 73 (1995).

163.   With Stephen A. Carlton (eds.): *Cybernetics of Cybernetics*, Second Edition, Future Systems Inc., Minneapolis, 523 pp. (1995).

164.   "Metaphysics of an Experimental Epistemologist." in *Brain Processes, Teories and Models*, Roberto Moreno-Diaz and Jose Mira-Mira (eds.), The MIT Press, Cambridge, Massachusetts, 3–16 (1995).

Index